Technology

BRAD AND TERRY THODE

Delmar
Publishers Inc.

DELMAR TECHNOLOGY SERIES

DELMAR STAFF
New Product Acquisitions: Mark W. Huth
Developmental Editor: Sandy Clark Gnirrep
Senior Project Editor: Christopher Chien
Production Supervisor: Teresa Luterbach
Senior Design Supervisor: Susan C. Mathews

For information, address Delmar Publishers Inc.
3 Columbia Circle Drive, Box 15015
Albany, New York 12203–5015

Printed in the United States of America
Published simultaneously in Canada
by Nelson Canada,
a division of The Thomson Corporation

10 9 8 7 6 5 4 3 2   XX 99 98 97 96 95 94

Library of Congress Cataloging-in-Publication Data:
Thode, Bradley R.
     Technology / Brad and Terry Thode.
         p.   cm.
     Includes index.
     ISBN 0-8273-5098-8
     1. Technology.   I. Thode, Terry.   II. Title.
T47.T55 1993
600—dc20

92–23972
CIP

CONTENTS

Introduction viii

Preface ix

Acknowledgements...xi

Using the Barcode-Driven Technology Laserdisc.............xiv

Chapter 1   **Getting Started in Technology 1**

What is Technology?
Technology - Growing Fast!
Solving Problems Step by Step
Research and Experimentation
Working in Groups
Being Creative

Chapter 2   **How Does Technology Affect Me? 28**

How Has Technology Changed Our World?
Technology and Other Subjects
Careers in Technology
Future Technology

Chapter 3   **Using Computers 50**

Computer Basics
Using Your Computer
Writing with a Computer
Drawing with a Computer
What's a Database?
Using Spreadsheets
Finding Information
The Future of Computer Technology

Chapter 4   **Inventing Things......................................78**

What is Innovation?
Getting Ideas
Who Invented That?
Getting a Patent

Chapter 5 : **Making Things.................................100**

Where Do We Get Resources?
Using Resources
How Products are Made
Is Our World Disposable?

Chapter 6 : **How Do Things Work? 114**

What is a System?
Designing Mechanical Systems
Designing Electronic Systems
Designing Fluid Systems
Designing Thermal Systems
Designing Chemical Systems
Putting Systems Together

Chapter 7 : **Designing Things 142**

Measuring Things
Designing Products for People
Choosing the Right Material
Making Models and Prototypes
Ecology of a Product

Chapter 8 : **What is Automation?........................170**

What about Automation?
Moving Materials
Robots
Working with Technology
The Factory of the Future

Chapter 9 : **How Does Business Work? 202**

What is a Company?
Mass Production
Total Quality Improvement
Packaging Products
Marketing and Advertising Products
Doing Business in the Future

Chapter 10 : **Building Things............................234**

What is Construction?
Design
Structural Design

Designing Communities
Energy Conservation
Building for the Future

Chapter 11   **Using Energy**............................**264**

Where Do We Get Energy?
How Can We Save Energy?
What is Alternative Energy?

Chapter 12   **Moving Things**............................**290**

What is Transportation?
Land Transportation
Water Transportation
Air Transportation
Future Transportation

Chapter 13   **Finding and Using Information**...............**326**

How Do We Communicate?
Writing and Drawing Instructions
Communicating with Pictures and Symbols
Electronic Communications
Communications in the Future

Chapter 14   **Space Technology**............................**360**

Space History
Space Physics
Living and Working in Space
Space Spinoffs and the Future

Chapter 15   **Biotechnology**............................**394**

You Are What You Eat
What Are Distillation and Fermentation?
Designing New Life: Bioengineering and Medicine
Bionics and the Future of Biotechnology

Chapter 16   **Technology and Your Future**...............**414**

Appendix   Table of Trigonometric Ratios   **423**

Glossary   **425**

Index   **431**

INTRODUCTION

The study of technology is a study of process that includes both technical and social processes. It has to do with designing, making, and doing things. It is enhanced by discoveries in science and shaped by the designs of engineers. Technology is the way that things are introduced into society.

This book, *TECHNOLOGY*, is designed to help people apply technology in the solution of major problems that face society. While complete in itself, the text is enhanced with technology such as bar-code-driven laserdiscs, computer software to support design brief activities, and videotapes. It is a book *about* technology to be used *with* technology to produce a dynamic learning experience for the student.

The content and activities featured in this book are the result of years of work by two of the most innovative, creative, practicing technology teachers today. They are national award winners for their excellence in education. Their vast experience has been captured in this publication to provide a form of education that will help citizens live optimally in an advanced technological society.

Kendall N. Starkweather, Ph.D.
Executive Director
International Technology Education Association

PREFACE

We live in an exciting age of rapid technological growth. Keeping up with this rapid growth doesn't have to be frustrating. Learning about technology is fun! Technology is putting knowledge to use. In this book, you will use technology to learn about technology. *Technology* will help you to become technologically literate. You will learn that it is important to use your hands and your mind to solve problems.

Organization

Technology offers students the opportunity to explore the fascinating world in which they live. The 16 chapters will provide an in-depth exploration of many different aspects of technology today and in the future.

Chapters 1 and 2 introduce students to technology and how it affects them. Chapter 3 deals with computer applications so students have some technology tools to use in later chapters. Chapters 4–7 cover inventing, designing, and producing things using technology systems.

In Chapters 8 and 9 students work with automation and modern business setups. Chapters 10–12 let students explore how products are built, energy systems, and how things are moved using technology.

Communication technology and space technology make up Chapters 13 and 14. The area of biotechnology is covered in Chapter 15. The final chapter is a synthesis for the student.

Special Features

- Every chapter has hands-on, minds-on activities that allow the students to get involved with technology. You use technology to learn about technology!
- Each chapter starts with a technology career and a list of major concepts covered in the chapter.
- Technofacts add interesting sidelights throughout the chapters.
- Boxed articles feature special interest materials to supplement the text information.
- Throughout the text, technology is integrated with other subjects such as math, science, language arts, etc. The inter-

disciplinary connections are indicated with a curriculum icon. The following icons will help you identify the cross curriculum connections:

Language Arts

Social Studies

Mathematics

Science

Art

▶ Solving problems systematically, individually as well as in groups, is emphasized throughout the text.

Using *Technology* to Learn About Technology

Laserdisc

This book is full of ideas, pictures, and activities related to technology. To bring these exciting concepts alive, the book can be used with a set of laserdiscs that further illustrate the concepts using video technology. As you read the book, you will notice a series of thick and thin lines in the margins. These are barcodes that let you find and watch video information quickly and easily. If you want to see and learn more about what you read, just pass a wand across the barcode. The laserdisc player will quickly find a video presentation that further explains the topic. It couldn't be easier to learn about the exciting world of technology that we live in.

Computer Software

The best way to learn something is to put your knowledge to use. There are over sixty technology activities in this book that will guide you through exciting ways to use technology. Computer technology can be used to help you solve the problems in the activities. The software can be used to help you with each step and keep track of your progress. In each activity, you will enter information into the computer. When you finish, you can make a hard copy (computer printout) of your work. Each of the activities you complete can be put together into your own book of technology.

No matter how you look at it, learning about technology is fun!

About the Authors

Brad and Terry Thode are a husband-and-wife technology teaching team with 37 years of combined teaching experience in K–12 technology education. Their dedication to helping other teachers implement technology education has been recognized at the local, state, and national levels. Their never-ending enthusiasm for teaching about technology and their love of learning have encouraged many teachers to re-think their curriculum, facility, and presentation techniques.

Brad Thode has been teaching technology education for over twenty years. He serves as a member of the International Technology Education Associations (ITEA) Advisory Council, and has conducted technology workshops in over 25 states. He is a past president of the Idaho Industrial Technology Education Association, Idaho's 1991 IBM and *Technology & Learning* Magazine Teacher of the Year, and the 1992 Idaho Technology Teacher of the Year. He had the ITEA Program of the Year in 1987, and he is president of Sun Valley Scientific, Inc. He has authored two other textbooks and numerous articles in national journals. He continues to challenge students to understand technology and its uses. Technology Education is exciting for Brad's students at Wood River Middle School in Hailey, Idaho.

Terry Thode is an elementary technology teacher whose honors include a 1991 Challenger Seven Fellowship, the 1990 National Foundation for Improvement in Education Learning Tomorrow Award, the 1990 Technology Education for Children Council (TECC) Educational Leadership Award, and the 1990 Presidential Award for Excellence in Science and Math Teaching. She is active in the ITEA, TECC, NSTA and other professional organizations. She has authored numerous articles in national and international journals and serves as vice president of Sun Valley Scientific, Inc. She currently teaches Technology Education K-6 grades at Hemingway Elementary School in Ketchum, Idaho. Her enthusiasm for using technology to teach about technology continues to grow exponentially.

Dedication

This book is dedicated to our daughter, Linly, whose love of learning and concern for the environment will help assure the appropriate use of technology in the future.

Acknowledgments

We would like to thank all of the people that have cooperated in various ways to make this book a reality. Their help and encouragement in this project is greatly appreciated.

Julie Caldwell, Hemingway School Technology Advisory Board, Idaho

Rob Campbell, Idaho State Division of Vocational Education

Jan & Keith Carter, Hemingway School Technology Advisory Board, Idaho

John Dominick, Hemingway School Technology Advisory Board, Idaho

Linly Ferris, University of California at Berkeley Boalt Law School

Ray Grosvenor, Wood River Middle School, Idaho

John Hickey, Chief Joseph Middle School, Montana

Phil Homer, Blaine County School District, Idaho

Tom Keck, Union 32 Jr./Sr. High School, Vermont

Bill Kenworthy, Sun Valley Scientific, Inc., Idaho

Marty Lukes, Power Engineers, Idaho

Barbara Morgan, NASA Teacher in Space Designee, Idaho

Doug Polette, Montana State University

David Rees, University of California at Berkeley Boalt Law School

Tom Smith, Boltz Junior High, Colorado

Kendall Starkweather, International Technology Education Association

Chuck Turner, Wood River Middle School, Idaho

Pete Van Der Meulen, Power Engineers, Idaho

Mick Williams, Blaine County School District, Idaho

Prior to writing *Technology,* the authors and the publisher met with a panel of educators in the field of Technology Education to discuss the planned text. Meeting participants brainstormed and discussed the text, its philosophy, and its activities for a number of days. The intelligence, wit, and humor of the panel members made the task a pleasure. The authors and Delmar would like to extend a sincere thanks to the following people for their input into the project from its inception:

Jeff Bush
Oakland Schools
Waterford, Michigan

Colleen Hill
California State University at Long Beach
Long Beach, California

Ted Hitchens
Rossville Middle School
Rossville, Georgia

Tom Smith
Boltz Junior High
Fort Collins, Colorado

Jerry Weddle
Roanoke County Schools
Salem, Virginia

We would also like to think those individuals who reviewed the manuscript and offered suggestions, feedback, and assistance:

Thomas Barrowman
Queensbury Middle School
Queensbury, New York

Jeff Bush
Oakland Schools
Waterford, Michigan

Rob Campbell
Idaho Division of Vocational Education
Boise, Idaho

Colleen Hill
University of California, Long Beach
Long Beach, California

Alan G. Horowitz
Clarkstown Central School District
Stony Point, New York

John K. Vandervelde
Amherst Middle School
Amherst, Virginia

Steven R. Ware
Dalton Junior High School
Dalton, Georgia

James Whitley
Dunbar Middle School
Crowley, Texas

We would like to express our deep appreciation to the Delmar staff for their hard work and creativity in making this text exciting and fun to read. Specifically, we would like to thank Sandy Clark and Mark Huth for their dedication and support.

Using the Barcode-Driven
Technology Laserdisc

The laserdisc included with the *Technology* textbook represents a major breakthrough in textbook publishing. The laserdisc provides a database of video presentations and simulations tied directly to specific segments of each textbook chapter. To this end, the textbook is *brought to life* through high-impact video presentations and simulations in which students are asked to offer advice, discuss, or complete a video-worksheet. Following many of these activities, the laserdisc provides feedback on concepts they have been exploring.

The *Technology Laserdisc* is designed for use on any laserdisc player with a barcode reader (employing the "LaserBarcode" standard). Many laserdisc players are "barcode-ready." Some require only a barcode reader (a hand-held wand used to swipe the barcode). Other laserdisc players without barcode capability can usually be converted for barcode utilization by factory-authorized service centers. Consult with your school's audiovisual specialist if your player does not have a barcode reader.

The barcode reader makes interaction between user and laserdisc easy and convenient. A simple swipe of any barcode printed in the text will instantaneously play the respective segment.

Disc Architecture

There are four double-sided laserdiscs (eight sides) available for the *Technology* text. The disc set is designed much like the textbook itself. Each of the eight sides contains chapters of interactive video material that directly corresponds to text chapters.

Disc Side	Text and Disc Chapters
1	1, 2
2	3, 4
3	5, 6
4	7, 8
5	9, 10
6	11, 12

Disc Side	Text and Disc Chapters
7	13
8	14, 15, 16

To begin viewing any chapter, simply play the corresponding disc side by swiping the barcode at the top of the textbook chapter's opening page. The barcode is identified as a *Chapter Opener*. (It is not necessary to initiate laserdisc play at the *Chapter Opener*.)

After viewing the *Chapter Opener* video segment, the video screen will fade to black and pause, waiting for the user to respond. As the teacher or student proceeds through the text they will come to subsequent barcodes which they can swipe to initiate the next video segment. At the end of a video segment, the user will be met with one of two alternatives. Either the screen will fade to black and the user will proceed to the next barcode, or the screen will pause and present a prompt that reads *swipe next barcode*. This prompt will occur most often when the primary video segment presents a question or scenario that initiates class or small group discussion or individual thought. The subsequent video segments that the user initiates at the *swipe next barcode* prompt provide feedback. For example, a segment might ask students to think about the ethical use of nuclear technology and pause with the *swipe next barcode* prompt. After the users have explored the topic to their satisfaction, they can swipe the next barcode and view feedback on the issue.

There are four varieties of interactive laserdisc segments:

Chapter Opener: The *Chapter Opener* is a "kickoff" presentation that accompanies each chapter. *Chapter Openers* typically provide an overview of the chapter contents with intriguing, high-impact video segments. Each *Chapter Opener* ends with one or two questions. These questions are designed to stimulate interest, thought, and questioning on the part of students by illustrating the relevance of the information presented in the forthcoming chapter.

Special Report—Technology: The *Special Report — Technology* is a video presentation that reveals new and anticipated technological developments. Questions for student discussion are often embedded in the *Special Report*.

Techno Teaser: *Techno Teasers* present a variety of interactive video simulations which draw students into video-prompted activities including group discussions and worksheet activities. Most *Techno Teasers* include barcode-access feedback; use these <u>after</u> the activity has been completed.

Techno Talk: *Techno Talk* includes a video report on technological topics that may be controversial or particularly thought-provoking. *Techno Talk* provides a base for highly creative thinking and discussion activities for individuals, groups, or the class as a whole.

During most of these segments, the video will periodically pause for presentation of questions and class discussion. These are prime opportunities for active involvement of the entire class or small groups. The video will freeze indefinitely to allow for ample discussion time. A prompt at the bottom right corner of the screen will advise users what to do once student discussion or activity is finished.

In addition to the above segment varieties, users may exercise other play options:

Pause and Freeze: In addition to pauses and freezes programmed on the laserdisc, teachers can pause or "freeze" the action at any time to highlight video of particular significance to the class, or interject a special discussion or activity. In this way, teachers can customize the laserdisc presentation to meet unique needs. A "pause," "still," or similarly labeled button is found on most players. To resume play, simply press "play," or a similarly labeled button. Use of this function will in no way hinder subsequent barcode function.

Repeating: At any time, a segment may be repeated by simply swiping its barcode again. However, it is important to always strike the *first* barcode if there is more than one for a single segment.

Programming: The laserdiscs can be programmed to run at Level III interactivity with a Macintosh or IBM (or clone) computer. Consult your school's computer specialists for programming details.

A Video Segment Directory: The disc package design enables the discs to be used as a "database index" or collection of randomly accessible interactive video presentations, simulations, and activities. Packaged with the discs is a *Technology Laserdisc Image Directory* with key descriptors, disc reference number, and respective barcodes that enable the discs to be used independently of the text. By simply playing the laserdisc side indicated and striking a barcode, an interactive video presentation can be instantaneously accessed by anyone. Students, for example, could conveniently bring a class report to life and generate discussion with a disc-based simulation. This feature may also be used by teachers who wish to assemble a specially designed interactive video experience for students who are interested in exploring a particular technology concept. In this case, teachers can compile an ordered list of barcode descriptors for the students to use.

Setting Up the Laserdisc Player

1 Connect the video-out port at the back of the laserdisc player to the video-in port of the television or monitor.

2 Connect the audio signal from the laserdisc player to the television or monitor.

Note: Many players and televisions are capable of a single cable connection that connects both video and audio simultaneously (such as "8-pin" or "RF" ports).

General Operating Instructions

1 Turn on the monitor/TV power switch.

2 Turn on the player's POWER switch.

3 Press the OPEN/CLOSE button. The disc table will be extended from the player.

4 Place the disc on the disc table with the side you wish to play facing up. Use only one laserdisc at a time. Take special care to align the disc inside the guides on the disc table.

5 Press the OPEN/CLOSE button. The disc table will close.

6 Some disc players require that you press a "play" button to begin use.

All directions for moving through the disc will appear on the monitor screen.

Using the Barcode Reader

1 Press and hold down the READ button on the barcode reader.

2 To read the barcode, continue to press the READ button while holding the scanner in a vertical position similar to holding a pencil. Move the scanner horizontally across the barcode from left to right starting from the outside edges. Be sure to swipe the entire barcode with the reader. A light, steady swipe of the barcode is all it takes. When the barcode is read, an electronic "beep" signal sounds. If the player doesn't respond, modify the angle at which you hold the barcode reader or your scanning speed and try again. After a few successful swipes you will have a feel for using the barcode reader.

3 If the barcode reader is hardwired to the player, instruction is sent to the player automatically after your swipe. In the wireless mode, aim the barcode reader at the player and press the SENDING key.

Caring for Laserdiscs

1 When loading or removing a disc, try not to touch its playing surfaces. Hold the disc by its edges.

2 Although fingerprints or other dirt on the disc will not directly affect the recorded signal, dirt on the disc will cause the brightness of the light reflected from the signal surface to be reduced. This may adversely affect sound and picture quality. If the disc is dirty, clean the disc with a soft, clean cloth before playing it. DO NOT CLEAN THIS DISC with record cleaning sprays or static prevention sprays. NEVER USE a cracked, scratched, or warped disc. This may damage the player or cause it to malfunction.

3 After using a disc, always remove it from the player and return it to its jacket. Store it vertically/upright, away from heat and humidity.

Follow other manufacturer's instructions for using the player, barcode reader, and laserdisc.

Instructional Tips

Using technology to teach technology is a natural, logical approach. It's also very exciting and helps students learn. When the laserdiscs are used in conjunction with the *Technology* text, subjects can come alive for students. After students learn the facts about an eximer laser, what better way to complete their learning than by *seeing* an eximer laser from reading the *Technology* text, in action and in an applied situation? The laserdisc segments provide students with the opportunity of seeing technology working in the real world and add an exciting video component to text study. While the discs were designed to enhance *Technology*, they can also be used as a stand-alone item. The laserdiscs and the accompanying *Technology Laserdisc Image Directory* can also add a relevant, visual component in any Technology Education class using any text.

With the Text: There are a number of ways that the laserdiscs can be used to enhance the learning that takes place in your classroom. While you will often use the laserdiscs to add video examples and relevance to your presentations, students also have the ability to access the laserdisc video segments themselves. Because barcodes have been incorporated directly into the text, students can use the laserdiscs individually or in small groups at a workstation, putting an exciting learning tool directly in the hands of students. If students, for example, are reading the text to learn about the system model, they can swipe the barcode in the margin of the text that corresponds to this topic. By swiping this barcode, students will learn what the system model is and see how it works in a variety of real-world applications.

Interactivity: One of the unique and educationally important features of the discs is their interactivity. With the interactive feature of the discs, students become active, engaged participants in the learning process. As students read their text and call up video segments that reinforce their reading with interesting relevant examples, the laserdisc segments usually pause with a question to engage students in worksheet activities, brainstorming, questioning, and discussion. These questions ask students to pull information together, synthesize, and use higher-level thinking skills to form their own <u>informed</u> ideas. When students have completed their activity, they will resume the laserdisc and receive interesting disc-based feedback.

With the Technology Laserdisc Image Directory: As a teacher, you can use the discs and the *Image Directory* to augment your presentation of material. If you're teaching ways in which industry is improving quality with Statistical Process Control (SPC), you could go to the *Image Directory* and select the video segment on SPC. After students have listened to your presentations, you can swipe a barcode and your class will be transported to a company to see how they actually use SPC in their business. Not only have you incorporated a video component, you've added relevance and real-world application to your class presentation. The *Image Directory* can also be used by students to research topics they are interested in learning more about. If students are working on a design brief to design a product with environmentally friendly packaging, they might use the database to look up this topic. The information then could be incorporated into their project.

Bringing Technology to Life: Clearly, there are a variety of creative ways you can incorporate the laserdiscs into your classroom — your vision and creativity are the only bounds. The ability to see technology in use in practical applications adds real-world relevance to Technology Education. The laserdiscs' unique interactive capabilities keep students involved and excited. Using laserdisc technology, the study of *Technology* comes to life!

Chapter 1

Getting Started in Technology

Things To Explore

When you finish this chapter, you will know that:

▶ Learning about technology is fast-paced, exciting, challenging, and fun!

▶ A technologically literate person can put technology to use to solve problems.

▶ Technology is changing and growing very quickly.

▶ Technology started over 2 million years ago.

▶ A problem-solving strategy can make it easier to solve any problem.

▶ Working in groups and brainstorming ideas are important skills.

▶ You can have more creative ideas if you open your mind to new things.

Chapter Opener
Getting Started

TechnoTerms

acronym	microchip
alloy	parameters
brainstorm	problem statement
bronze	research and development (R&D)
components	
cycle	retrieving
design brief	scientific method
evaluate	strategy
experimenting	tech
explore	technologically literate
exponential rate of change	
knowledge base	testing

(Courtesy of Thiokol Corporation.)

Careers in Technology

Millions of people all over the world work in technology-related careers. Technicians use problem-solving steps in jobs that produce or transport products, communicate messages, and research ideas. There is a good chance that the career you choose will involve technology in some way. Here, technicians are preparing the inside of one of the solid rocket boosters used to put the Space Shuttle into orbit around the earth. What kind of job do you think you would like?

1

Ask students to give their definition of technology.

TECHNOFACT

Introduction

In this technology textbook, you will find many interesting facts related to technology, called "technofacts." Every technofact will give you something to think about that relates to our high-tech world.

Technofact 1

Primitive people once washed themselves with ashes and water. If that wasn't strange enough, they would add animal grease or oils to the ashes! This sounds very messy, but it wasn't as strange as you might think. The soap we use today is very similar chemically to the ashes, water, and grease used thousands of years ago! Today scientists and technicians continue to improve the process of making soap. Maybe primitive people weren't as primitive as we think.

The idea of being technologically literate should be presented as a possible key to a successful future career and to being an informed citizen.

What Is Technology?

You live in a "high-tech" world. **Tech** (pronounced tek) is short for *technology*. Is technology robots, satellites, lasers, and computers? Or is it tools such as saws and hammers? All of these are products of technology, but technology is a lot more! Technology is a combination of people like you, your ideas, and the tools you will use to solve problems. It involves both thinking and doing. Technology is fast-paced, exciting, challenging, and fun! As you learn about technology you will be:

▶ Using knowledge from science, math, and other subjects to solve problems.

▶ Designing, inventing, and making things using your creative ideas.

▶ Building things such as products, houses, bridges, and devices that people have created to make life easier.

You can see that technology has many different definitions. Most definitions agree that technology is the use of knowledge, tools, and resources to help people. You will probably come up with your own definition of technology after working with it for a while.

Our high-tech world is very exciting. Technology is a combination of your ideas and the use of tools to solve problems. (Courtesy of Ball Aerospace.)

Because technology deals with people and the environment, you need to know how technology affects you. A person who understands the effects of technology is **technologically literate**. If you are technologically literate, you will be able to make decisions about your future and technology based on facts. As a technologically literate person, you will be able to:

▶ See how technology has changed through time.

- ▶ Think through a problem and come up with an answer.
- ▶ Decide whether a technology is good or bad for people or for the environment.
- ▶ Understand the newest uses of technology.
- ▶ Use the tools of technology to solve problems.

The effects of technology are not always good for society or for the environment. Some advancements in technology have caused environmental problems such as acid rain. Other technologies are being developed to help solve those problems. Let's find out more about technology.

Divide the chalk board into two areas. Have the class brainstorm good and bad effects of technology on our world.

Pulling Together The Global Puzzle

Earth is a planet of extraordinary complexity. Even so, until recent times the commonly held belief was that global change occurred on a geological time scale measured in tens of thousands of years. But that belief is changing.

Now, scientists are observing more rapid changes than previously thought possible. Consider that holes in the ozone layer have been detected over both poles.

In addition to the normal rhythms of nature, more recent man-made perturbations—expanding population, increased consumption of resources, spreading pollution—are accelerating global change. But exactly what is the interplay among contributors to change? Are all causes and effects even identified?

Answers to such questions remain fuzzy. Until uncertainties are resolved scientists cannot confidently predict long-term consequences. They don't fully understand linkages between what happens in the environment, the oceans and the terrestrial ecosystem. Nor can they accurately determine outcomes of actions aimed at mediating the changes. Ambiguities abound.

Technical advances, particularly in space sensors and computer modeling, may help answer these questions. Technology has already brought about a shift in how the earth is viewed.

Earth observation from space ➔

Data from Hughes-designed SSM/I microwave sensor, shown above, is transformed to color-coded images indicating seasonal changes in the Northern Hemisphere during 1988. In January, a blanket of snow covers most of North America and Asia. By July, vegetation prevails; tropical air that had entered the Gulf of Mexico a month earlier is moving up the eastern seaboard. October brings snow cover, an expanding polar ice pack and Typhoon Ruby, which killed more than 300 in the Philippines, churning the South China Sea.

Technology affects the whole world. The effects have been good and bad. One problem that was created by technology is the shrinking of the ozone layer in the atmosphere. Hopefully, technology will help us to solve the problem, too. (Courtesy of G. M. Hughes Electronics Corporation.)

Technology—Growing Fast!

Technology is changing faster and faster all the time. As the population grows, more people are adding more new ideas and inventing more new tools. When these people combine their ideas, we have even more new machines and tools. Over 90 percent of all technologies we have today were invented in the last 25 or 30 years. That means that technology is causing lots of change very fast.

Technology is causing rapid changes in our everyday lives. For example, there are now smart highways, where radar, satellites, and video cameras help police keep traffic moving when there is an accident. (Courtesy of G. M. Hughes Electronics Corporation.)

Discuss the different levels of technology of various cultures in the world today.

When did technology start? Many people think that technology started in the 1700s when factories began to make things. But technology really began many thousands of years ago. Prehistoric people used simple tools such as clubs and axes made of stone to work on different materials.

One way to organize history is to divide the past into time periods. These periods are based on the kinds of materials people used. This

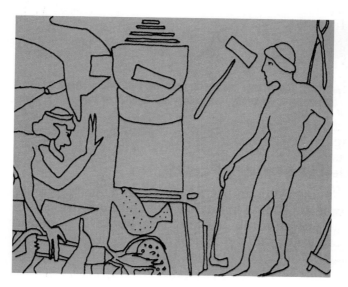

Prehistoric people had a technology very different from ours.

A world history chart can be useful in helping students visualize a time line of events. A laminated chart lets you easily mark and erase important events in technological history.

method is useful because people develop new technologies to meet their changing needs. But they don't throw away the old technology; they build on it. Let's take a brief look at technology through the ages.

▶ *The Stone Age* (2,000,000 B.C. to 3000 B.C.)

During the Stone Age, prehistoric people used tools made mostly of stone, animal bones, and wood. The tools were important to them as weapons or for gathering food. Prehistoric people also discovered uses for fire.

▶ *The Bronze Age* (3000 B.C. to 1200 B.C.)

During the Bronze Age, people learned how to mix copper with tin to make a stronger metal called **bronze**. Bronze is an **alloy**. An alloy is made when two or more metals are mixed together. Bronze was stronger than either copper or tin, and it could be used to make better tools and weapons.

This would be a good time to work with the social studies department to see what multidisciplinary topics can be reinforced or expanded.

▶ *The Iron Age* (1200 B.C. to A.D. 500)

People started to make their tools from iron instead of bronze. They preferred iron because it was harder and held a cutting edge better than bronze. Iron ore was also easy to find and less expensive. With the use of iron, more tools developed faster. Iron is still used in industry today.

▶ *The Pre-Industrial Revolution* (A.D. 500 to 1750)

There were very few changes in science and technology during the first part of this time period. People around the world did not advance from one stage of technology to the next at the same pace. People in some parts of the world moved ahead, while others stayed the same. Some people remained hunters and gatherers while others became farmers and herders. For example, the Chinese were ahead of other people in technology because they had gunpowder, the compass, and movable type for printing. Today there are a few isolated places where people are still living in the Stone Age.

Interested students should be encouraged to research the uses of steam power and the burning of coal. A model steam engine can spark the imagination of students.

During the second part of the pre-Industrial Revolution, technology and science became important. For example, steam was first put to use in 1712 by Thomas Newcomen. He made a steam engine to pump water out of coal mines. Products were made in homes rather than factories during this period. Several important scientific instruments such as the microscope were invented. This is when scientists started using the **scientific method** to find answers.

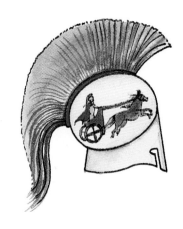

▶ *Industrial Revolution* (1750–1900)

Beginning about 1750 many inventions brought changes that affected all of society. Before this time products were made by craftspeople using their own tools in their work-

The scientific method is commonly introduced in science classes at this age. This would be a good time to correlate the scientific method with technology and set the stage for the steps of problem solving, which come later in this chapter.

The Scientific Method

The process of scientific problem solving is called the **scientific method**. The scientific method was developed by the Italian scientist Galileo Galilei (1564–1642) and the English philosopher Francis Bacon (1561–1626). This method is used as a general rule and is not strictly followed in its purest form. The five steps of the scientific method are:

1 Recognize the problem.

2 Form a **hypothesis** (an educated guess) about the correct answer.

3 Make a prediction.

4 Experiment to test your prediction.

5 Make a general rule that will apply, or organize the hypothesis, prediction, and results of your experiment.

Scientists and technologists face the same traps in trying to solve problems. Everyone has a set of ideas about how the world works. These ideas often block new ideas or restrict our thinking to one path. Scientists must keep an open mind about their investigations; this is important because their results have to be based on scientific facts. Scientists can't let their results be changed by what they think the results should be. Another important reason for keeping an open mind is that many scientific discoveries happen by accident!

All scientists have a few very important things in common. They have investigating minds, they are willing to perform experiments carefully, and they don't let themselves be influenced by anything but proven facts. Their experiments have to be done so that if someone else performs the exact same experiment, he or she will get the same results.

shops. With new inventions and machines, people set up factories that could produce things cheaper and faster.

Before reading this section, ask the students to name the age that we live in now.

▶ *Recent History* (1900–Today)

The early 1900s started a period of very rapid growth in technology. Recent history is divided into ages that describe the technology of each period just as prehistoric times are described by the materials people used. Some of the recent ages include the Air Age, the Atomic Age, the Jet Age, the Space Age, and the Information Age. The ages overlap and build on each other. For example, some of the principles of flight used by Orville and Wilbur Wright, the inventors of the first successful airplane in 1903, were important to scientists in the Jet Age and the Space Age.

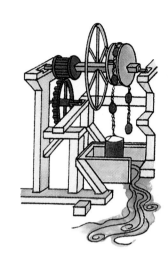

As technology has changed, so have the ways people live. In earlier times, people lived in an agricultural society where they needed tools to live off the land. Then, during the industrial period many machines were invented that changed the ways in which products were made. Many people moved away from farms and worked in factories. Today we are in an Information Age, where skills such as finding and using information are important.

Many people moved from farming to factory work in the Industrial Revolution. Children often worked 10 to 14 hours a day. (Courtesy of The Smithsonian Institution.)

(Courtesy of The Smithsonian Institution.)

(Courtesy of G. M. Hughes Electronics Corporation.)

Today we live in the Information Age. Computers are often used to sort and find information. (Courtesy of G. M. Hughes Electronics Corporation.)

General Safety Rules

Throughout this book, you will be doing many activities that require you to use safety equipment and conduct yourself safely. These general safety rules should be followed at all times:

1 Follow the teacher's instructions. Do not fool around in the technology lab area.

2 Pay close attention to what you are doing at all times. Do not let others distract you while you are using a machine.

3 Always wear eye protection. Special eye protection may be needed for some activities such as using a laser, welding, or using chemicals. Ask your teacher for help.

4 Never use any tool or machine without a demonstration by your teacher. Do not follow the advice of a friend. Ask the teacher for help.

5 Be careful not to wear loose clothing, jewelry, or other items that could get caught in a machine.

6 Always use the guards on each machine. Keep hands and fingers away from all moving parts.

7 Keep the work area clean.

8 Report all injuries to your teacher at once.

9 Know the safety rules related to a specific machine before you turn it on. Remember to ask for help if you are not sure.

10 Put warning signs on things that are hot and could cause burns. Warn others of the hazards of laser light.

11 Do not use electric tools near flammable liquids or gases. Store oily rags in a proper container. Know where the nearest fire extinguisher is and how to use it.

THINGS TO SEE AND DO:
Prehistoric Communication

Special Report
Beginnings & Growth

Introduction:
Many people think prehistoric people were not very smart, but they were able to use the technology of their day very well. In this activity, you will try to communicate as Stone Age people might have.

Design Brief:
You will need a partner for this activity. Use only natural materials that were available to prehistoric people. Communicate any one of the following messages using symbols only.

You might want to make up your own messages.

- ➤ **DANGER! FLOOD AREA**
- ➤ **GOOD FOOD**
- ➤ **GOOD HUNTING AHEAD**
- ➤ **KEEP OUT**

Materials and Equipment Needed:
Rock, wood, animal bones, plants, or other natural materials that were available to Stone Age people. You may *not* use modern-day materials or tools.

Procedure:

1 Work with your partner, and select a message to communicate.

2 Gather materials needed.

3 Communicate the message using symbols only.

Early people communicated with each other using drawings. What do you think these people are trying to communicate?

You can use technology in this activity by sharing the message with students in other schools. You might take pictures, shoot videos, and so on.

Students may enjoy other examples of exponential rate of change. For example, if a sheet of paper is .005 inches thick, how thick would the paper be if you folded it 50 times? You can have the students use calculators.

Evaluation:

Share your message with students in another class. Do they understand what the message says? How long would your message last if you left it outside? How would nature's elements affect it?

Technology touches almost every part of your life today. With technology comes change—change in the way people do things, change in the way machines work, and change in how we think. Changes in technology took place very slowly before the Industrial Revolution. Now, with so many new inventions and improvements on older inventions, change occurs faster than ever. Some people say our **knowledge base** (all the facts known to people today) is doubling every two to three years. When something changes, and keeps changing faster and faster, we call it an **exponential rate of change**.

For example, if someone gave you one penny on your birthday and told you that the amount would double every day for a year, you might be disappointed at first. Think of it this way. On the first day you would have only one cent. On the second day you would have double that amount: two cents. On the third day you would double the two cents. Now you would have four cents. That doesn't sound like much, does it? By the end of 30 days, however, if you continued to double the amount each day, you would have over $5 million—$5,368,709.12 to be exact. That is how your one-cent investment grew at an exponential rate. You

TECHNOFACT

Technofact 2

When you look at stars, you are really looking back in time hundreds, thousands, or even millions of years! The light that we see shining from stars takes time to travel to earth. Even though light travels very fast (186,282 miles per second), stars are very far away. You probably have seen the North Star, for example. The light you are seeing is really 680 years old! It is strange to think that the light you see when you look at the North Star left the star during medieval times.

EXPONENTIAL GROWTH

Technology is growing at an exponential rate. It is doubling every few years. To make it even faster, the time it takes to double is getting even shorter!

If you were given one penny on your birthday and twice as much everyday after, how much money would you have at the end of one year?

.01   .02   .04   .08   .16   .32   .64   1.28

Technology is growing at an exponential rate. It is doubling every few years. To make it even faster, the time it takes to double is getting even shorter.

can imagine how much knowledge there is now and how the knowledge base is growing. Because the knowledge base is so large and is growing so rapidly, it is important for you to learn where and how to find information.

THINGS TO SEE AND DO: *Making a 3-D Model of Exponential Growth*

Special Report
Exponential Growth

Introduction:
Exponential growth is easy to understand if you make a model of it. Graphs are often used to model mathematical information (numbers). In this activity, you will build a three-dimensional (3-D) model showing the exponential growth of the knowledge base.

Ask the students to brainstorm other possible materials that could be recycled in this activity. You might save time by providing students with the needed materials. In this way, this activity can be completed in less than one hour.

Design Brief:
Working in groups of three or four students, design and make a 3-D model that shows the exponential growth of the knowledge base. Use only recycled materials such as empty milk cartons, soda cans, or scrap paper.

Materials and Equipment Needed:
Recycled packaging or other materials, scissors, tape, glue, markers.

Procedure:
1. List possible materials you can use for the 3-D graph.
2. Make a sketch of how your graph will look.
3. Gather materials, and build your model.

Evaluation:
Share your group model with other groups. How many doublings could you use before the model became too big?

Make a display in your school hallway of the best 3-D graphs.

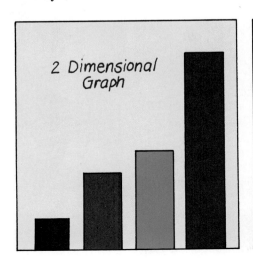

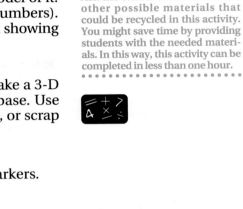

Three-dimensional graphs can make it easier to understand information. Computers can be used to make 3-D graphics colorful and interesting. (Courtesy of G. M. Hughes Electronics Corporation.)

Every student should be able to list the steps in problem solving. The steps in problem solving will be reinforced throughout this book.

The term "design brief" will be used in each activity in this book.

Solving Problems Step by Step

Have you ever had a problem that you wanted to solve but you didn't know where to start? Many people give up before they even try. If you have steps to follow in trying to solve a problem, getting an answer will be easier. A step-by-step plan for solving problems is called a problem-solving **strategy**. Once you learn a problem-solving strategy, you can use it to solve all kinds of problems. If you really learn how to use it, it will help you throughout life. Any problem in school, at home, or with your friends can be tackled using a problem-solving strategy.

Problems aren't always bad or complicated. For example, a problem can be as simple as deciding whether to eat a chocolate chip cookie or an oatmeal cookie. Most of the time, the problems you will be solving in technology are more complex (complicated). All problem-solving strategies have some common parts. Usually, the first step in solving a problem is being able to say, or state, exactly what the problem is in your own words. The definition of the problem is sometimes called the **problem statement**. A more detailed problem statement is called a **design brief**. Here are some examples of problem statements used in technology.

▶ Design a solar-powered vehicle.
▶ Build a bridge to span a river.
▶ Produce a television show.
▶ Program a robot to move objects.

If the problem statement is not very **specific** (exact) then you might not know when you have found a solution. For example, in the problem statement "Design a solar-powered vehicle" you might come up with a design that worked. However, its range might be too limited, it might be unsafe, or it might be too expensive to be practical. A design brief gives specific details of the problem. The design brief for the same solar-powered vehicle problem might include:

Using a problem-solving strategy is helping these students to find a solution by exploring many different ideas. (Courtesy of U. S. Department of Energy.)

Solar panels help to solve the problem of using fossil fuels like gasoline. Each solar cell changes sunlight into electricity. The energy produced by the sun can charge batteries to power electric vehicles. (Courtesy of U. S. Department of Energy.)

Getting across wide rivers has always been a problem to solve. (Courtesy of San Francisco Convention & Visitors Bureau.)

The problem of producing products quickly in a hazardous workplace is often solved by using a robot. (Courtesy of Motoman.)

Producing TV shows that inform, educate, and entertain people is a problem. (Courtesy of Turner Broadcasting System.)

TECHNOFACT

Technofact 3

Did you know that the highest wind speed ever recorded on earth was 231 miles per hour? That is faster than Indy race cars! The high-speed wind was recorded in 1934 atop Mount Washington in New Hampshire. This incredible wind was a gust during a storm. The extremely high speed resulted from storm winds being funneled over Mount Washington by weather and geography.

Design Brief: Solar-powered Vehicle

Design a solar-powered vehicle that will safely carry a minimum of two passengers at least 100 miles without recharging. The design budget for the project is $10,000. The vehicle must meet the federal safety standards for use on public highways.

As you can see, a design brief gives you a lot more information than a problem statement does. The specific requirements for a problem are called **parameters**. Parameters might limit the amount of money you can spend to solve the problem, the amount of time, or the kinds of materials you can use.

Once the problem is identified, you need to **explore** ideas and gather information related to your problem. Ideas can come from many sources, as you can see in the chart. You need to use as many different resources as possible. You also need to be sure your information is current. You will find a great deal of information on most topics. You will need to **evaluate** the information and decide whether or not it is useful in solving your problem. A simple question to ask yourself is, "Is this particular information 'need to know' or is it just 'nice to know'?" Often other people will be working with you on the problem. Working in groups and sharing ideas are important parts of problem solving.

Now that you have all the ideas, you have to do something with them. You **select** what you think is the best idea to try first. Many times your first idea will not work, and you must choose a different idea that might work better.

The fun part is **testing** your idea or solution. During the test you should review the parameters of the problem so that you can stay on track. Testing usually requires taking measurements or accurate notes about what's happening.

The most important part of the entire problem-solving strategy is **evaluating** what happened during the testing. Sometimes you want so much for your idea to work that you do not look at the results carefully.

Saving our world is a problem we all need to work on. Solar-powered vehicles can help to save our environment. They produce far less pollution and are quieter than the cars we have today. (Courtesy of G. M. Hughes Electronics Corporation.)

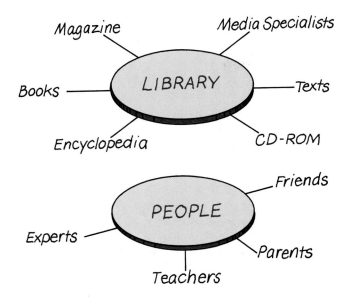

You may find ideas in many places. (Photo courtesy of Jeppeson Sanderson, Inc.)

Remember that there may be many possible solutions to a problem. Just because your idea did not work, it doesn't mean you failed—it just means you learned what doesn't work. That is a big step toward solving the problem!

The chances of your first idea being totally perfect are slim. You can use the successful parts of your first idea along with some changes to create a second idea. You will need to test this new idea and evaluate it too. You might go through this **cycle** (a loop) many times before you reach your final solution. Let's review the basic steps of any problem-solving strategy:

PROBLEM SOLVING LOOP

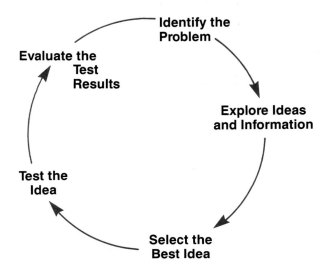

The steps in this problem-solving loop can be used to help you solve problems by learning from past experiences and mistakes.

- ▶ Identify the problem.
- ▶ Explore ideas and information.
- ▶ Select the best idea.
- ▶ Test the idea.
- ▶ Evaluate the test results.

Using an **acronym** to represent the steps might make it easier for you to remember a problem-solving strategy. An acronym is a word formed from the initial letters of the words in a phrase. The acronym SEARCH uses the main problem-solving steps and adds a few extras.

S tate the problem in your own words

E xplore all ideas and information possible.

A pply the best idea and test it. If it doesn't work . . .

R etry the idea or

C hange your idea and retry it

H ook what you have learned to real-life problems

Acronyms can help you remember steps in a process. The acronym SEARCH, for example, was developed by a group of students to help them remember the problem-solving steps.

Techno Teasers
The One-Armed Drummer?

Techno Teasers
Answer Segment

THINGS TO SEE AND DO:
Problem-Solving Acronym

Introduction:
In this activity, you will design your own acronym to help you remember how to solve problems step by step.

Design Brief:
Think of a four- or five-step procedure for solving a problem. You should use a key word for each step. The steps should fit the main parts of any problem-solving strategy.

Materials:
- ▶ Thesaurus

Advanced classes might use an electronic thesaurus on the computer.

Procedure:
❶ List a key word for each problem-solving step.

❷ Using a thesaurus, find at least four **synonyms** (words that have the same meaning) for each word.

❸ Think of an acronym that uses any of the words you found in the problem-solving order.

❹ Make a diagram illustrating your acronym steps.

❺ Share your acronym and diagrams with the class.

Problem-solving Reminder:

Did you use the problem-solving strategy to make your own problem-solving acronym?

Research and Experimentation

In the problem-solving strategy, you spent a lot of time exploring ideas. Companies are always looking for new ideas, too. That's because every product starts as an idea. Someone may suggest an idea or invent something to solve a particular problem. Many companies have special **research and development (R&D)** departments. The people who work in R&D look for new ideas, and then they develop them into products. Of course, not every idea works, as you discovered in the problem-solving unit. Only one out of every few ideas actually works.

People working in research and development departments are always looking for new ideas. This researcher is taking soil samples to help find ways to improve our environment. (Courtesy of British Petroleum.)

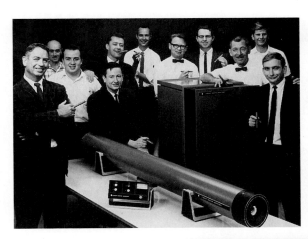

A research and development team worked together to make this first carbon dioxide laser in 1966. Since that time, many new uses for lasers have been found. (Courtesy of Coherent, Inc.)

A microchip is a tiny piece of silicon that has thousands of electronic parts in microscopic size. The microchip makes everyday items like digital watches, calculators, and computers smaller and less expensive. (Courtesy of G. M. Hughes Electronics Corporation.)

A small research library in the classroom will provide ready access to needed resources. Have students send for information on technology topics and start a file of the materials received.

Research helps us gather information. There are two kinds of research—**basic research** and **applied research**. **Basic research** is gathering information that helps us understand things about the world around us. Sometimes the information cannot be used right away. For example, people learned how to make a laser in the 1960s, but they did not know what to do with it. Today we use lasers for many things, from supermarket pricing to surgery.

Applied research is the kind of research done to solve a particular problem. The information can be used right away. For example, people wanted to make computers smaller and more portable. To do that they needed to put more **components** (parts) in a smaller space. The **microchip** (a tiny silicon chip containing thousands of electronic circuits) was invented to solve this problem.

Use a hammer and chisel to break open the dual inline package that surrounds a microchip. Use a microscope or magnifying glass to view the intricate circuitry of the chip.

Basic and applied research use the same two methods to get information: retrieving and experimenting. **Retrieving** is gathering already-known information from sources such as books, people, and computers. The information is part of the knowledge base. Using this kind of information saves time because you do not have to start from scratch every time.

THINGS TO SEE AND DO:
Technology Research

Introduction:

Relate the activities to possible future careers. For example, what kind of occupation would involve this kind of activity?

Your job requires you to research information on various technology topics. You will need to get information on three topics from companies all over the country. To find the companies' names and phone numbers, you will use resources in the library or from your teacher.

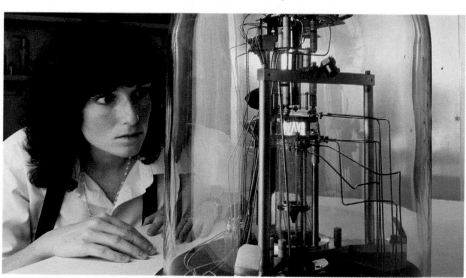

You might choose to research lasers or fiber optics. New uses for lasers are being found every day. Try to find the most recent information you can. (Courtesy of G. M. Hughes Electronics Corporation.)

Design Brief:

Research three technology topics of your choice. Find information on the topics and on companies that make products or do research related to your topics. Request information on your topics either by phoning or writing to the companies.

Procedure:

1 Choose any three of the following technology topics:

- Fiber optics
- Composite materials
- Conveyor systems
- Pneumatics
- Soldering
- Super-conductors
- Robotics
- Fasteners
- Hydraulics
- Plastics
- Lubricants
- Welding
- Computer controls
- Any technology-related topic you think of

Encourage students to ask their parents, friends, and relatives where they can get information on their topic.

2 Research your topics using the resources in your library or those available from your teacher. Record information about the resources you used in the chart below.

Topic 1: _____ Company Name: _____

Phone Number: (800) _____ Resource _____, Page _____

Topic 2: _____ Company Name: _____

Phone Number: (800) _____ Resource _____, Page _____

Topic 3: _____ Company Name: _____

Phone Number: (800) _____ Resource _____, Page _____

3 Get your teacher's permission to call each company (use "800" phone numbers if possible), or write a letter to request information on their products. Ask them to send their catalog or product information to you at the school address. Don't forget to say "please" and "thank you;" good manners are appreciated in business.

Contact local businesses or public libraries. Ask them about donating recent editions of the *Thomas Register* to you. Many companies update their set every year and discard the old version.

Evaluation:

Which of your three topics were easy to research? How many companies sent you information on their products? Share the information you received with the members of your class. If you owned a business, what would you do to help students learn more about technology?

Research is also done by **experimenting** to find out how things work. This knowledge can then be used to change or improve technology that already exists. You are not forced to "re-invent the wheel" each time you want to begin a project. If you wanted to design a bicycle, for example, you would not have to invent the wheel. Somebody already did that for you. But you might just want to change the *shape* of the wheel.

THINGS TO SEE AND DO:
Lumpy Wheel

Introduction:

Have you ever heard someone say "you don't have to re-invent the wheel?" People often use this saying to mean you do not have to start from the beginning to solve a problem. Most inventions are just other inventions used in a different way. In this activity you will see that even though you don't have to reinvent the wheel, that doesn't mean you shouldn't try.

Design Brief:

Using drawing tools, draw and build a lumpy wheel that will work better than you might think.

1.

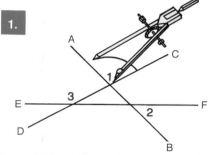

Draw 3 lines that cross. The angles are not important. Start with the point of your compass at point 1. Draw an arc (part of a circle) as shown.

2.

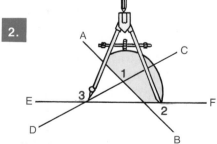

Now put the compass point at number 3. Adjust the compass to continue the arc as shown.

3.

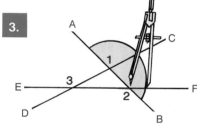

Put the compass at point number 2 to continue your ''wheel.''

4.

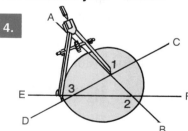

Using points 1, 3, and 2 as corners, finish the drawing as illustrated. If you were careful, you should end-up with a smooth connection to your starting point.

Do you think this shape would make a good wheel? You might be surprised.

Use the same shape that results in step #4 of instructions.

The lumpy wheel

Materials:
- Cardboard
- Paper
- Pencil
- Wood dowel rod
- Wood glue

Equipment:
- Drawing compass
- Ruler
- Scissors

Procedure:

1 Follow the directions shown in the drawing to make your lumpy wheel design.

2 Cut out the design, and trace it eight times onto thick cardboard.

3 Cut out the cardboard "wheels," and glue them together in pairs. You will end up with four thick wheels.

4 Sharpen the ends of a 4-inch-long by ¼-inch-thick wood dowel rod in a pencil sharpener.

5 Carefully poke the pointed ends of the dowel axles into the middle of your lumpy wheels. Your finished wheels should look like the lumpy wheels in the illustration.

Evaluation:

It is very simple to test your wheels. First, try to predict what would happen if you rolled a flat board over your wheels. The wheels are not round, so you might think that the board would go up and down as the lumps come around. Try it and see what happens.

Experimenting is like the testing part of your problem-solving strategy. If the idea doesn't solve the problem, you try other ideas. Researchers have to also. They might try many ideas until the problem is solved. You will learn more about research and development in Chapter 5 titled *Making Things*.

Working in Groups

Problems are often easier to solve with the help of other people. Knowing how to work in a group is an important skill that will help you in school and later in a job. The old saying "two heads are better than one" is really true. When you are working to solve a problem, you need as many different ideas as possible. Having people with different experiences and backgrounds in your group gives you more information to explore. As you move through the problem-solving strategy, you can put each person's talents to good use. Some of us are better at tossing out new ideas. Others are better at putting ideas into action. You will have a chance to do both as part of a group.

Many people judge an idea as soon as they hear it. Many good ideas are lost because someone puts the idea down before it has a chance to be discussed. When you **brainstorm** ideas, you list as many ideas as possible without people saying anything good or bad about any of them.

TECHNOFACT

Technofact 4
Did you ever wonder how much snow could fall at one time? The greatest snowfall on record happened in 1921. The town of Silver Lake, Colorado, had an incredible snow storm that left 6 feet, 4 inches of snow on the ground in only 24 hours!

Advanced classes could be challenged to do this activity on the computer using CAD software.

Many activities in this book encourage cooperative learning strategies. Every effort should be made to ensure that all students within a group are participating.

A space science class working together on a rocketry project shares ideas to find a solution. Brainstorming ideas brings out a lot of ideas for people to use. (Courtesy of NASA.)

Everyone in the group has a chance to contribute to the idea bank. From that idea bank your group selects the best idea.

The abilities to work both independently and as part of a group are important to you in solving problems.

Special Report
Odyssey of the Mind

THINGS TO SEE AND DO: *Working in Groups to Solve Problems*

Introduction:

In this activity, you will work in groups using a step-by-step procedure to solve a problem. Note: you may choose any of the problem statements at the end of the activity.

Design Brief:

Design and build a model of a skyscraper structure. Your goal is to build the tallest structure using only the materials listed below. The finished structure must support the weight of this book without falling down. You will have only one class period to complete your group structure.

Materials:

You will be limited to the following materials:

- Straws
- Paper clips

Equipment:

- Pliers
- Scissors

Procedure:

❶ Brainstorm at least four or five designs with your group.

❷ Sketch all the designs for the structure. Evaluate the designs, and as a group select one to construct.

❸ Use scissors to cut the straws to the lengths needed for your structure.

❹ Bend the paper clips as needed to attach the straws according to your design.

❺ Measure the height of your structure from its base to the top edge. Write the height of your test structure on your sketch.

❻ Test your structure by balancing this book on top.

❼ Evaluate your structure with your group. What were the **constraints** (things that held you back) of this problem? For example, if your group had had more time or different materials, would your structure have been better?

Problem-solving Strategy Reminder:

Did your group try different ideas if the first one didn't work?

Being Creative

Sometimes people make excuses for not doing things by saying, "I'm just not creative." But everyone can be creative. You just have to learn how. First, to be a creative thinker you need to use facts, feelings, experiences, knowledge, and some basic concepts.

You're much more likely to find something original if you try different things in different ways. Many good ideas were discovered when someone looked in unusual places, found some ideas, and then applied the ideas to their own field. For example, military designers in World War I borrowed from the style used by Pablo Picasso, a famous artist, to make more effective camouflage patterns for guns and tanks. In another case, an "unbreakable code" used by World War II military leaders was based on the Navajo language.

You might have to break your usual rules for thinking or create some new ones when trying to make an idea work. You need to ask "what-if" questions. By doing this, you allow yourself to question the possible, the impractical, and even the impossible. For example, what if there were a pill that would make anyone smarter? Who should take it? How many would you have to take to be really smart? Would you want everyone to have the same degree of "smartness"?

To improve your creativity even more, take a closer look at things you usually take for granted. For example, who says a wheel has to be round? You found out that it doesn't in the lumpy wheel activity. Part of being creative is changing patterns, looking at things differently, and

Alternate materials for this activity might include popsicle sticks and hot melted glue, or toothpicks, pipe cleaners, and so on.

TECHNOFACT

Technofact 5

Only a few hundred of the thousands of human-made objects in space today are considered working spacecraft. What's left is "space junk." On the moon you will still find two golf balls, an astronaut pin, a stereo camera, a television camera on a tripod, various geology tools, and even an armrest. Close to twenty tons of junk were left on the moon after the Apollo landings!

There's no such thing as "can't" in these activities. All ideas should be accepted. Students should not be allowed to give up without trying.

Make a list on the blackboard of ways students could increase their creativity.

Creativity was a very important ingredient in the making of the space shuttle. (Courtesy of NASA.)

experimenting. Remember, if your idea doesn't work, it doesn't mean you've failed. It just means you learned what doesn't work, and then you can try another idea.

THINGS TO SEE AND DO:
Design a Flying Hospital

Introduction:
Advances in technology can have a great positive impact on the world if they are made available to all people. Doctors use technology to save the lives of sick or injured people. In this activity, you will use your creative ability to design a flying hospital. If medical equipment and people could be put in an airplane, people all over the world might be helped.

Design Brief:
You are working as a designer for an aerospace company. You have been given the job of designing a passenger jet as a mobile hospital for emergency surgery. The Boeing 747 should include at least the following specific areas:

▶ Seating for four doctors and four nurses
▶ Operating room
▶ Patient examination and x-ray room
▶ Preparation room
▶ Galley (meal preparation area)

You may remind students that a Boeing 747 has two levels or allow them to discover that fact independently. Two levels would require two separate floor plans.

- Recovery area for patients
- Restroom
- Cockpit
- Storage for equipment and supplies
- Laboratory

Materials:

In addition to imagination and creativity, you will need:

- Paper
- Pencil

Equipment:

- (Optional) Computer with graphics software.

Procedure:

1. Sketch an outline of an airplane onto a blank sheet of paper.

2. Sketch your idea for the flying hospital on the tracing. (Using graphics software, draw the areas listed in the design brief. Move the areas around on the computer until you are happy with your design.)

The areas on your floorplan should be numbered and keyed to explain your ideas.

Thanks to creative thinking, liquid crystal technology has many familiar uses, from digital watches and pocket calculators to heat-sensing products. The color of the liquid crystals changes with temperature. (Courtesy of NASA.)

Liquid crystal battery testers change color to show if the battery is strong or weak. (Courtesy of NASA.)

How many emergency exits are available on each student design? How much space is used as passage or hallway space?

TECHNOFACT

Technofact 6

About a century ago skyscraper construction was made possible through the use of steel framing. Then the invention of the elevator made the buildings usable! The world's tallest self-supporting structure of any kind, anywhere, is a TV and radio transmission tower in Toronto, Ontario, Canada. The CN Tower is more than 1,815 feet tall.

Evaluation:

1 Would sketching freehand or drawing on a computer make it easier for you to complete this assignment? Why?

2 List and describe three other ways that medical technology could be provided to remote places on earth.

3 For extra credit, make a similar design for a space station hospital.

Summary

Technology is rapidly changing the way your world works. Most of the products of technology such as lasers, robots, and satellites were invented to meet the needs of people. The technology of the Stone Age people met their needs just as computers fit today's high-tech world. Technology is a combination of knowledge, ideas, and resources put to work for us. It is easier to solve problems that involve technology if you have a problem-solving strategy and if you work with others to find the best solution. Creative thinking is important when you are developing new technologies and evaluating their uses.

Challengers:

❶ Design and test an earthquake alarm. The materials may include a simple, battery-operated electric buzzer. Come up with and test ideas for other methods that would wake up a sleeping person in an earthquake.

❷ Design and build a model of a permanent space station that would be in lunar orbit (in orbit around the moon). Your design should provide for eight people to live and do experiments for up to six months without resupply. Make your model out of recycled materials (cans, boxes) painted to look like your design. Suspend the finished model from the ceiling of your classroom using fishing line.

❸ Design and test a device that will protect a raw egg from breaking when it is dropped from the second story of a building.

❹ Design and make a model of a school locker that would be useful to a wheelchair-bound student. Your design should provide storage for coats, books, pens, pencils, notebooks, and so on.

❺ Practice opening your mind by looking at ordinary things in a new way. Turn a potato or an apple into something else. How many different uses can you find for each one?

❻ You ordered 10,000 Popsicle sticks for your factory because you are sure someone creative can find something useful to do with them. What other uses can you find for the sticks besides making Popsicles?

❼ Asking "what-if" questions helps open your mind. Think about this one. "What if people had 12 fingers instead of ten?" Write down your ideas.

❽ Make a collage of cave paintings. Underneath each painting, write what you think its message is. Research cave paintings and pictographs in your library.

❾ Find a new way to show exponential growth. Make a graph or a model to show your results.

❿ Working in small groups, come up with a list of ten products you think will be useful in the future.

Techno Talk
Technology Flops or Catalysts

Techno Talk
Answer Segment

Techno Talk
Problem Solving in Groups

Techno Talk
Answer Segment

See Teacher's Resource Guide.

Chapter 2

How Does Technology Affect Me?

Things to Explore

When you finish this chapter, you will know that:

▶ Technology has made our lives easier and more comfortable.

▶ Technology is changing the way we play, learn, and work.

▶ The impact of technology on our world depends on how it is used.

▶ Technology is growing at a faster and faster rate.

▶ Science and math are important in technology problem solving.

▶ Changing technology requires people to be adaptable.

Chapter Opener
How Technology Affects Me

TechnoTerms

adaptable	molecule
audio	nucleus
clip art	on-the-job training
consensus	repel
electron	retail
flowchart	sound effects (SFX)
friction	theory
icon	video
impact	video effects (EFX)

Careers in Technology

Technology is a part of our everyday lives. Most careers today and in the future will require you to understand technology. People grow accustomed to technology and start to depend on it. Most of us watch television every day. Do you ever stop to think about how the TV show is made or how it gets to your home? Thousands of people work in the communications industry to bring you up-to-date information or entertainment.

How Has Technology Changed Our World?

Encourage students to bring in examples of music that they associate with their parents' or grandparents' era. Discuss how musical instruments and production techniques have changed.

In Chapter 1, you learned how fast technology is growing. This rapid growth of technology has caused our society to change rapidly, too. Ask your parents or grandparents what games or music they liked when they were your age. They probably played board games and listened to radio, while you might prefer video games, tapes, and CDs!

Changes in technology make a lot of different things possible. Before low-cost calculators were available, people used slide rules and adding machines to do math problems. Now anyone can use calculators without special training, and anyone can afford to buy one. Before 1961, no one had ever seen the whole Earth. Now, thanks to new space technology, you know what the Earth as well as Jupiter and the other outer planets looks like.

Dig out your old slide rule and show students how math used to be done!

No one had ever seen the entire Earth at one time until 1961. Today we can view the Earth and watch environmental changes as they happen using satellite technology. (Courtesy of NASA.)

TECHNOFACT

Technofact 7

A search-and-rescue-satellite named *NOAA/-Sarsat* orbits the earth from the pole to pole about every 1½ hours. *Sarsat* picks up distress signals from planes and ships. It then relays (sends) them to ground stations.

Technology has also made it possible for us to view planets close up. This view of Saturn's rings was sent back to Earth from the unmanned satellite *Voyager 1*. *Voyager 1* was 11 million miles from Earth when it took this picture. (Courtesy of NASA.)

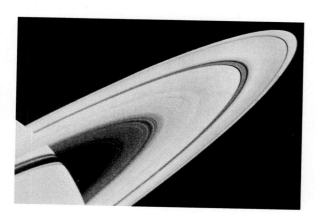

In your classroom, you might be sitting in chairs and desks designed by computer to fit your size. You might even use computers or electric word-processing typewriters to help you with your schoolwork. Teachers now teach with video lessons instead of movies in the classroom. Even the equipment you use in gym classes or football has been improved by technology.

Technology has changed the way we work and the way we play. People see some of these changes as being good and some as being bad. Some people just don't like any change at all. But one thing is for sure; change keeps happening. We can't always predict accurately how a change will affect us or our world. We can only be sure that everything changes eventually. Your job is to constantly evaluate how technology can be used with the most benefits to people and the environment.

Some questions you can ask about a technology's impact on you and your environment are:

- ▶ Does the new technology require more or less energy or natural resources than an existing technology?
- ▶ Will it damage the environment, for example, by not being biodegradable?
- ▶ Is it easier to use, or does it save time?
- ▶ Does it require special training to use?
- ▶ Does it put people out of work?
- ▶ Is there a real need for this technology?
- ▶ Is the technology safe?
- ▶ Is this an appropriate use of technology?

Use these questions to evaluate the following technology topics: robotics, fiber optics, genetic engineering, calculators, and computers.

For example, what if someone invented an electric banana peeler? How would you evaluate this product using the questions above? Evaluating the use of an electric banana peeler at home might be different from evaluating its use at an ice cream parlor that specializes in banana splits. Sometimes people invent a technology they think is needed for a certain area but forget to evaluate its **impact**, or its effects, on other areas such as the environment or you.

Many technologies give us waste products that are not easy to get rid of. These products cause a big storage problem. Soon the second highest point in the eastern United States will be a huge pile of garbage! (Courtesy of Waste Management, inc.)

The same questions about the impact of a product can be asked about the steps in making a product. The steps can become very complicated. There is an easy way to keep track of the steps. **Flowcharts** are used to illustrate the steps using symbols. Each symbol represents a special step in the process of producing a product.

Flowcharts

Sometimes the steps in a process can get very complicated. Symbols can be used to show the steps and make it easier to understand the process. For example, graphic symbols represent various steps in the process.

▶ Transportation: Moving an object from one place to another

▶ Operation: Usually, using machines to change the shape of a product

▶ Delay: Waiting for the next step

▶ Inspection: Checking the quality of a product

▶ Storage: Putting the product in stock

A flowchart can show where a process might be stopped because an important step couldn't be completed. Flowcharts help in the planning process to try to meet deadlines.

This sample flowchart shows the steps in the process of making a peanut butter sandwich. Can you find any steps that are missing?

Get Peanut Butter

Get Plate

Open Jar

Get Knife

Get Bread

Apply Peanut Butter to Bread

Inspect Coverage

Eat Your Peanut Butter Sandwich!

A flowchart uses graphic symbols to show a series of steps in a process.

THINGS TO SEE AND DO:
Life Cycle of a Product

Special Report
Lifecycle of a Product

Introduction:

Most of us buy, use, and throw away products every day. We seldom think about where they came from or how they were made. In this activity, you will make a picture flowchart, or major steps in a process, of how a product is made.

Design Brief:

Pick a product from the list below (or better yet, think of your own), and make a flowchart of how it was made and how it is used and disposed of.

- Book
- Hamburger
- Pen
- Bracelet
- Rubber band
- Paper clip

- Coins/money
- Videotape
- Running shoe
- Candy bar
- Insect repellent
- Think of your own

Contact the school media center or public library for old magazines that could be used in this activity. See Teacher's Resource Guide for further ideas.

Materials:

- Posterboard
- Glue
- Markers
- Old magazines

Equipment:

- Optional: Computer with graphics software
- Scissors
- Computer with clip art images

Procedure:

1 Work in groups of three or four. Choose a product from the list, or think of your own.

2 Research how the product was made. Use the resources in your library, or find a resource person to talk to or call.

3 Sketch a flowchart of the life of your product. Remember, a flowchart shows the major steps in the process from its beginning to its end.

4 Pick one of the following ways to illustrate your flowchart:

- Glue pictures cut from old magazines to posterboard. Finish the flowchart using markers.
- Draw your own illustrations on the posterboard.
- Use computer graphics software and **clip art** (already-drawn pictures saved on a disk) to make your flowchart.

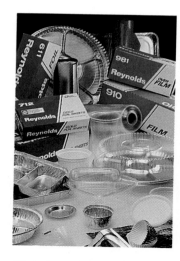

All products and even their packages have to be made from raw materials. (Courtesy of Reynolds Metals Co.)

Aluminum is easy to form into useful products. Here aluminum is being rolled into a thin sheet we know as aluminum foil. (Courtesy of Reynolds Metals Co.)

Aluminum cans are made quickly thanks to automation. (Courtesy of Reynolds Metals Co.)

Aluminum is a material that is easy to recycle; you save energy and the environment at the same time. (Courtesy of Reynolds Metals Co.)

Evaluation:

Share your flowchart with the class. In your group presentation, discuss these topics:

▶ Would your parents have been able to buy the product when they were your age? Would your grandparents?

▶ What is the impact of the product on the environment?

▶ What suggestions to change the product would your group give the maker of the product?

Technology and Other Subjects

Technology touches everything. We depend on it to make our lives more comfortable. In your everyday life, you often take technology for granted. For example, common items such as your toothbrush and your shoes are very different from those of the past because of developments in plastics technology. Toothbrushes were once made of hog bristles instead of plastic. Today's running shoes, made of plastics, are very different in weight from the kangaroo leather running shoes of earlier days.

Technology is not only part of your world but it is part of school, too. Think of the technologies in your classroom today. Your desks and chairs were designed to fit you instead of a third grader. Many of you have access to VCRs, televisions, laserdisc players, and computers. But the most important way technology reaches you is through your school subjects. Students sometimes lose sight of the fact that school subjects are supposed to prepare them to be good members of our society. Our high-tech world requires you to understand what technology is and how it works. Let's see how technology plays a part in subjects such as science, math, social studies, and language arts.

Most people think that technology is related only to science and math. While this connection is easy to see, technology is just as much a part of social studies and other subjects. You will work with technology in connection with other subjects in the following activities.

Technology and Science: Technology and science are closely tied together, but they are different. Science usually gives you the **theories** (ideas about how nature works), while technology lets you use your knowledge and resources to solve different problems.

Life science and technology can be combined to produce food in a fish farm. (Courtesy of British Petroleum.)

Make a hallway display of the finished flowcharts.

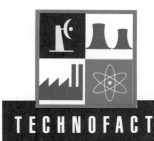

TECHNOFACT

Technofact 8

Satellites provide a lot of information for us. Cartographers (mapmakers) use satellite information to map remote areas they can't get to easily. Geologists like those using GIS can find deposits of metals and oil. *Landsat V* is a satellite that orbits the earth from pole to pole every 100 minutes. The instruments aboard *Landsat V* send information to computers in ground stations that build images of the earth's surface for us.

Have the students describe how different their daily routine would have been 100 years ago.

Discuss each student's classes in school and how they prepare them for a successful future.

SCIENCE THINGS TO SEE AND DO: Can You Touch an Atom?

Introduction:

Most people have the wrong idea of what atoms are. You know from your science class that all materials are made up of atoms, but did you know that atoms are mostly empty space? In this activity you will sketch illustrations that show the size relationships of atoms.

Design Brief:

Everything in our world is made up of atoms and combinations of atoms called **molecules**. In this activity, you will visualize the real size of atoms and the parts of atoms.

You know from studying science that **electrons** (one of the three basic parts of an atom) spin around the **nucleus**, or the center of, an atom. You also know that electrons have a negative electrical charge. The negative charges of electrons spinning around atoms keep the atoms from touching each other. Like charges **repel**, or push away from each other, so even though it feels like the atoms are touching, they really aren't! They are pushing away just like when you try to hold the north poles of two magnets together.

Procedure:

❶ First, let's think about how small an atom really is. Divide a piece of paper into three equal spaces. Number the spaces 1, 2, and 3. In each space you will make a sketch that shows the size of atoms to scale.

❷ In space 1, make a sketch that illustrates the following fact: If a baseball were enlarged to the size of the earth, its atoms would be the size of marbles.

❸ In space 2, make a sketch that illustrates the following fact: If a single atom were enlarged to the size of a fourteen-story building (140 feet tall), the nucleus would be the size of a grain of salt.

❹ In space 3, make a sketch that shows the atoms of your finger as it "touches" the top of your desk.

Evaluation:

❶ From what you have learned, can you really touch an atom? Explain your answer.

❷ What is in the space between the nucleus and the electrons of an atom?

Technology and Social Studies: People use history as a way of charting or planning the present and the future. Technology definitely has changed with the times. Some technologies no longer exist because there isn't a use for them today. Other technologies have changed to bet

Draw the traditional model of an atom, showing the nucleus, on the blackboard. Discuss the fact that this model is not accurate. Real atoms are mostly empty space and not nearly as neatly organized as the model shows.

Explain to the students that the atoms are not really touching. They are really repelling each other.

(Courtesy of G. M. Hughes Electronics Corporation.)

New technology often comes from improvements in old technology. The study of history can provide clues to help today's technology.(Courtesy of The Bettman Archive.)

ter meet our needs. At each step in time, different technologies were important for what they could do to help us.

SOCIAL STUDIES THINGS TO SEE AND DO: *Technology in History*

Introduction:

Advancements in technology have given us the quality of life we enjoy today. It is easy to see how the speed of technology has increased when you chart it on a type of graph called a *timeline*. In this activity, you will organize and chart some of the events in history that have led to the high level of technology we have in our world today.

Design Brief:

Technology is sometimes divided into job-related groups called *career clusters*. Four common career clusters are: communication, construction, manufacturing, and transportation. In this activity, you will sort events in history into one of the four clusters and make a timeline.

Materials:
▶ Adding machine paper

▶ Markers or pens

Equipment:
▶ Meter stick

Techno Teasers
Technology Growth by Cluster

Techno Teasers
Answer Segment

Contact the social studies department in your school and ask the teachers to discuss the technology available during different time periods.

Note that technology encompasses many more topics than four career clusters.

Procedure:

❶ Work in groups of four. Measure and cut 1m of adding machine paper.

❷ Use a meter stick to draw one line on the paper for each of the four clusters. Label each line.

❸ You will use a scale of 10 cm = 1000 years. Mark your timeline starting at 8000 B.C.

❹ Chart each of the following events in history on the appropriate line. Use circled numbers to represent each event.

- 3500 B.C. Writing first used by Sumerians
- 3000 B.C. Egyptians made first book
- 1500 B.C. Pulleys and simple machines
- A.D. 1045 Movable type used in printing
- A.D. 1450 Printing press
- A.D. 1725 Steam engine developed
- A.D. 1835 Morse code/telegraph
- A.D. 1876 Telephone invented
- A.D. 1892 Reinforced concrete
- A.D. 1906 Radio developed
- A.D. 1926 Television invented
- A.D. 1926 Liquid-fueled rocket
- A.D. 1940 FM radio
- A.D. 1946 ENIAC computer
- A.D. 1950 Semiconductors invented
- A.D. 1957 Sputnik—First artificial satellite
- A.D. 1960 First laser
- A.D. 1961 First man in space
- A.D. 1966 First soft landing on the moon by Luna 9.
- A.D. 1969 Neil Armstrong becomes first man on the moon.
- A.D. 1977 The Apple II starts the personal computer industry.
- A.D. 1977 Fiber-optic cable first used in commercial communication networking by AT&T.
- A.D. 1977 MRI (Magnetic Resonance Imaging) first used by doctors.

As an alternative to working in small groups, the entire class could work on one large timeline made on a roll of butcher paper or newsprint.

Challenge students to research and add technology events of their own.

- A.D. 1978 The 5¼ -inch disk becomes the standard format for storage of computer data.

- A.D. 1981 First flight of a reusable spacecraft: U. S. Space Shuttle *Columbia*.

- A.D. 1982 Sale of synthetic insulin, the first drug manufactured by recombinant DNA.

- A.D. 1985 British Antarctic survey team discovers a hole in the ozone layer.

- A.D. 1986 Karl Muller and Beorge Bednorz discover a ceramic material that is able to superconduct at 30 degrees Kelvin for a new high temperature.

- A.D. 1988 The U. S. Patent Office approves a patent for a genetically altered mouse.

- A.D. 1988 A voice-operated typewriter that recognizes dictated words.

- A.D. 1988 The world's first public MAGLEV system goes into operation in West Berlin.

- A.D. 1990 A new line of biodegradable plastics is developed by ICI Americans Inc.

- A.D. 1991 First permanent seawater desalination plant starts to serve residents of Catalina Island in California.

Evaluation:

How does your timeline show the rapid growth of technology? Explain.

Technology and Language Arts: Being able to communicate with others is a very important skill that you work with in all your subjects. Usually you think of it only with language arts or reading, but in all your courses you need to be able to let your teachers and others know what your ideas are. Technology lets you use many different ways to communicate with others.

LANGUAGE ARTS THINGS TO SEE AND DO: *Writing Scripts*

Introduction:

If you are like many people, you see or hear dozens of commercial messages each day on television or radio. Most people don't think twice about how commercials are made. In this activity, you will use language arts skills and technology to produce your own commercial. You will work in groups to write and produce a **video** (TV, something seen) or **audio** (radio, something heard) commercial for an imaginary product.

An alternate activity could include a game show, news broadcast, or documentary video.

Advancements in technology would come to a grinding halt without the ability to communicate in words and pictures. Writing and reading skills are an important part of technology. (Courtesy of International Business Machines Corporation.)

....
Commercial sound effects are available on compact disks or audio tapes.
....

Materials:
- Audio or video blank tapes
- Props

Equipment:
- Video camera
- VCR
- Television
- Audio tape recorder
- Sound effects equipment
- Computer (optional)

Procedure:

1 Work in groups of four or five students. Elect someone in your group to be the director. The director will organize the production of your commercial.

2 Brainstorm an imaginary product. Make a sample of your product to be used as a prop if you are making a video commercial. You might make a sample label that could be glued to a box, for example.

3 Write the script for your commercial. You may include everyone in your group, but you will need one person to operate the camera or the audio recorder. Your group might need to use **sound effects (SFX)** or **video effects (EFX)**. All actions, sounds, or dialogue (talking) must be a part of your script. Use the following format:

SOUND	VIDEO	DIALOGUE	
LOUD ROCK MUSIC	FADE-IN: TWO GIRLS TALKING	MARY:	ARE YOU GOING TO THE DANCE SATURDAY NIGHT?
		KEISHA:	NO, HECTOR NEVER ASKED ME.
PHONE RINGS	ZOOM-IN: KEISHA ON PHONE	KEISHA:	OH, HI HECTOR, WE WERE JUST TALKING ABOUT YOU.

A script should explain what you hear and see. This format will help you organize your production.

....
Play the commercial for other classes in the school.
....

4 Gather any other props you need, and plan to bring any special clothes or costumes for your rehearsal and taping. The director of your group should schedule the use of the video or audio equipment with the teacher.

5 Rehearse and revise your production so that it lasts exactly 60 seconds. Record the final version.

6 Play your finished commercial for the class.

Evaluation:

Evaluate the production for the following: sound, video, theme, effectiveness in selling a product. If possible, visit a local TV or radio station. Ask questions about how commercials are made.

Technology and Math: Math and technology work well together. Advances in technology have produced hand calculators and computers.

Techno Teasers
Video Production

They are useful because they can make many math computations faster and more accurately than a person usually can. You still have to know what math operations—addition, subtraction, multiplication, and division—to use in solving a problem. You also must know how to enter the information correctly into the calculator or computer. Let's see what you can do with this activity.

Techno Teasers
Answer Segment

Math plays an important role in the engineering of almost every product or structure.
Math is often used to solve problems related to technology.
(Courtesy of British Petroleum.)

TECHNOFACT

Technofact 9
In 1991, it was estimated that over 25 percent of all American homes had at least one personal computer. What do you think that percentage is today?

MATH THINGS TO SEE AND DO:
Hamburger Math

Introduction:
The fast-food industry has used technology to help make food as efficiently as possible. Every fast-food restaurant must try to keep costs down to be competitive.

Design Brief:
Your fast-food company plans to sell 2 billion hamburgers. As business manager, you will need to solve some of the following problems.

- Beef, 113.5 g
- Ketchup, 2.1 mL
- Mustard, 1.5 mL
- Salt, .19 g
- Mayonnaise, 2.76 mL

Materials:
- Paper
- Pencil

Equipment:
- Calculator or
- Computer and spreadsheet software

After this activity, students could develop their own, similar, design brief; for example, how many pieces of pepperoni would it take to make one million pizzas?

Procedure:
Answer the following questions using a calculator or computer spreadsheet.

1 If the average cow yields 175 kg of ground beef, how many cows will be needed for you to reach your 2-billion-hamburger goal?

② If a tank truck holds 10 m³ (cubic meters), how many truck-loads of ketchup, mustard, and mayonnaise will be needed?

③ How many tons of salt should be ordered? (1 lb = 454 g)

④ If your company sold hamburgers for $1.49 and they cost $1.19 to make, what would your annual profit be?

Evaluation:

Did your company make enough profit to stay in business? Would your place of business survive in your own hometown? Why or why not?

Technology and Health/P.E.: Many of the new developments in technology have been made for the field of health. We want and need to keep healthy and fit. Technology has come up with new products that help us do that.

HEALTH/P.E. THINGS TO SEE AND DO:
The Brain Strain

Introduction:

It is important to exercise both your muscles and your brain. Why not do both at the same time? In this activity, you will design an exercise and study cell where you can do homework while exercising your muscles.

Design Brief:

Work in groups of two and three to design and sketch a combination exercise machine and study area. The exercise equipment might be similar to a stationary bicycle, a stair-stepping machine, or anything you can create. The study area should include a place to write, a light, and a place for a computer. The design must let you exercise and study at the same time!

Materials:
▶ Paper
▶ Pencil
▶ Catalogs

Equipment:
▶ Computer with graphics software (optional)
▶ Exercise machine (optional)

Procedure:

❶ Brainstorm possible solutions to the problem with your group members.

❷ Evaluate each idea and come to a **consensus** (agreement) on a practical design.

❸ Use old catalogs or ads to find pictures of the equipment you would like to put into your exercise-study cell.

❹ Design the exercise-study cell to be safe, quiet, and easy to use.

5 Make a sketch or a computer drawing of your ideas. Think of a name and estimate a **retail** (store) price.

6 Present your product idea to the class.

Evaluation:
Evaluate each product by asking the following questions:

1 Does the exercise-study cell adjust to fit people of different sizes?

2 Could the exercise-study cell be used by a physically-challenged person? Explain.

3 Does the design include "pinch points" where you or small children might become caught or injured?

4 Have the following items been considered in the design?

- Fire resistance
- Low cost
- Environmental impact

5 Would you buy such a product? Explain why or why not.

You can probably think of many other ways technology affects your life and your studies. To help you see connections between technology and other subjects, **icons** (small pictures), will appear next to some paragraphs. Studying technology can also lead you to some exciting careers.

Careers in Technology

Have you thought about a job or career you want to try? Most people will actually end up changing jobs several times during their working life. What's important is that you start thinking about what skills you will need to enter the job market and what interests you have related to your future career. Throughout this book, you will be introduced to many careers that deal with technology. Maybe one of them will interest you.

Just as technology itself is always changing, so will the jobs of the future. One of the most important abilities for you to develop is being **adaptable.** Being adaptable means you can change your work skills without difficulty, or it means you have a variety of skills that you are good at using. Because things are changing so rapidly, schools cannot prepare you for every job there is or will be. Therefore, companies often require **on-the-job training**. You learn the main skills needed to do a specific job as you are doing the job. This specialized training builds on the basic skills (reading, math, and solving problems) you learned in school. For example, an employee working in a computer assembly factory will constantly need retraining as new circuit boards are developed. The new training will add to the skills the employee already has.

Two important things you can be learning now for any future job are dependability and punctuality. Employers depend on their employees

Careers in technology give you a chance to work in fields such as energy, transportation, communications, biotechnology, aerospace, and many others. (Courtesy of British Petroleum.)

TECHNOFACT

Technofact 10
Hershey Foods Corp-oration has developed and patented a heat-resistant milk chocolate bar. The special chocolate bars are supposed to hold their shape at temperatures of up to 140° F.

Reinforce the idea that the jobs of the future will require employees to be very adaptable.

Discuss the fact that many companies have to retrain their employees to do the basic skills of reading, math, calculation, and problem solving.

Careers in Technology— Technology Teacher

Technology teachers teach students about their world. Tying all the subjects that are learned in school together makes learning fun for students and teaching fun for the teacher! In your technology class, the teacher tries to make learning and exploration a "hands-on, minds-on" experience. Technology teachers help students develop the skills they need to solve real-world problems. They also teach students how to evalutate the effects of technology, both good and bad.

Teaching technology never gets boring or dull. As new technologies are discovered they can be brought into the classroom as new activities. Technology teachers use computers, robots, lasers, satellite communications devices, and other high-tech equipment to help you learn about technology.

The best part of being a technology teacher is helping students plan for a successful future through an understanding of technology.

Being a technology teacher is exciting. You have the opportunity to work with students to help them be technoliterate. You also can help students plan their future.

to be at work on time. You can start developing these good habits by completing your homework on time and getting to class on time!

Your future in a technology career will be exciting. No one knows exactly what the jobs of the future will be. We do know that being adaptable, dependable, and punctual will be important in any job, now and in the future.

Future Technology

How will technology shape your future? It's hard to say for sure, but you can be sure that future technologies will make your life easier or more productive. That is why technology is here. Future technologies should also bring good changes that fit the environment and our needs. You may hear words such as *genetic engineering* or *biotechnology, voice recognition, superconductivity, magnetic levitation,* and *artificial intelligence.* These are all areas where technology is being put to use for the future. Some of them used to be considered science fiction. Now they are real. Here are brief descriptions of a few future technologies. You will have a chance to explore them more in future chapters.

▶ *Genetic Engineering or Biotechnology*

The ability to design or redesign life forms. Developments in biotechnology or genetic engineering can lead to better production of food, prevention of disease, and improved medicines.

Biotechnology combines life sciences with technology. Here NASA researchers are experimenting with growing lettuce in space to provide food for long space flights or lunar (moon) colonies. (Courtesy of NASA.)

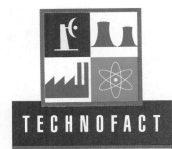

Inexpensive superconductivity demonstration kits are available from science and technology suppliers. Students will enjoy this demonstration.

An activity allowing students to experiment with Maglev trains will come in Chapter 12.

▶ *Voice Recognition*

A computer's ability to react to a person's voice. Instead of using a keyboard, you will give commands by voice.

▶ *Superconductor*

A material that loses all resistance to electricity, usually at low temperatures. This property is unusual because all materials have some resistance. Anything that uses electricity could benefit from this technology.

Special Report
Excimer Lasers

Careers in Technology— Medical Researcher

Medical researchers might work on a project such as laser angioplasty (pronounced An-gee-oh-plas-tee). This medical procedure uses a laser to vaporize (change from a solid to a gas) fatty deposits called plaque found in your arteries. Researchers are working with this process as an alternative to bypass surgery, in which blood vessels near the heart are replaced.

Until now the problem has been that lasers have been too hot for angioplasty. They might damage human tissue. However, one group of researchers has developed a "cool" type of laser called an excimer laser that doesn't damage the walls of the blood vessels. A catheter (a tiny tube) carries the laser light source through fiber optics. Watching from a video display screen, the physician finds the plaque buildup and quickly vaporizes it with the laser.

Medical research and technology offer many exciting careers.

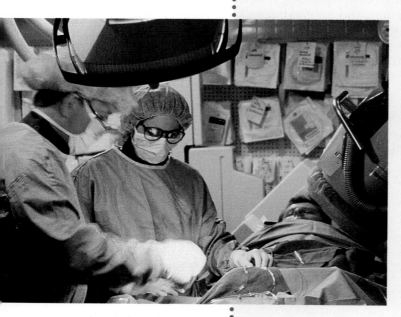

One possible career in technology involves medical research. In surgery, laser light is carried in a fiber-optic bundle into a person's artery. (Courtesy of NASA.)

Here the physician is holding a fiber-optic cable that he will use to fire short laser bursts to remove built-up plaque (fat) deposits. (Courtesy of NASA.)

Materials that have no resistance to electricity are called superconductors. Here a magnet is levitating (floating) above a superconductor made by an eighth-grade girl. The superconductor must be cooled to −300° F for it to have no resistance to electricity. Here liquid nitrogen is used to cool the material. (Courtesy of G.M. Hughes Electronics Corporation.)

T E C H N O F A C T

Technofact 13
Setting up dominoes is fun unless something knocks them over. The most dominoes anyone has ever set up at any one time is around 1½ million. Some students in Holland spent a long time setting them up, but it took only 1 hour for all of them to fall after the first push.

▶ *Magnetic Levitation*

Trains that float above a magnetic field instead of using wheels on a track. Levitating trains reduces the **friction** made when two materials rub against each other. Levitation also makes it possible for trains to go over 300 miles per hour safely.

▶ *Artificial Intelligence*

Programming a computer to be able to reason through a problem as well as any person. The computer would then be able to recognize problem situations and make the right decision. This means that the computer would be able to figure out what it needs to know on its own instead of being told!

Advancements in one field often lead to advancements in another field. Your school subjects are related in much the same way. You need math skills to solve science problems, and you need language arts skills to record your findings. Technology can be used to tie all your other subjects together. Throughout your study of future technology, you'll find ideas that interest and challenge us.

Summary

Technology has changed the way we work and the way we play. We see some of the changes as good and some as bad. Part of your job as a member of society is to evaluate technology's effects on you and your environment. Since technology has an impact on everything you do, it is related to the subjects you are studying in school like math, science, social studies, and language arts. Careers involving technology are exciting and offer you new challenges. Future technologies are being developed according to the products we want and think will benefit us most.

Thermal energy imaging can be used by doctors to see where stress may be a problem during exercise or sports. (Courtesy of G. M. Hughes Electronics Corporation.)

Careers in Technology— Geographic Information Systems Analyst

This analyst is using a computer and special software that uses data from satellites to forecast the amount of water available to farmers. Geographic Information Service (GIS) computer workstations can combine information to form a new picture of our world.

Geographic Information Systems can be used to predict and to visualize millions of pieces of information that would otherwise be impossible to understand. The analyst uses the system to help local, state, and federal governments plan land management, fire and safety, rapid transit system, real estate development, and other activities.

Utility companies and contractors can use GIS to plan the route of power transmission lines or pipelines. This technology can save time and money. It can also route utilities so that they don't disturb the environment. The job of a GIS analyst is important in environmental issues.

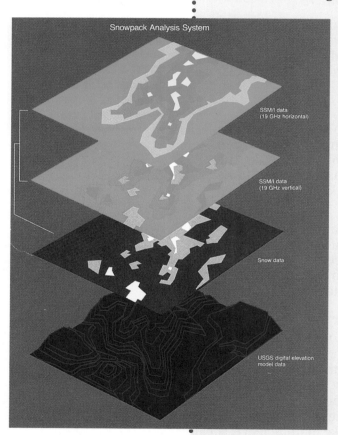

Snowpack Analysis System

SSM/I data
(19 GHz horizontal)

SSM/I data
(19 GHz vertical)

Snow data

USGS digital elevation
model data

Geographic systems technology combines graphic data into one computer image. Here the snowpack is presented as a 3-D computer model. Using this information, analysts can forecast the rate of snowmelt and runoff paths. (Courtesy of G. M. Hughes Electronics Corporation.)

A geographic information systems analyst uses a computer to make a three-dimensional (3-D) model of land masses. (Courtesy of G. M. Hughes Electronics Corporation.)

Challengers:

❶ Research the role of technology in sports:
- What causes a curve ball in baseball?
- Why do golf balls have dimples?
- Why are tennis balls fuzzy?
- What is a photo finish?

❷ Place a plastic bag and a paper bag outside in the same area. Observe them over a long period of time. Write what you predict will happen. Write the actual results.

❸ Choose any product to evaluate using the seven suggested questions at the beginning of this chapter.

❹ List five ways in which technology and science are alike and five ways in which they are different.

❺ Research ten math formulas that could be used in technology. Explain how they could be used.

❻ How would your life be different if electricity cost 100 times as much as it does today?

❼ Research space junk in earth orbit. Write a paragraph telling what you think we should do about the problem.

❽ List ten products you think should be improved to help the environment. Tell what improvement you would make.

❾ List ten inventions that led to the development of another invention.

❿ Interview someone in a technology-related career. Share the results of your interview with the class.

See Teacher's Resource Guide.

Techno Talk
Technology & You

Techno Teasers
Answer Segment

Chapter 3

Using Computers

Chapter Opener
Using Computers

Things to Explore

When you finish this chapter, you will know that:

▶ Computers are valuable tools used to help solve technology-related problems.

▶ The miniaturization of electronic components has made computers better and less expensive.

▶ The mother board has a central processing unit microchip that serves as the brain of the computer.

▶ The electronic parts of a computer system are called hardware. The programming code that computers use is called software.

▶ Computer memory can be stored in a microchip, on a disk, or on a tape or CD.

▶ Computer applications are programs that make it easy to do things like word processing and computer-aided design.

▶ A modem can make it easy to get information from other computers.

TechnoTerms

application	microchip
ASCII	modem
baud rate	mother board
binary system	peripheral
bit	printed circuit (PC) board
byte	
central processing unit (CPU)	program
	random access memory (RAM)
components	
computer-aided design (CAD)	read only memory (ROM)
data	software
database	word processing
hardware	

Careers in Technology

Many people think of computers when they hear the word *technology*. Computers are valuable tools used to help solve problems in technology and to run our world. There is a very good chance that your job in the future will require you to use a computer. Today, researchers use very fast computers to help design our future. Here, a person is working on a system that links engineering and manufacturing data. The system makes production processes faster. This reduces the time it takes new products to get to market. Computers are used to design cars, aircraft, houses, machines, and spacecraft that will travel to Mars.

TECHNOFACT

Technofact 14

In Walt Disney's Magic Kingdom, computers run the show from raising curtains to opening exit doors. Computers make sure that every recorded sound matches the movements of animated figures. If something goes wrong with a ride or special attraction, the computers immediately shut it down.

Predicting the weather accurately can help save lives. Computers analyze satellite information to produce weather maps that can warn of tornadoes or hurricanes. (Courtesy of British Petroleum.)

Computer Basics

Computers are important tools for solving technology problems. A few major uses of computers include writing, drawing, organizing information, calculating numbers, and finding information. In this chapter and throughout your technology study, the computer will be put to use in a variety of ways to help you solve problems. Just like any other tool, it is meant to make your job easier. The computer can calculate problems using very large and very small numbers quickly and accurately. It can find information rapidly, and, in some cases, even organize it for you.

Word processing. (Courtesy of British Petroleum.)

Finding and using information. (Courtesy of G. M. Hughes Electronics Corporation.)

Computers are used in many different ways. Here are just a few uses of computers in technology. (Computer graphics, Courtesy of Battelle, Inc.)

Did you know that many events in your life are controlled by a computer? Many of the routine things that happen every day are controlled by computers you do not see. For example, city traffic lights are monitored by computers. Weather forecasters use computers to analyze information received from satellites above the earth. Telephone networks across the nation are also controlled by computers. In some hospitals, computers are used to draw precise diagrams for doctors to use in performing heart surgery. Even modern cars have computer devices to control the engine.

Many inventions and technological developments led to the first computer. The computers of the 1950s and 1960s were huge machines that took up whole rooms. Because of changes in technology, today a small desktop computer can do the same jobs that those older computers did. The early computers were used by government and big corporations. It wasn't until 1977 that the first home computers became available. Since then, the power of the computer has increased many times, while its size has decreased. Today you can have a computer that is more powerful than the computers of 40 years ago and could fit on a wristwatch!

Computers have changed a lot in the past few years. New computers are smaller, less expensive, and much more powerful than older models. (Courtesy of International Business Machines Corporation.)

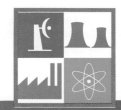

T E C H N O F A C T

Technofact 15

Silicon, the best ingredient for making computer chips today, also turns out to be a great material for making mechanical parts. At microscopic sizes, silicon is lighter than aluminum and stronger than steel. Researchers can build micromachines (very small motors, gears, lights, and so on) using the technology developed to make computer chips. Micromachines are so small they can easily fit into one strand of human hair. A round silicon wafer just 4 inches across can hold thousands of micromachines. Micromachines are being used today to monitor automobile brakes and to trigger the air bags in crashes.

The technology that led to smaller and more powerful computers was the **microchip** or **integrated circuit (IC)**, a combination of hundreds of tiny electronic parts on a silicon chip. Electronic parts together form circuits. Electronic parts such as transistors and resistors are called **components**. The combination of microscopic components and circuits together forms the microchip or *i*ntegrated *c*ircuit (IC). Computers use ICs to store information, or **data**.

Thousands of electronic components are shrunk to fit onto a tiny silicon chip. The patterns on the table above have been reduced so that hundreds can fit onto a round piece of silicon called a wafer. The finished microchip is about as big as the period at the end of this sentence. (Courtesy of G. M. Hughes Electronics Corporation.)

Special Report
A Bag of Chips

Even the most complicated computer uses only the number zero and one. The numbering system that uses only zero and one is called the **binary system**. Using this simple system, the computer can process information quickly.

THINGS TO SEE AND DO:
Counting in Base Two or Binary

Design Brief:

You are going to practice computing in binary just as a computer does.

Procedure:

❶ With a partner, make a place value chart for binary like the one shown in the example. Count to 100 in binary, filling in the chart as you go.

❷ Adding two numbers in binary works the same way it does in the decimal system, except that you have to move to new columns more often. Try this problem:

```
  10000111
+  1111000
─────────
```

The Binary Number System

Work with the math department to reinforce the use of the binary system. Advanced students might be challenged to learn the hexadecimal system.

The number system you generally use in everyday life is called the **decimal system**. It has ten digits from 0 to 9. You can use these ten digits to write numbers larger than 9 by using a tens column, a hundreds column, a thousands column, and so on. In other words, each place to the left of the ones place increases in value by a power of 10. For example, 256 means 2 hundreds (10^2) + 5 tens (10^1) + 6 ones.

Computers understand only one language: the binary code, or base 2 system. It has two digits, 0 and 1, that stand for "off" and "on" electronic signals. In the binary system, each place to the left of the ones place increases in value by a power of 2. In order to write numbers larger than 1 in binary, you use a twos column, a fours column, an eights column, a sixteens column, and so on. In the chart, the decimal number 5 would be 101 in base 2. It would be 1 group of 2^2 + 0 groups of 2^1 + 1 in the ones column.

Four Systems with Different Bases

DECIMAL			BINARY Base 2						OCTAGONAL Base 8			HEXADECIMAL Base 16		
PLACE	PLACE	PLACE	PLACE	PLACE	PLACE	PLACE	PLACE	PLACE	PLACE	PLACE	PLACE	PLACE	PLACE	PLACE
100	10	1	32	16	8	4	2	1	64	8	1	256	16	1
		0						0			0			0
		1						1			1			1
		2					1	0			2			2
		3					1	1			3			3
		4				1	0	0			4			4
		5				1	0	1			5			5
		6				1	1	0			6			6
		7				1	1	1			7			7
		8			1	0	0	0		1	0			8
		9			1	0	0	1		1	1			9
	1	0			1	0	1	0		1	2			A
	1	1			1	0	1	1		1	3			B
	1	2			1	1	0	0		1	4			C
	1	3			1	1	0	1		1	5			D
	1	4			1	1	1	0		1	6			E
	1	5			1	1	1	1		1	7			F
	1	6		1	0	0	0	0		2	0		1	0

Binary numbers

Bits and Bytes

The smallest unit of information used by computers is called a **bit**. The word *bit* is an acronym for *b*inary dig*it*. Eight bits togther are called a **byte** (pronounced bite). Computers often process information one byte at a time.

Why is this important? Computers use a standard code to represent letters, numbers, and punctuation. A code of zeros and ones is used so that computers can communicate with each other. When you hit a key on a computer keyboard, you use one byte of memory and send a binary code. For example, if you typed TECH, the binary code would look like this:

T   01010100

E   01000101

C   01000011

H   01001000

Bits and bytes

The computer code in the United States is called **ASCII** (as-key). ASCII stands for American Standard Code for Information Interchange.

External computer interface cards are available that show one byte of information using eight LEDs. Students can easily see which bit is on or off.

Have students research the ASCII code. Have them find the decimal equivalents of letters and symbols.

❸ Now try to change these binary numbers to base 10:

11111 = _____

1011 = _____

10101 = _____

❹ Now change these base 10 numbers to binary:

45 = _____

120 = _____

198 = _____

Evaluation:

Is it easier to count in tens or twos? Explain.

Challenge:

How can the computer handle such large amounts of information when it uses only the binary code? Research octal (base 8) and hexadecimal (base 16) numbering systems to see how they are used by computers.

You need to know something about the parts of your computer system. The electronic parts of a computer system are called **hardware**. One part of the hardware, the **central processing unit (CPU)** is the brain of the computer. It is a special kind of integrated circuit that along with others are mounted on a **printed circuit (PC) board**. Some computers have many printed circuit boards. The main circuit board that contains the CPU and connectors to other parts of the computer is called the **mother board**.

Other basic hardware parts of a computer system might include a monitor, a mouse, a floppy disk drive, a hard drive, and a keyboard.

TECHNOFACT

Technofact 16

The first large digital computer was ENIAC (electronic numerical integrator and computer). It was 100 feet long and weighed 30 tons. Besides taking up an entire room because it was so large, it was very expensive. Today, the same computing power can be found on an inexpensive microchip smaller than your fingertip.

Open the case of a personal computer to show the mother board, CPU, other printed circuit boards, electronic components, power supply, and so on.

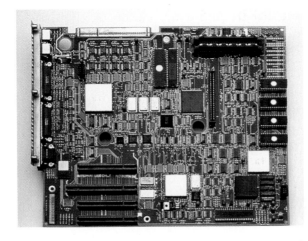

Printed circuit (PC) boards have many electronic components and integrated circuits soldered onto them. The main PC board in a computer is called the mother board. (Courtesy of International Business Machines Corporation.)

Techno Talk
Debugging the Eniac

Techno Talk
Answer Segment

Computer Memory

Information can be stored and used by computers in different ways. You know that information can be stored magnetically on computer disks, for example. Computer disks come in different sizes.

Computer memory is called either **RAM** or **ROM**. RAM is an acronym for *random access memory*. RAM memory is used up by the computer programs and your documents. ROM is an acronym for *read only memory*. ROM is built into the computer's electronic circuits on a microchip. ROM gives the computer operating instructions that can be used (read), but not changed by the user. An important difference between RAM and ROM is that each time you boot (turn on) a computer, it reads the operating instructions built into the ROM. As you type or draw on a computer, you use RAM memory. If the power is cut off or you turn off the computer, the RAM memory is erased. When you restart the computer, the RAM is gone but the ROM is still there. To keep from losing the work you do on a computer, save it to a disk frequently.

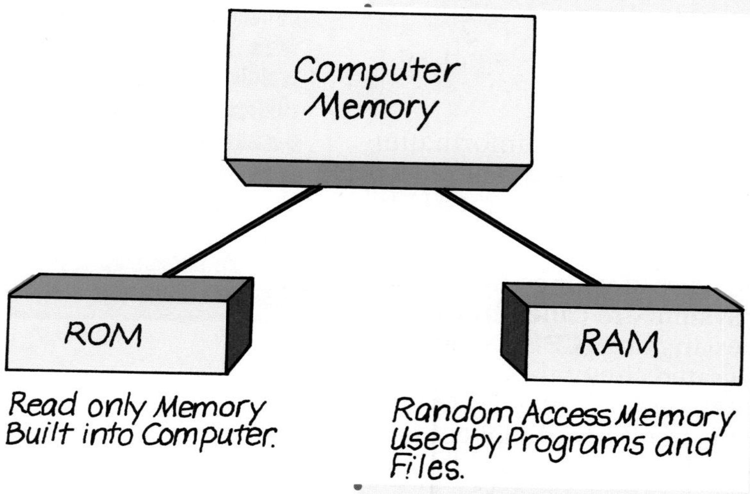

Challenge students to make a display of the different sizes of floppy disks ranging from the old 8-inch disks to 3½-inch size. Research the memory storage capacity of each format size.

Computer Memory

ROM
Read only Memory
Built into Computer.

RAM
Random Access Memory
Used by Programs and Files.

ROM and RAM

Other devices that can be connected to your computer system are called **peripherals.** Peripheral devices you might use include the following:

▶ *Printer:* A machine that puts text (words) and graphics (pictures) on paper.

▶ *Modem:* A device that lets your computer communicate with another computer over the telephone line.

▶ *Scanner:* A machine that copies text and graphics from paper to the computer.

▶ *Digitizer:* A device that changes video images into computer images.

▶ *MIDI (musicial instrument digital interface):* A device that lets you put music into a computer from an electronic music keyboard.

▶ *Joystick:* A device that changes hand movements into actions on a computer screen.

Have students make labels for all of the computer equipment. An engraver or hand lettering device can be used to make labels that last. If all of the equipment in the technology lab is labeled in this way, just being in the room can be a learning experience.

A

Things that connect to computers to provide input and output are called peripherals. There are hundreds of different peripherals that can be used; here are just a few. (Figure A Courtesy of Apple Computer, Inc./ Julie Chase and John Greenleigh; Figure B Courtesy of International Business Machines Corporation.)

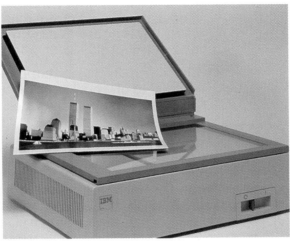

B

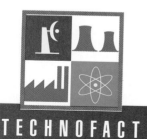

TECHNOFACT

Technofact 17

An ARU (audio-response unit) is the computer voice that lets you do banking by telephone. When you call the bank, ARU is the voice that asks you to "please enter your account number" or "please hold." Its computer is programmed to take you through an entire step-by-step banking transaction.

Computer hardware doesn't work by itself. You need to use some kind of **software**. Software is the coded instructions that a computer uses. Software is most often stored magnetically on floppy disks. Software used for a specific purpose is called an **application,** or **program**. Applications or programs are used to make **documents**. Applications might be word processing, databases, spreadsheets, and so on. When software is used by the computer, it uses RAM memory space.

New uses for computers, computer software, and computer peripherals are being developed every day. Chances are good that your future job will depend on the use of a computer and some of these peripherals.

Using Your Computer

No matter what kind of computer you are using, there are some basic steps to follow.

❶ **Boot your computer.** Booting your computer means to turn it on. Each computer system has a special way of starting. In some systems you have to put the floppy disk in before turning it on. In others you have to turn the computer on first and then put the floppy disk into the drive. If you have a hard disk drive, you will not have to put in any floppy disk to start an application. Be sure you know which way is correct for your computer.

❷ **Run an application.** The computer will read the information that lets you use the computer to do word processing, graphics, etc. Some software applications have a tutorial program that helps you learn how to use the application.

❸ **Create a document.** Using that application, type in information or create a drawing.

❹ **Save your document frequently on a data disk.** It's frustrating to work on the computer and lose what you've done due to a power failure or striking the wrong key. It's easy to prevent that from happening by saving your work every few minutes. Your program can be saved on a hard disk or a floppy disk.

❺ **Shutdown.** Each computer has a special way of being shut off to prevent loss of information. Follow the right procedure for your computer.

As you learn more about computers, you will find abbreviations used to describe computer terms. Often the abbreviations are acronyms. An acronym is a word made from the beginning (or sometimes ending) letters of words, as you learned in Chapter 1. As you use computers, you will want to print what you have on the screen. A funny sounding acronym to do that is called **WYSIWYG** (wizzy-wig). It stands for "*what you see is what you get.*" You will find tons of acronyms in this chapter. Now let's look at some real-life uses of your computer.

Acronyms You Should Know

Computer technology is sometimes confusing because many acronyms and abbreviations are used. Here is a list to help you.

bit = *b*inary dig*it*

byte = eight bits

CAD = *c*omputer-*a*ided *d*esign

CAM = *c*omputer-*a*ided *m*anufacturing

CD-ROM = *c*ompact *d*isk *r*ead *o*nly *m*emory

CIM = *c*omputer-*i*ntegrated *m*anufacturing

CPU = *c*entral *p*rocessing *u*nit

DTP = *d*esk*t*op *p*ublishing

IC = *i*ntegrated *c*ircuit

kilobyte (K) = 1024 (2^{10}) bytes

Mb = *m*ega*b*yte (1024 K)

MIDI = *m*usical *i*nstrument *d*igital *i*nterface

modem = *mo*dulator/*dem*odulator

PC = *p*rinted *c*ircuit

RAM = *r*andom *a*ccess *m*emory

ROM = *r*ead *o*nly *m*emory

WYSIWYG = *w*hat *y*ou *s*ee *i*s *w*hat *y*ou *g*et

Challenge a group of students to make a set of flash cards with the acronym on one side and its definition on the other. Set up a "technology pursuit" game using the cards.

Techno Teasers
Processing Words

Techno Teasers
Answer Segment

Word-processing software lets you choose many different styles, fonts, and sizes of letters for your document.

Writing with a Computer

Writing with a computer is called **word processing**. Word processing is quickly replacing typing in the business world. It is faster, easier, and more efficient to use a word processor than it is to use a typewriter. Imagine you had typed a 60-page report and then found that you had skipped a paragraph on page 3. Correcting that mistake would change the page setup all the way through the report. If you were using an old-style typewriter, you would have to retype the report completely. On a word processor, you would simply fix the error on the computer screen before the report was even printed. Think of the time you just saved!

Type Styles

This is plain text

This is bold text

This is italic text

This is outline text

Type Sizes

This is 10 point

This is 14 point

This is 24 point

Type Font

This is the Helvetica font

This is the Courier font

This is the Park Avenue font

Word processing on a computer can vastly improve the appearance and quality of student writing. The process is slowed painfully by lack of keyboarding skills. There are many computer programs that are designed to help students learn to type. Students should be encouraged to learn to type at an early age.

Word-processing software lets you choose different **fonts** or kinds of **characters** (letters). It lets you change the size of the characters and the style of print. It even lets you change between single and double spacing easily. Most word-processing software even places page numbers, tabs, and paragraphs where you tell it to.

The computer software also can point out spelling errors. Some word-processing software packages also have an electronic **thesaurus**. A thesaurus gives you **synonyms** (words that have the same meaning) to help you in your writing.

THINGS TO SEE AND DO: *Word Processing*

Introduction:

Using a computer to do any kind of writing can save a lot of time. Using a computer as a type of advanced typewriter is called word processing. Word-processing software can help you become a better writer by making it easy to rearrange words or sentences. It can even check your spelling and grammar. In this activity you will use a computer and word-processing software to make and print a document.

Design Brief:

Choose any one of the following topics or think of your own idea. Use a computer and word-processing software to make and print a document about your topic. You might do any of the following:

- Write a letter to a company requesting information about their products.
- Write a resume. A resume is a list of important facts about yourself. You will give your resumes to employers when you apply for a job.
- Write your own definitions to the key terms in this chapter.
- Write a program for an event at your school such as a band concert or football or basketball game.
- Write a book report for another class.
- Write an article for the school newspaper or a letter to parents.
- Write a video or audio script.
- Or think of your own idea.

> Work with language arts teachers to reinforce written communication skills. Ask other teachers to require students to complete some assignments using word processing.

Equipment:

- Computer with word-processing software
- Printer

Procedure:

1. Boot a computer with word-processing software.

2. Type your document. Use one of the topics suggested or get your teacher's approval on your own idea.

3. Learn how to rearrange words and sentences using the word-processing software.

4. Experiment with changing type sizes, styles, and fonts.

5. Check the spacing and grammar of your document using computer software.

6. Have a friend **proofread** (check over) your work.

7. Save your document to a data disk.

8. Print your document.

Evaluation:

1. What was the easiest part of doing word processing?

2. Time yourself for one minute as you type. How many words per minute did you type?

Students should be challenged to use the features of the software. Encourage students to find out how to incorporate graphics, for example.

Inexpensive desktop publishing software is available for every brand of personal computer. Some DTP software may require the use of a laser printer. Check the requirements of the software and compare it to the hardware you have available.

A

B
Many different brands and kinds of computers are being used today. Here are just a few. (Figure A Courtesy of Apple Computer, Inc./Will Mosgrove; Figure B Courtesy of International Business Machines Corporation.)

Special Report
Computers Make the News

Challenge:

Ask your teacher, or use the **documentation** (instructions) for the word-processing software, to find out how to do the following: set tabs, change line spacing, make columns, set or change margins, and so on.

The software you use is designed to work on the specific model or brand of computer you are using. All computer brands have at least one word-processing software program. The major applications listed earlier are also available for any brand of computer.

After you write the text with word-processing software, you can then organize it using desktop publishing software. **Desktop publishing** (DTP) lets you put text and graphics together to make a report, newsletter, or newspaper. Desktop publishing has changed the way printed materials, from business cards to books, are published. It is especially exciting because now anyone who has learned to use a computer can do something that before only trained professionals such as typesetters and printers could do.

Desktop publishing software lets you put together text and graphics to make reports, newspapers, or newsletters. (Courtesy of International Business Machines Corporation.)

The manuscript for this book was put together with word-processing and graphics software on a **laptop** (portable) computer in a camper! It was then reorganized with desktop publishing software before the final printing. At any stage it was easy to change information without redoing the entire chapter or book.

Drawing with a Computer

The old saying "a picture is worth a thousand words" is really true. Whenever you can add pictures or graphics to your words it makes reading much more exciting and fun. Graphics also help you understand the meanings of words.

There are graphics software programs for every computer brand. These programs allow you to make drawings using a mouse, a keyboard or other device. You don't have to be an expert artist to use graphics software. You can use predrawn images called **clip art** available on floppy disks or CD-ROM disks.

Computer-aided design (CAD) software used on a computer makes it easier to draw and then change your drawing if necessary. Being able to change part of your drawing gives you the same advantage as changing text in word processing. You don't have to start over!

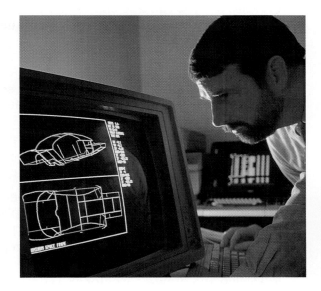

CAD is an acronym for computer-aided design. Computers can be used to make drawings of products or parts quickly and easily. (Courtesy of Reynolds Metals Co.)

T E C H N O F A C T

Technofact 18

Did you know doctors now use computer graphics or imaging to explore the human body? PET (positron emission tomography) makes images of the brain showing where tasks such as looking, listening, or thinking occur. CT (computer tomography) scans the human body in seconds and can turn that information into an image in less than a minute. Even better, a CT scan of your hand, for example, can show your fingers and muscles in motion!

The images on a computer screen are made of tiny rectangles called **pixels**. The word **pixel** stands for "picture element." Each pixel can have its color changed or can be erased using CAD software. The computer software controls the location and color of every pixel on the screen. A common computer screen might be 512 pixels wide and 342 pixels high. That makes 175,104 little rectangles on your screen! On a black-and-white screen, these pixels are turned on or off in patterns that make up a picture. Color images are more complicated because the computer program has to control hundreds or thousands of color pixels as well as black and white. Imagine having to change each of those pixels by hand!

CAD software also lets you see an object in three dimensions (3-D). Seeing all three dimensions gives designers, architects, and engineers a big advantage. For example, architects may design a **floor plan** for a house. A floor plan shows the house from above, looking down on the house as if it had no roof. It can be difficult to really see in your mind how the house will look when you have only a flat drawing of it. With CAD software, you can see the rooms on the computer monitor as if you were standing inside or outside them.

CAD drawings can be produced on a printer or a **plotter**. A plotter is a peripheral that uses ink pens to draw the picture.

THINGS TO SEE AND DO:
Computer-Aided Design (CAD)

Special Report
Drawing & Designing

Introduction:

Even if you can't draw a straight line with a ruler and a pencil, you can make very complicated drawings easily with a computer. Computer-aided design (CAD) or graphics software makes it easy to draw lines and shapes. The best part about drawing with a computer is that it is very easy to change or modify your drawing. Not only can you erase easily, but you can also change the size or shape of things you have drawn using CAD or graphics software. In this activity you will make a drawing with CAD or graphics software.

Design Brief:

Make a drawing using CAD or graphics software and a computer. You might choose one of the following ideas for your drawing or think of your own:

▶ Design and draw a school **logo**. A logo is a symbol that represents a product or company or in this case your school.

▶ Design and draw a floor plan for a house or school of the future.

▶ Design and draw a sign that reminds people to recycle waste materials such as aluminum, glass, or plastics.

▶ Design and draw a cover for a report. This cover might include a drawing of something related to science, social studies, math, or art.

▶ Design and draw a cover for a CD or tape.

▶ Or better yet, think of your own!

Equipment:

▶ Computer with CAD or graphics software
▶ Printer or plotter

Procedure:

1 Boot a computer with CAD or graphics software.

2 Start your design using one of the suggested topics or get your teacher's permission to do your own.

3 Experiment with erasing, moving, and resizing part of your design until you are happy with the way it looks.

4 Ask a friend to check your drawing. Look for places that will make the drawing better. Watch for overlapping corners, rough "free-hand" curves, and so on.

5 Make any final corrections to your drawing.

6 Save your drawing.

7 Print or plot your design.

Evaluation:

❶ Do you think that great works of art will ever be made on a computer? Explain.

❷ Is it easier to change a drawing on paper or on a computer? Explain.

Challenge 1:

Use the documentation (instructions) for the software to find out how to do the following:

▶ Make a scale drawing
▶ Duplicate objects
▶ Use clip art images with your design

Many of the drawings created with CAD software are combined with text using desktop publishing software. For example, you may design a graph to show the number of products sold by your company. CAD software can be used to make the graph. When combined with word-processed text, it can make a business report or letter.

Challenge 2:

Combine a word-processing document with computer graphics using desktop publishing software. You might make a class newsletter or an announcement to advertise a band concert, athletic event, or other school activity.

Challenge students to draw a cartoon for the school newspaper using graphics software.

CAD software is useful in another area. When a CAD drawing is sent directly from the computer to automated machines that make parts, it is called **CAM,** or *c*omputer *ai*ded *m*anufacturing.

A computer used mostly for CAD is called a **CAD station**. CAD stations can be **networked** (hooked together) to a main computer that controls an entire production line. Using computers in this way to produce entire products is called *c*omputer *i*ntegrated *m*anufacturing, or **CIM**.

The combination of CAD, CAM, and CIM makes it possible to produce products quickly and inexpensively.

CAM is an acronym for computer-aided manufacturing. Computers can be used to control machines to accurately make parts for products. (Courtesy of Cincinnati Milacron.)

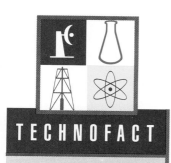

People use databases to organize and find information on many subjects such as people, products, and historical events. Using a computer and database software makes it very easy to search for and sort information, and saves many hours of work. (Courtesy of Deere & Company.)

Ask the class to brainstorm possible uses of a database.

What's a Database?

Data is another way of saying "information." A **database** is a computer application that lets you organize and find information quickly and easily. Databases are sometimes called "electronic file cabinets." This means you can use your computer to find information that would normally be stored in a file cabinet. For example, your school has a file for every student, including you! Most likely these files are stored in a file cabinet near the office. If the school needed to send a letter to just the parents of seventh-grade girls taking technology classes, the school secretary would first have to find the files for all seventh graders. Then the secretary would have to sort out all the girls and then go through each girl's file to see if she is in technology. This process could take a long time! The computer database application would do it in a matter of seconds.

To start a database, you need to make fields, records, and files. These are the major parts of most databases.

▶ **Field:** One part of the data that helps describe the record. For example, the person's first name would be one of many fields that would make up a record for that person.

▶ **Record:** All of the fields put together. For example, the database record of a person might include fields such as first name, last name, and phone number.

▶ **File:** A group of records. For example, all the records of the students in the seventh grade.

In the school records example, a file might be all seventh graders. A record would be the information on each seventh-grade student. The fields would include first name, last name, parents' names, address, class schedule, and so on.

In everyday life, databases are used by companies to keep track of information on people. Credit information, hobbies, sports interests, and travel preferences are some kinds of data stored about you and your family on computers. This kind of database sometimes leads to "junk mail" being sent to your home. Some people consider this use of computers as an invasion of their privacy. What do you think?

THINGS TO SEE AND DO: *Making a Database*

Introduction:

Data is another word for information. You have learned that the amount of information in the world is called the knowledge base. You also know that, thanks to technology, the knowledge base is doubling every few years. That's a lot of data! In this activity you will learn how to keep track of data using a computer program called a database. The advantage of using a database for storing information is that it can be found and organized quickly on a computer. You can think of a database as an electronic file cabinet. In this activity you will design and use a database to organize and sort information.

Design Brief:

Design and create a database that can be used to organize and find information. Choose any of the following topics, or think of your own.

- Events in technology
- Space exploration
- Major inventions
- Instructional videos
- Technology-related magazine articles
- Think of your own idea

Equipment:

- Computer with database software
- Printer

> The database portion of an integrated software package is all you need to complete most information gathering activities. Expensive and complicated dedicated database software is not necessary.

Procedure:

1 Work in groups of three or four. Choose one of the ideas in the design brief, or get your teacher's permission to do your own idea.

2 Design the fields that would make up a database record for your topic.

3 Show your database design to your teacher. Research your topic, and enter the data into a database application on a computer.

4 Save your data for future use.

5 Try to use your database to sort and find information on a topic. Challenge other students in other groups to use your database to do research.

Evaluation:

1 List ten ways that stores or companies might use a database.

2 Do you think that a database with information about you should be given to companies, or should it be kept confidential (secret)? Explain.

Challenge:

Use the documentation (instructions) for the database software to do the following:

- Sort alphabetically
- Sort numerically
- Create a report
- Edit (change) data already entered

Using Spreadsheets

Just as databases keep track of information on a computer, spreadsheets keep track of numbers. The computer's ability to calculate numbers is called **number crunching** because it can do it so fast! Spreadsheets are often used to make budgets. Suppose you wanted to know how long it would take you to save enough to buy a stereo system. Your budget might look something like this:

Column

Row {

Month	Income	Expenditures	Balance
September	$22.50	$10.35	$12.15
October	$25.00	$ 8.50	$16.50
November	$42.25	$ 7.25	$35.00
December	$65.00	$23.00	$42.00
		Cash Available	$105.65

Cell

A spreadsheet is often used to make a budget to keep track of money.

In the sample spreadsheet, the earnings, expenses and savings are in vertical **columns**. The data for each month are in horizontal **rows**. All spreadsheets are made up of columns, rows, and cells. Usually rows are numbered, and columns are given letter names. Each space in a column or row is called a **cell**. Each cell can be identified by a combination of its column letter and row number. The cell in the upper left corner would be A1.

The advantage of spreadsheets is that they let you answer "what-if" questions. For example, you might ask, "What if I earned twice as much in January?" How would that change the total? To find the answer, you would change the amount in the January earnings cell. The computer would make the changes instantly using a formula, and then would show you the answer. A formula might be as simple as adding numbers in a row or column, or it can be more complicated. It could be an engineering calculation. You can easily add your own formula to a computer spreadsheet.

THINGS TO SEE AND DO:
Working with a Spreadsheet

Introduction:

Whenever a lot of numbers need to be added or combined in any way, a computer spreadsheet can help. Spreadsheets are often used to keep track of money. Budgets can be made to automatically calculate how much money is left (balance) after each **expenditure** (expense). In this activity, you will design and use a spreadsheet that will help you make math calculations.

Design Brief:

Make a spreadsheet that will do any of the following:

▶ Budget money to be spent on presents for family members and friends.

▶ Budget money earned for odd jobs, allowances, and expenses.

▶ Keep track of products made by a company.

▶ Or think of your own idea.

Equipment:

▶ Computer with spreadsheet software

▶ Printer

The spreadsheet portion of an integrated software package is sufficient for this activity. Again, expensive dedicated spreadsheet applications are not necessary.

Procedure:

1 Work with a partner. Pick one of the topics listed, or think of your own. If you are using your own idea, get help from your teacher.

2 Design how you will make your spreadsheet on paper first.

3 Boot a computer with spreadsheet software.

4 Enter the column and row headings that you need.

5 Enter the numbers in the correct cells.

6 With the help of your teacher, enter the formulas that you need to add, subtract, multiply, or divide your numbers.

7 Change the number in one of the cells to see how it affects the other cells.

8 Save your document.

9 Make a hard copy (printout) of your spreadsheet.

Evaluation:

1 Think of five different uses for a spreadsheet in the real world.

2 What advantage is there in using a spreadsheet instead of a calculator?

Challenge 1:

Design and make a spreadsheet that will do one of the following jobs.

▶ Calculate gas mileage from automobile trips

▶ Keep track of expenses for your household

▶ Organize information related to sports statistics in football, basketball, or any sport you choose.

Challenge 2:

Use the documentation (instructions) for the spreadsheet software to find out about other math calculations that can be made besides addition, subtraction, multiplication, and division.

Finding Information

You live in an information age. You've already learned how fast the knowledge base is growing. You also know it is impossible for any one person to know everything. A smart person isn't necessarily someone who knows everything. A smart person knows where to find the information he or she needs!

Being able to find information when you need it is important. This security guard uses the computer to watch different areas. She can easily monitor many different activities at one time. (Courtesy of G. M. Hughes Electronics Corporation.)

If you do not have a CD-ROM drive in your classroom, check with the media specialist in your school or the public library.

A computer can be used to **access** (find) information on many topics. You can access information in several ways. Personal computers can be interfaced, or attached to, a **CD-ROM** (compact disk-read only memory) drive. An entire electronic encyclopedia, for example, can be accessed through a **compact disk** (CD). A compact disk can hold 1,300 times as much information as a floppy disk. That's a lot of information at your fingertips! Besides that, it accesses information in a matter of seconds. If you needed information on crash testing of cars, by the time you pulled the encyclopedia from the shelf, the computer would have that information for you. In addition, the computer lists cross-references that allow you to tie the information with other subjects. Sometimes not all the information is available on CDs but many electronic encyclopedias also include a **bibliography** (list of resources) so you can explore further.

A CD-ROM disk can hold an entire set of encyclopedias. Finding information quickly is easy using a computer with a CD-ROM drive. (Courtesy of International Business Machines Corporation.)

Another way to access information from the outside world is by hooking the computer to a **modem** and a telephone line. A modem is an electronic device that lets computers share information. Attaching a modem to your computer lets you communicate with any other computer that has a modem. *Modem* is another acronym. It stands for *m*odulator-*d*emodulator.

Two computers can communicate with each other only if one computer can **decode** (understand) the information received from the other computer. When you receive a file over a modem it is called **downloading**. When you send a file over the modem to another computer it is called **uploading**. Communications software on each computer sets up a **protocol** (special settings) to allow information transfer. One important protocol is how fast information is sent on the telephone line. This is called the **baud rate**. Most common modems operate from 1,200 to 9,600 baud. The faster the baud rate, the less time you have to spend using the telephone line. The baud rate is important because if you're communicating long-distance the cost can add up quickly!

Using your computer and modem, you can access information stored in huge mainframe computers. Information services are companies that sell information to you. Did you ever think about paying for information like you would for another product such as a hamburger? A rapidly growing use of computers is for accessing information. Whenever a topic needs to be researched, an information service can save valuable time. It gives you bibliographies and **abstracts** (brief summaries) of articles in magazines as well as books. Abstracts let you evaluate whether that resource will be useful to you.

- -
A modem can open a new world of communication to technology students. A virus-detecting program should be used to prevent your files from being infected.
- -

- -
Remind students to save telephone time by transferring pre-made files on the modem rather than typing while on-line.
- -

THINGS TO SEE AND DO:
Finding Information

Introduction:
Computers can "talk" to each other over telephone lines using a modem. By connecting a modem to a computer, you can find information that is stored in large computers, called mainframes. Any brand of computer can send or receive information from any other computer. In this activity, you will get information from another computer using a modem.

Design Brief:
Search and find information using a computer, modem, and communication software.

A dedicated phone line for the modem is best; it will prevent an interruption of data transfer if someone else picks up a remote receiver.

Equipment:
▶ Computer with communication software
▶ Modem
▶ Phone line
▶ Optional: Printer

Procedure:
❶ Working in groups of four or five, get a list of information services from your teacher.

❷ Research the type of information available from the information service you have selected.

❸ With your teacher's permission, boot the communication software. Dial the appropriate phone number. Note: You must have your teacher's permission!

❹ Quickly search and download (retrieve) information. Do not take the time to read all of the information received. You can read the data later.

You should plan the modem activity to prevent wasted long distance telephone charges.

❺ **Log off** (hang up) the service. Be sure you are not still connected to the information service. Remember, every minute you are on the phone line costs money!

❻ Read through the information you received.

❼ Optional: You may save your information and print it out.

Evaluation:
❶ List ten information topics you might access using a modem.

❷ How could you save money by using a modem?

❸ Why would banks or stores use modems? Explain.

Challenge 1:
Exchange information with other schools using a modem.

Challenge 2:
Access education-related information services such as NASA's SpaceLink.

The Future of Computer Technology

In 1950, there were fewer than 150 computers in the world. In 1970, there were 50,000. Today, companies make 50,000 computers every day! A huge mainframe computer that cost millions of dollars twenty years ago can now be purchased as a portable computer for less than $5,000. Advancements in computer technology and electronics will make future computers even more powerful and less expensive.

Because technology is changing so fast, it is hard to predict what computers of the future will be like. Here are a few possibilites:

▶ Voice-activated computers without keyboards
▶ Computers that operate with laser light instead of electronic chips
▶ Computers that react to your eye movements
▶ Superconducting, superfast supercomputers
▶ Mind-reading computers
▶ Computers that can think and learn on their own

Research is being conducted on each of these areas today. For example, your home or car of the future will be controlled by computers. Parts of them already are.

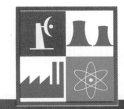

TECHNOFACT

Technofact 20
Computers are making highways "smart" by helping control traffic in crowded areas. The computers receive information from video cameras. Then the information can be composed into a computer traffic map that tells where major problems such as accidents or hazardous-material spills are located on the highway.

Have students brainstorm features they would like to see in future computers.

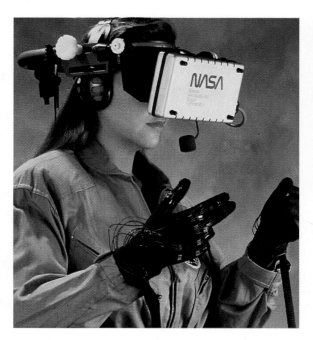

Using 3-D computer graphics this scientist feels as if she is an active part of a real environment. In the future, surgeons hope to use this technology to do telesurgery, where surgery is performed by a robot whose movements are guided by a surgeon on earth. (Courtesy of NASA.)

New technologies will change the way you do things, even how you play games. Right now you can purchase airplane tickets through your own computer or order items from catalogs. (Courtesy of AT&T/Bell Laboratories.)

Techno Talk
Future of Computers

Techno Talk
Computers from Worlds Beyond

TECHNOFACT

Technofact 21

Computer memory can sometimes be confusing. It may help to know the following terms:

bit = binary digit: smallest unit of information

byte = eight bits: amount of memory used to press one key

kilobyte (K)

= 1024 (2^{10}) bytes: often used to show the size of a document

Megabyte (Mb)

= 1024 K: large unit of memory often used to give the size of disk storage space or RAM memory

Summary

Computers are often used for writing, drawing, calculating numbers, organizing information, and finding information. Our daily lives often depend on how computers are used. The first computers were huge and expensive. Today you can have a computer on your wristwatch. Microchips, or integrated circuits, reduced the size and price of computers and other electronic products. Computer systems are made up of hardware and software. The electronic parts make up the hardware. The coded information that runs the hardware is called software.

Sometimes computers are used as powerful electronic typewriters. This is called word processing. Computers also use computer-aided design (CAD) software to design products or advertisements. Computer graphics, or pictures, can be added to words in desktop publishing. Information can be organized and used easily in a computer database. Numbers can be organized, calculated, and changed easily in a spreadsheet application. Future computers might respond to your voice so you won't have to type, and they will be faster and less expensive.

Challengers:

1 Explain how computers are used to forecast weather or control traffic.

2 Research ways computers affect your everyday life. Explain your findings.

3 Make an acronym that will remind you of the basic steps to follow in using a computer.

4 Make a display of electronic components, and label each part.

5 Using desktop publishing software, make a school newspaper with text and computer graphics.

6 Use a word-processing application to write a long paper or report. Include a cover and title page made from a graphics application.

7 Make a database of everyone in the class.

8 Find out how businesses in your area use computers.

9 With the help of your teacher, arrange a long-distance problem-solving activity with another school using a modem.

10 Find out how your school library uses computers. Make suggestions on other uses you think would help you and other students.

See Teacher's Resource Guide.

Chapter 4

Inventing Things

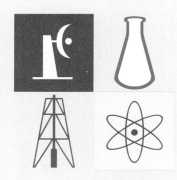

Things to Explore

When you finish this chapter, you will know that:

▶ New ideas or different methods of solving problems are called innovations.

▶ Innovations and inventions can be challenging and fun to design.

▶ Problem-solving steps can help in the invention of new products.

▶ Visualizing new ideas in your mind can help you make them in reality.

▶ Prototypes are the first attempts or models for new products.

▶ Patents protect ideas from being copied without permission.

TechnoTerms

feedback
format
innovation
NASA
patent
physically challenged
prototype
specifications

(Courtesy of Cyprus Minerals Co.)

Careers in Technology

Choosing a career is one of the most important decisions you will make in your life. A job that is challenging, fun, and rewarding can make your life enjoyable and profitable. Whatever career you choose, it is important to do your best. The technician shown here is really getting into his work! As you complete the activities in this book, think about the jobs that might require you to do the same thing or something similar. Talk with people and watch them at work. Ask yourself if you would be happy doing their job. It's not too early to start thinking about your career right now.

••••••••••••••••••••••••
Ask students to give examples of
serendipity in their lives.
••••••••••••••••••••••••

What Is Innovation?

Innovation is the creation of new ideas or devices or different approaches to doing something. Innovation happens in many ways. In Chapter 1, you learned some steps you can use to be more creative. Most innovations are a result of creative thinking. Sometimes valuable innovations happen by accident, or **serendipity**. Some examples of serendipitous innovations include Teflon (a nonstick surface), safety glass for car windshields, and even the process for making breakfast cereals like corn flakes! In those cases, inventors were looking for something else and accidentally came up with a different idea or product. At other times innovation happens because people or companies work toward a goal using a combination of skill, creativity, and knowledge. Because innovation makes you think about "new" things, it is usually tied closely to change.

T E C H N O F A C T

Technofact 22

Teflon, a trade name for a special plastic, is used for items from nonstick frying pans to spacesuits to artificial heart valves. The discovery of Teflon is an example of serendipity. The waxy white powder was accidentally discovered by a chemist, Dr. Roy Plunkett. It has some remarkable properties. It is not changed by strong acids or by heat, and nothing can dissolve it. It is also extremely slippery, so nothing sticks to it. Because it is one of the few things the body doesn't reject, it can be useful for artificial heart valves, tracheas, teeth, corneas, and substitute bones for the nose, chin, skull, knee, and hip joints. Teflon is used for the outer skin of space suits. It is also used for heat shields and some fuel tanks on space vehicles.

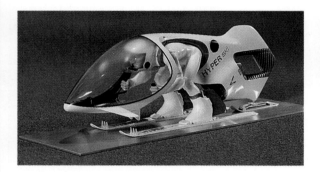

The Hyper Ski is a very innovative prototype, or model, of a snow ski machine powered by twin jet turbines. (Courtesy of GM World of Motion.)

The Lean Machine is similar in size and weight to a traditional motorcycle, but it can travel up to 200 miles per gallon. In this prototype the rider is inside a protective fiberglass pod. (Courtesy of GM World of Motion.)

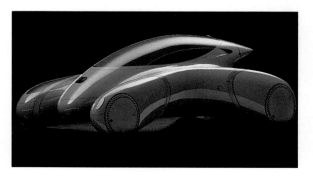

This is a prototype design of a futuristic streamlined car for two people. Anyone can design a dream car from their imagination, but innovators can carry the design through to meet the needs of people. (Courtesy of GM World of Motion.)

Change is happening all around us all the time. People who are innovators are excited about making changes. In business, change might be a different way to make the same product, or it could be an entirely new product. **Trends**, or current needs that people have, also determine what changes companies will make in their products. Many companies have R&D (research and development) departments whose job is to find out what people want now and what they will want in the future. By studying trends, R&D departments can provide the products you want to make your life better.

R&D departments design products to benefit physically challenged people. This is a computerized wheelchair with three wheels around one axle. It makes it easier to go up and down stairs. (Courtesy of Battelle, Inc.)

R&D departments are trying to develop ways to protect the ozone layer around the earth. Here different materials are being tested as a propellant in spray cans. These do not have chlorofluorocarbons (CFCs). (Courtesy of Battelle, Inc.)

Innovation is not limited to one area either. Innovation happens in every field from aerospace to education. For example, aerospace engineers working for **NASA** (National Aeronautics and Space Administration) started out in the 1960s with a space vehicle that could hold only one person without much movement. The astronauts were not very comfortable and could not stay in space for long periods of time without some movement. That meant changes and new innovations were needed. The result is the Space Shuttle, where several astronauts exercise, eat, and move freely inside the spacecraft. What do you think the next changes in space travel will be? Why do engineers need to continually make changes?

Early space vehicles were really cramped for even one person. The Space Shuttle lets up to eight people move freely. (Courtesy of NASA.)

Have students make a list of innovative products that they use.

Special Report
Physically Challenged

One reason is that what worked for us yesterday might not work for us today. That's where innovation and creativity are important. Sometimes we simply want different things such as new car styles, different exercise equipment, or new foods. At other times we need change for a specific reason. For example, today we don't know what to do with our nuclear waste products. Up to now, we haven't found a way to store the materials safely for long periods of time. If someone like you could think of an innovative way to use or recycle the materials, it would really benefit our society.

It's easy to see how some innovations have made our lives easier. You've probably seen old movies where people had to crank the engine to get a car started. Thanks to innovation, all you have to do today is turn the key. Maybe someday all you'll do is talk to the car's computer!

A   B

Old cars were kind of "cranky" compared with today's models. (Figure A Courtesy of the Henry Ford Museum, Dearborn, MI. Figure B photo by Doug Persons.)

If you don't like to tie your shoelaces or untie the knots in them, you can use elastic shoelaces. The stretchy laces make your shoes like slip-ons. It's a great innovation for young children and the **physically challenged** (those with physical handicaps).

Innovation not only makes our lives easier but it can save lives, too. In many cities and towns in the United States, you simply dial 911 to get emergency help. You might even be using a cellular phone or a cordless one to make that call.

Ask students if they ever thought about a person inventing a video game.

If you don't like to tie your shoes or to untie knots, try elastic laces. They make your shoes into "slip-ons." (Photo by Mike Hemberger.)

Many innovative ideas go unnoticed by most people. For example, today many parts of cars are made of plastics to reduce weight and to prevent rusting. Many of your clothes are made of new kinds of plastics that don't feel or look any different from the materials that were used before. In the aerospace industry, adhesives (glues) are used instead of rivets to hold airplane wings together.

Other innovative ideas, such as the development of the ruby laser, had to wait until people found a practical way to make and use them. Innovation needs some knowledge combined with creativity. There are lots of people out there with knowledge, but they don't know how to use it creatively. Being innovative means you can do something new with your knowledge and experience. It also means you can recognize when something happens accidentally or through serendipity, even if you're not exactly sure what it is!

Noland Bushnell is an example of a really innovative person. In the 1970s he created "Pong," the first interactive electronic table tennis game, because he was bored with just "watching" television. He wanted to play with the television and have the television respond. His innovation was the start of video games as you know them.

Sometimes people, like Rube Goldberg, like innovation just for the pure fun of it. Goldberg invented complicated, funny ways to do simple jobs. Innovative ideas like these usually don't become an actual product or service. How practical an idea is, how economical it is, and how it is marketed often detemine whether you and I ever see it.

Airplane wings are commonly fastened together with rivets. Today adhesives or glues are being used to make the assembly stronger and easier to produce. Above: Courtesy of G. M. Hughes Electronics Corporation; Below: Courtesy of Grumman.)

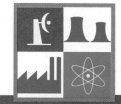

TECHNOFACT

Technofact 24
How would you like to rent a special one-of-a-kind Cadillac limousine available in Beverly Hills, California? It seats twelve and has a satellite dish, large-screen television, VCR, microwave oven, freezer, refrigerator, pinball machine, personal computer, telescope, and even a library. If you want it to, the roof can slide back, the seats fold up, and there is a hideaway hot tub for you to use!

Noland Bushnell's invention of "Pong" in the 1970s led to today's video games. (Courtesy of Nintendo of America Inc.)

Techno Teasers
The 25-Step Pencil Sharpener

Techno Teasers
Answer Segment

Mad Machines

In the early 1900s, a talented engineer and cartoonist made funny drawings of incredibly complicated machines designed to do simple tasks. In his newspaper cartoons, Reuben Lucius Goldberg included things such as leaking hot water bottles, birds, pulleys, strings, kites, and many other common items in his silly drawings. Today we call any machine that is made too complicated for a simple job a "Rube Goldberg."

Today, we remember Mr. Goldberg for his incredible imagination, creativity, and one other thing. The highest award that a cartoonist can get is call a Reuben in Mr. Goldberg's honor. This is the cartoonist's version of an Oscar.

Rube Goldberg is famous for his silly and complicated ways to do simple things. (Courtesy of King Features.)

THINGS TO SEE AND DO:
A Rube Goldberg Invention

Introduction:
Sometimes designing a silly way to do something can spark ideas about how to do it in a practical way. Whenever a machine is made too complicated to do a simple job, it is called a Rube Goldberg. Mr. Goldberg would be happy to know that his cartoons still entertain people today. In this activity, you will design your own "Rube Goldberg" machine to do a simple job.

Design Brief:
Design and draw a "Rube Goldberg" type machine that will do one of the following simple jobs:

- Turn on a television
- Turn the page in a book
- Water a house plant
- Untie your shoe
- Feed your dog
- Press and shape cookie dough
- Hammer a nail
- Fry an egg
- Turn on a light

Your "machine" must include at least one example of each of the six simple machines: pulley, wedge, wheel and axle, lever, inclined plane, and screw. You must have at least twelve different steps.

Students may also think of their own task for their Rube Goldberg machine.

Materials:
- Pencil
- Paper

Equipment:
Optional:
- Computer with graphics software
- Printer
- Scissors
- Computer with clip-art software

Your Rube Goldberg invention might look something like this.

TECHNOFACT

Technofact 25

In Hong Kong, the latest microwave technology is being used to dry hair at beauty salons. It takes only 90 seconds to totally dry an average head of hair. But hairdressers have to be careful. If they forget to remove any hair curlers or bobby pins, they will explode or melt when the microwave hair dryer is on!

Fads come and go. (Photo by Doug Persons.)

Procedure:

1. Work in groups of three or four. Choose one of the tasks listed above, or think of your own simple job.

2. Make a sketch of each of the six simple machines.

3. Cut out each of the simple machines and arrange them on paper in the order your group chooses.

4. Make a rough sketch of your design. Discuss your design with other students to see if they can follow the steps.

5. Refine your design, and make a finished cartoon on paper or computer.

6. Show your design to students in a lower grade to see if they can understand what your machine is designed to do.

7. Write a step-by-step list of how your machine works.

Evaluation:

1. Why are cartoons of this type funny?

2. List five parts of real machines that are examples of simple machines.

3. How could a computer be used to make cartoons?

Challenge 1:

With your teacher's help, use parts of old machines and recycled materials to build a working model of your Rube Goldberg machine.

Challenge 2:

Can you think of a way to have your machine do two things at the same time? Make a quick sketch to show your idea.

Getting Ideas

Where do ideas for inventions start? There really isn't a specific set of steps to follow. Lots of innovations are a result of looking at old ideas in a new way. For example, Johann Gutenberg combined two unconnected ideas, the coin punch and the wine press, to create a new product—the printing press and movable type. Other inventors set out to solve a specific problem. The search for cures for diseases like AIDS is ongoing in medical research laboratories. People also come up with ideas for inventions by serendipity, or just being in the right place at the right time. Edward Jenner's smallpox vaccine was a result of serendipity. It has since saved millions of people from a horrible disease.

Sometimes innovations result in **fads**, or temporary ideas. Many fads disappear quickly because people enjoyed them for the moment but did not find a real need for the product or idea. What are some fads of today?

One thing that all inventors have in common is thinking, or using their brain. As recently as 500 years ago, most people did not even know

Serendipity—Safety Glass

The accidental discovery of safety glass happened when it was most needed! The automobile with glass windshields had just been invented. Glass, of course, has been around a long time. Even the Romans used it for windows in their buildings. But it wasn't until the car was invented that glass windows became a real safety problem for drivers and passengers.

A French chemist, Edouard Benedictus, invented safety glass in 1903. Here's how it happened. One day he dropped a glass flask on a hard floor. The flask shattered, but he was surprised that the tiny pieces of glass didn't fly apart. When he looked at the flask closely he saw that a film on the inside was holding the shattered pieces together. Benedictus has been working on an experiment using a special plastic solution called collodion (cellulose nitrate made from cotton and nitric acid) that had been in that flask. When the collodion evaporated, it left that special film. At that time, Benedictus put a note on the shattered flask but then moved on to his own project.

The invention of safety glass has saved many people from injury. (Photo by Doug Persons.)

Later the next week, after reading two stories in the paper about people injured by glass windshields, he dashed to his laboratory to find that flask. He spent the next night figuring out how to use a coating to protect glass from shattering. The name "triplex" was used to describe the new safety glass. It was made of three layers with two sheets of glass as the outside layers. The inside layer was cellulouse nitrate. Benedictus did not take out a patent until 1909 for his safety glass. Even then, it wasn't until the 1920s that layered (**laminated**) windshields became standard parts of all U.S. automobiles. Before Benedictus's safety glass was used in windshields, it was being used to make the lenses of World War I gas masks.

Since then, there have been other changes in safety windshields. One is tempered glass. It doesn't have a plastic layer. It shatters into very small pieces, and so it is not so dangerous. It is used in most side and back windows of cars.

In the late 1980s, a plastic layer was added to the inside of the automobile windshield. In tests, the plastic coating called the antilacerative shield prevented people from getting any cuts.

Challenge students to make a large chart of the brain and label the functions of the left and right hemispheres.

T E C H N O F A C T

Technofact 26

With powerful computer systems and very sophisticated graphics, people can create virtual (not real) environments where your dreams can come true. If you think you'd like to fly, you would like the new HiCycle innovation by Autodesk. If you pedal a bicycle fast enough, you can "take off" and "fly" above a graphic landscape on the computer screen. You aren't really flying, of course, but the goggles you wear give you a three-dimensional (3-D) effect and you think you are!

where their brain was located in the body. If you didn't know where your brain was located, where would you think most thinking and emotion took place? Many people thought it was in the heart or stomach areas, because that's where you feel pain. Even today, after years of research into how people think, it is estimated we still know less than one percent of all there is to know about the brain.

Scientists have discovered that there are really two very different sides to the human brain. The left and right sides of your brain work in different ways. They are linked together by a very complicated network of nerves called the **corpus collosum.** In most people activities such as thinking logically, dealing with numbers, reasoning, and language skills are located in the left brain. The right side of your brain deals with imagination or creativity, music, rhythm, art, and daydreaming. It has been found that great inventors such as Leonardo da Vinci and Albert Einstein used both halves of the brain at the same time. Most of us don't do that as much as we could!

This front view of a human brain shows the different kinds of mental activity in each half. Are you using both sides of your brain as much as you could?

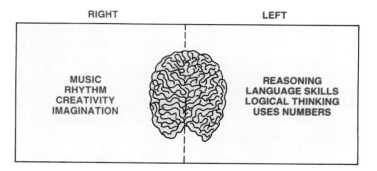

Both Leonardo da Vinci and Albert Einstein could use both halves of the brain at the same time. Do you think that ability is a mark of a great inventor and thinker? (Courtesy of The Bettman Archive.)

Inventors are also good at **visualizing**, or picturing ideas, in their minds. Often they can see a different way to do something that everyone else has missed. Or they can look at an everyday object and find new uses for it. For example, look at a regular pencil and then visualize it being used for something else besides writing.

Innovations make your world exciting. Besides, you're used to having new things happen. Can you imagine a world where nothing changed? What you have to remember is that invention combines ideas with some technological information. Inventors have to try new ideas, change old ideas, and look for new ways to solve today's problems. The key is knowing when, or if, to move from a **prototype** (model) to the real thing!

THINGS TO SEE AND DO:
Product Design

Techno Teasers
Mail by Rocket

Techno Teasers
Answer Segment

Techno Teasers
Old Ideas—New Ones

Techno Teasers
Answer Segment

Introduction:
New products or inventions are often a combination of old ideas put together in a new way. Inventors must be able to communicate their ideas in words and drawings. In this activity, you will invent new tools that someday may be as common as a pair of scissors.

Design Brief:
Invent four possible new tools that are a combination of two or more existing tools. You may combine some of the tools shown, or think of your own such as a spatula, egg beater, or spoon.

Some tools that perform several functions are available today. Challenge the students to think of new combinations.

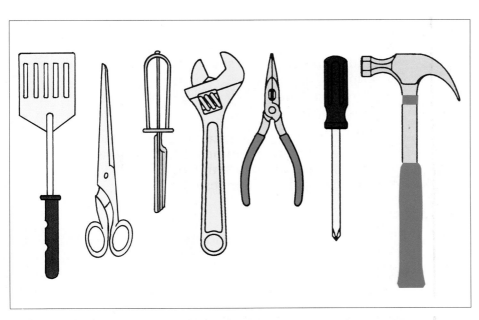

Choose two or more of these tools to combine into a new tool, or think of other possible combinations.

Materials:

- Pencil
- Paper

Equipment:

Optional:
- Computer
- Graphics/CAD software
- Clip art
- Word-processing software
- Printer
- Tape recorder
- Drawing tools
- Sample tools

Procedure:

1 Work with a partner. Brainstorm some possible ideas for a new tool.

2 Choose four of your ideas that you think have the most **potential** (ability to succeed).

3 Make some rough sketches of your four tools. Optional: Use a computer with graphics software to make your drawings.

4 Refine your ideas, and show them to your parents, relatives, teachers, or friends. Listen to the reactions of others. This is called **feedback**.

5 Redesign your ideas with the feedback in mind.

6 Make a finished drawing of all four of your ideas. Optional: Use computer with graphics or CAD software to make your finished drawing.

7 Choose one of your ideas that you like best. Write or use the word processor on the computer to make a script for a radio commerical to sell your product. Include in your script:

- Product name
- Cost
- Possible uses
- Where to buy it

8 Optional: Record your commercial on an audio tape recorder. Play your commercial for the class. Ask the class to sketch what they think your tool looks like without showing them your design.

Encourage students to think how the tool could be used safely for each of its functions.

Evaluation:

1 How many different tool ideas were you able to brainstorm?

2 What was the hardest part of the assignment? What was the easiest? Explain your answers.

3 Did you consider whether your tool design was safe? Are there dangerous pinch points, or sharp edges on your design?

Challenge 1:
Produce a video commercial for your product.

Challenge 2:
Find someone in your community who has designed a product. Ask that person to talk to your class, or interview her or him on the phone.

Challenge 3:
Think of other possible combinations of products. Some you might consider are:

▶ Athletic equipment
▶ Camping gear
▶ Food products
▶ Home appliances
▶ Biking, boating, skiing or other sports equipment

Who Invented That?

Invention and innovation have a long history. Many ancient discoveries and inventions that have been improved on over hundreds of years can't be credited to any one inventor. For instance, we don't know who discovered fire or invented the wheel. Thousands of inventions had a great effect on society. Many others didn't. You probably are aware of how television has affected our society. But you might not be aware that

Special Report
X-30 Space Plane

Have students brainstorm names of famous inventors and what they invented.

TECHNOFACT

Technofact 27

Imagine flying from New York to Tokyo traveling as fast as 25 times the speed of sound (1,100 feet per second)? It would take you only 2½ hours to make that trip in the National Aerospace Plane called the X-30, funded by NASA and the Department of Defense. Unlike the Space Shuttle, the X-30 can climb directly into orbit without booster rockets. It can cruise at hypersonic (greater than the speed of sound) speeds and then land like an airplane!

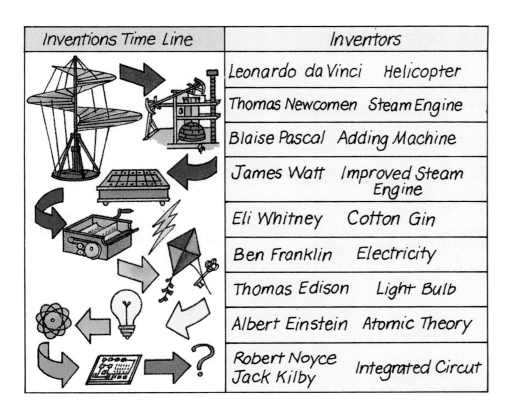

Inventions Time Line	Inventors	
	Leonardo da Vinci	Helicopter
	Thomas Newcomen	Steam Engine
	Blaise Pascal	Adding Machine
	James Watt	Improved Steam Engine
	Eli Whitney	Cotton Gin
	Ben Franklin	Electricity
	Thomas Edison	Light Bulb
	Albert Einstein	Atomic Theory
	Robert Noyce Jack Kilby	Integrated Circut

"Who invented what?"

there is an invention called an automatic hat-tipper. It was invented at a time when most men wore hats, and it was considered polite to tip your hat to every lady you passed. As the technology of the time changes and peoples' needs change, we expect invention and innovation to change also and to make our lives better.

THINGS TO SEE AND DO:
Invention Database

Introduction:

You have learned that some inventions can't be credited to just one person. With advancements in writing, better records were kept about events in history. Today we can research hundreds and even thousands of inventions to trace their development and find out who had the original idea. In this activity, you will make a computer database of inventions and inventors. It will make research easier.

Have other classes add to the database. It may be possible to share your database with classes in other schools.

Design Brief:

Design and create a database of inventions and inventors. The **format** (layout) of your database might look like this:

Materials:
- 3″ × 5″ file cards
- Research materials

Equipment:
- Computer with database software

DATABASE FIELDS:

Invention _____

Inventor _____

Date _____

Category _____

Abstract _____

Reference _____

SAMPLES:

Light Bulb

Thomas Edison

When Did it Happen?

Biotechnology
Construction
Communication
Computers
Machines
Processes
Transportation

Short Description of the Invention.

Where Did You Find It?

Invention database format.

Procedure:

1 Design the records and fields of your database so that everyone in the class is using the same format.

2 Each person in the class should research three inventions. You might use resources in your own classroom or the library.

3 Record your three inventions on a 3″ × 5″ file card using the format shown on page 92.

4 When you have finished your research, enter the data (information) into the computer database.

5 Put your file cards in a file box together with all of the cards from the rest of your classmates.

6 When everyone has finished, work as a class to sort the file cards according to the dates of the inventions.

7 Sort the computer database according to date also.

8 Combine the records made by other classes with your class's database.

9 Use the combined invention database to search for various inventions, inventors, and categories.

Evaluation:

1 Which took longer, sorting the cards by hand or using the computer database?

2 How else could you use the computer database to sort the inventions?

3 How could your database help other students find information in the future?

Challenge:

Make an invention timeline by tacking the file cards to a bulletin board in order. Make drawings or sketches to illustrate each invention.

Getting a Patent

A **patent** is a special government license that protects your idea or invention. When you patent your invention, anyone who wants to use your idea must get your permission or pay you. A patent protects an invention for 17 years. Anyone can patent an invention, but getting a patent takes time and money. You must first prove your invention is new or the first of its kind. This means going back through the thousands of patents filed at the Patent Office to be sure no one else has an existing patent for the same invention. Then you have to provide written plans and sketches that show how your invention works. Often you have to make a prototype, or model. The government keeps these plans on file while your patent is current.

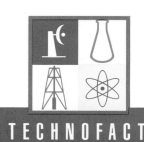

TECHNOFACT

Technofact 28

In the 1890s, the director of the United States Patent Office asked Congress to close the office. He said that all the inventions that ever could be made were patented and that the Patent Office was no longer needed. Don't you wish he were here today to see what has happened? Over 90 percent of today's technologies have been invented in the last 30 years!

Have students find the patent number on various products.

There are five types of patents that you can apply for:

1 **Design patent:** A drawing and protection of the general appearance and how an invention looks to the eye. An example of a design patent might be the outside package for an invention.

2 **Structure patent:** A mechanical patent that shows how the idea works, its history, and detailed plans.

3 **Process/method/system patent:** Protects the way a group of materials or parts work together. These usually have specific drawings and flowcharts.

4 **Combination of materials patent:** Descriptions of materials and how they are mixed and used.

5 **Living cell patent:** Covers new developments in biotechnology and biochemistry.

Every patented product will have a patent number on it. If you see the words "**patent applied for**" or "**patent pending**" on a product, it means the inventor has applied for a patent. Sometimes companies will start making a product before they have their patent. Because there are so many inventions all the time, it is important to protect inventions

Patents are important. They protect the inventor. (Courtesy of Canon USA Inc.)

(Courtesy of Sony Corporation of America.)

with a patent. You can imagine how disappointed you would be if you spent a great deal of time and money on an invention idea only to find out someone else already had a patent on a similar project. Getting a patent is an important protection step for anyone with an invention.

THINGS TO SEE AND DO:
Patenting Your Ideas

Introduction:

Your special ideas may be worthy of a patent. Not every idea or device can be patented, however.

Things That Can Be Patented:

▶ Processes or machines
▶ A method of manufacturing
▶ A new material or life form

Things That Can't Be Patented:

▶ Discovery of scientific principles
▶ A naturally occurring material
▶ A way of doing business

In this activity, you will apply for a patent to protect your ideas.

Design Brief:

Design, test, and patent a device that will protect a raw egg from breaking. Your packaging device or method will be tested by dropping the egg from a 15-foot height to the ground. At least one-half of the egg must be visible at all times during the test.

Materials:

You may use any materials you like for the package, but they must be safe to drop from a second-story window. For example, you might use recycled packaging materials such as plastic, foam, or paper. Other supplies you will need might include string, tape, rubber bands, and scissors.

Discourage students from dropping live chickens!

Procedure:

❶ Work in groups of two to four students for each invention team. Your teacher will assign one group to be the "Patent Office."

❷ The invention teams will start to brainstorm ideas on how to meet the design brief **specifications**, or requirements.

❸ The students working in the "Patent Office" should write down the specific requirements for the design brief. They should then design a patent application form. Here is a sample you can use or change to fit your needs.

Official Patent Application
Form 1-a

• Patent Office •

INSTRUCTIONS:

1. Complete the questions and drawings on this form.
2. Submit the form in person to the Patent Office.
3. You will be notified by a Patent officer if your patent was approved or denied.

Patent Office Use Only:
Patent application date:___/___/___
Patent application time:_____a.m–p.m.
Patent approval date:___/___/___
Patent approval time:_____a.m–p.m.
Patent number:_____

1. Complete the following information for each of the invention team members.

Invention Team Member 1:
• PLEASE PRINT •

NAME:_____ ____ _____
              LAST   M.I.   FIRST

Invention Team Member 2:
• PLEASE PRINT •

NAME:_____ ____ _____
              LAST   M.I.   FIRST

Invention Team Member 3:
• PLEASE PRINT •

NAME:_____ ____ _____
              LAST   M.I.   FIRST

Invention Team Member 4:
• PLEASE PRINT •

NAME:_____ ____ _____
              LAST   M.I.   FIRST

2. Describe your invention in the space provided below. Be specific.

_____
_____
_____
_____
_____
_____
_____
_____
_____
_____

3. List the quantities and types of materials used in your invention.

QUAN	MATERIAL	QUAN	MATERIAL
_____	_____	_____	_____
_____	_____	_____	_____
_____	_____	_____	_____
_____	_____	_____	_____
_____	_____	_____	_____

4. Make a detailed drawing of your invention. Make the drawing on a separate sheet of paper and attach it to this form. Be sure to clearly draw and label each part.

Patent application form

④ Each invention team should decide on a design and complete the patent application.

⑤ If another team had an idea similar to yours, your patent will be denied. If this happens, your team should choose a different design.

⑥ If your team is granted a patent, it will be given a number. Make your egg-protection package according to your design. Be sure to put your patent number on your finished package.

⑦ When all the teams have finished, test each egg package by dropping it from a 15-foot height.

⑧ Clean up any mess!

Evaluation:

① If your design survives the first test, try dropping it for a longer distance first. Then try tossing it down instead of just dropping it. What did it take to crack your egg?

② Did the patent office evaluate your patent application fairly? Explain.

③ List five different packaging materials, and rate them from 1 to 5 (best is 1).

Monitor the students working in the patent office. Be sure they are treating each application fairly.

Challenge 1:

Contact a technology class in another school in another state or in a foreign country. Design and make a package to send a light bulb. Send the package to that school with instructions to drop it from a second-story window before opening it. Then they should test the bulb to see if it still works. If it does, they should design a different package from yours and send it back to you to retest. See how many trips the light bulb can survive.

A Christmas tree ornament or other delicate object may also be used.

Challenge 2:

Find out if anyone in your community has a patent on a product. Invite the person to talk to your class about his or her invention.

Summary

Innovation is a new idea or approach to doing something. When innovation happens by accident, it is called serendipity. R&D (research and development) departments in companies are always trying to find new innovations. Innovations make our lives easier. Some innovations do not last; these are called fads.

Some of the great innovators in history were able to use both sides of their brains at the same time. The left side of the brain controls mostly thinking, numbers, and language skills. The right side of the brain deals with creativity, imagination, music, and daydreaming.

Inventors apply for patents or licenses that protect their ideas from anyone else using them. Patents are difficult to get. Special steps must be followed to get a patent issued for your invention. One of the major steps is to prove that your idea is new and the only one of its kind. Some companies produce a product even if they do not have their patent yet. Usually their product will be marked with "patent pending" or "patent applied for." All patented inventions have a special patent number.

Patented products have a special patent number. Some products waiting for patent numbers will be marked "patent applied for" or "patent pending." (Photo by Mike Hemberger.)

Challengers:

1 Invent a new way to hold loose papers together.

2 List ten different uses for a plastic spoon.

3 Invent a new eating utensil.

4 Invent an automatic pet feeder (either Rube Goldberg style or a practical one).

5 Make a list of ten products and their patent numbers.

6 Make a model of your favorite invention.

7 Invent a new board game.

8 Design a new package that is environmentally safe for your favorite breakfast cereal.

9 List how many things you can do with a newspaper besides read it.

10 Invent something that would help people who are physically challenged.

See Teacher's Resource Guide

Special Report
Riblets

Chapter 5

Making Things

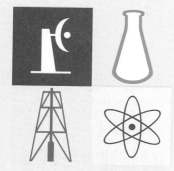

Things to Explore

When you finish this chapter, you will know that:

▶ Resources are used to produce the products we use.

▶ Some resources are not renewable. They cannot be replaced once they have been used up.

▶ Conserving resources and energy helps to make a better future for the next generation.

▶ Processes are used to change raw materials into finished products.

▶ It is necessary to understand and follow safety rules when using hand and power tools.

▶ Everyone can help to conserve energy and recycle materials.

Chapter Opener
Making Things

TechnoTerms

arc welding
closed system
compression
  molding
conserve
decompose
electroplating
extrusion
fossil fuels
geothermal
grinding
injection molding
interest

milling
nonrenewable
oxyacetylene welding
process
quality control
recycling
renewable
resources
shaping
spray painting
thermoforming
turning

Kintigh Station.
(Courtesy of New York State
Electric and Gas.)

Careers in Technology

Products and raw materials such as steel and lumber are all a part of the manufacturing industry. Even with automation and advancements in technology, it still takes thousands of people to produce the products we use and throw away every day. The control room shown here is used to monitor the production of steel used to make products such as cars and refrigerators. The operator can control millions of dollars worth of equipment with computerized precision. The steel produced in this way can be machined or formed exactly to the shape desired by the customer. If you had the training, do you think you could take on the responsibilities of operating this equipment?

Ask students to make a list of the resources needed to make a product.

Where Do We Get Resources?

Before you can make anything you will need **resources**. Resources can be anything that is used in the production of a product. What are some of the resources you can use, and where do you find them? Your resource list might include:

▶ *People*: Technology has been created by people. People have used their ideas and knowledge to invent and build products that meet their needs. Some companies hire people for their ideas and skills just as other companies buy raw materials for their products. You might need to contact people skilled in a particular field who can help you with your ideas and designs. With the proper training and education, you can be a resource person for others, too.

▶ *Machines*: Since the invention of the wheel and other simple machines, people have used their intelligence to create the many complex machines we rely on today. We rely on machines as a resource to help us do work, and through technology those machines often run automatically; we don't even have to think about them.

People are important resources. It is a great feeling to be able to help someone solve a problem. How could you be a resource to help others? (Courtesy of Motorola.)

Information is a very valuable resource. The amount of information today is enormous. Complex problems can sometimes require the use of a supercomputer like this one to quickly calculate and use data. (Courtesy of Martin Marietta Corporation.)

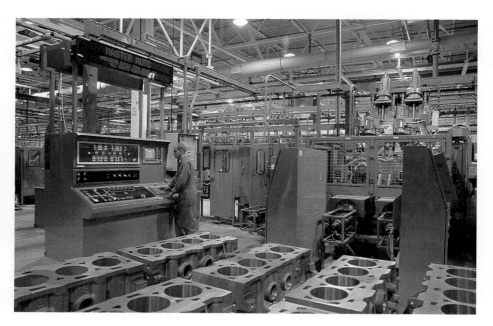

Machines are resources that help people by extending their muscle power to do things that would be impossible otherwise. Today, many machines operate automatically or with very little help from people. (Courtesy of The Ingersoll Milling Machine Company.)

▶ *Information*: You found out in Chapter 1 that the amount of information in the world is growing rapidly. Because there is so much information there are companies that gather, organize, and sell the information. People use this information to design, produce, and sell products.

▶ *Raw Materials:* You probably know that natural resources include water, land, minerals, **fossil fuels** (remains of dead vegetation and animals such as oil, coal, natural gas), and timber. But did you know that some of these resources are **nonrenewable**? Nonrenewable means that once the resource has been used up, it is gone forever! For example, the United States is highly dependent on oil. We use almost 16 million barrels of oil daily. The world's supply of oil will not last forever.

▶ *Energy:* Energy is used to make things and to transport products. It is also used to heat, cool, and light the buildings that we live and work in. Like raw materials, some sources of energy are limited. **Renewable** (resources that can be replaced) energy sources include solar energy, **geothermal** (heat from the earth) energy, and nuclear energy.

▶ *Money:* Even if you had all the other resources available to you, it still takes money to start a business. Money to start new businesses is often borrowed from banks. Loans must be paid back as well as any **interest** on the loan. The lender charges you additional money for letting you use its money. This charge is called interest. Even after a company is making a profit (money left over after all bills are paid), it continues to pay for the other resources it uses.

▶ *Time:* Many people consider time a resource because it takes time to make a product. In many instances, however, time is a factor that determines which other resources you might use and how you might use them. If you have to produce a product in a short time, you might choose only materials that are readily available and people who are already trained for that job.

Bauxite is a raw material used to make aluminum. Mechanical, chemical, and electrical systems are used to change this raw material into the metal we use every day. (Courtesy of ALCAN.)

Nuclear energy can produce much of the electricity we need. The problem is what to do with the radioactive waste products. (Courtesy of Tennessee Valley Authority/Ron Schmitt.)

Natural gas is sent through pipelines all over the country. Natural gas is a great resource, but it is not renewable. (Courtesy of Columbia Gas System, Inc.)

Energy is a resource that can come in many forms. Here coal is taken from the ground to be burned to produce electrical energy. (Courtesy of Coastal Corporation.)

TECHNOFACT

Technofact 29

Do you know what destroys more ozone than CFCs (chlorofluoro-carbons) do? It's nitrous oxide, more commonly called laughing gas. The amount of nitrous oxide in the atmosphere has been steadily increasing. Researchers once thought most of the increase in nitrous oxide was coming from the burning of fossil fuels and tropical forests. Guess what? Now researchers think it is coming from the manufacture of nylon products. The main problem is that once nitrous oxide enters the atmosphere it takes about 150 years for it to break down. So some companies that manufacture nylon are trying to either destroy the nitrous oxide or trap and recycle it before it escapes to the atmosphere.

Using Resources

Early people relied on muscle power to survive and to make everything. You learned in Chapter 1 that as technology grew and changed, people were able to extend their muscle resources by using machines. The energy of moving water, steam, oil, and the atom have been added to the list of resources that people can use to make things. Can you name other resources we use now to make products?

Whether you choose a resource often depends on its availability. For example, if you were going to start a company that needed a great deal of electricity, you would try to find an area in the country where electricity is available and inexpensive.

As you design your product you should pick material resources that best fit your product. Paper clips made of pure silver sound great, but they would be too heavy and too expensive.

You also want to use resources wisely and not waste them. The cost of energy used to make and transport products is increasing. If energy can be **conserved** (saved) in the manufacturing process, we can save energy and reduce the cost of products.

Some nonrenewable resources such as aluminum can be used appropriately through **recycling** (using them again). Aluminum requires a great deal of electricity to produce the first time. By recycling, you can save not only the mineral resource but energy as well.

Energy can be conserved in many ways. One way is to recycle materials such as aluminum. (Courtesy of ALCAN.)

Recycled aluminum may end up as another soda can or as a sports car! (Courtesy of ALCAN.)

THINGS TO SEE AND DO:
Setting Up a Recycling Center

Techno Teasers
The Energy Song

Techno Teasers
Answer Segment

Introduction:

Recycling materials such as aluminum, glass, paper, and batteries can help to save our environment. You and your school can help to make a difference. Just think of all of the paper and aluminum cans that are discarded every day. If your class can help to make recycling easy and convenient, more people are likely to pitch in and help. If students and teachers get into the habit of recycling at school, they are more likely to do it at home too. You can do a lot to save the earth; here's how.

Design Brief:

Design and make a recycling center for your school or classroom. The center should provide containers that are clearly marked for the materials to be recycled. Your design should also include posters that help people understand the importance of recycling and instructions on how to use your recycling center.

Materials:
- Posterboard
- Markers
- Garbage cans
- Plastic bags

Equipment:
Optional: ▶ Computer with graphics software

If recycling is already being done at your school, have students research where the recycled materials are used.

Procedure:

1 Work as a large group. Brainstorm ideas about setting up a recycling center in your school. Ask the principal of your school to give your class advice on the best way to set up your recycling center.

2 Decide how you would like to have your center work and its location. If your school is very large, you might consider having two or more centers or a small center for each classroom.

3 Make a plan of who will be responsible for each detail of your center. Some of the responsibilities might include:

- Who will empty the containers when they are full?
- Where will the materials be taken for bulk recycling?
- How will you safely handle broken glass or heavy paper?
- Which teachers can you get to help?
- Who will make the posters to encourage people to use the recycling center?

4 When your class has planned the recycling project, advertise in the student newspaper, bulletin boards, and any other place that students and teachers will notice.

Ask students if they are willing to change their expectations of products. Would they be happy to use recycled paper that isn't as white as first-generation paper?

Special Report
Making Things with Processes

Techno Teasers
Processes—Exercise Segment

Techno Teasers
Answer Segment

5 Start your recycling project with a special announcement from the principal or cooperating teacher. Encourage other students to help with the project.

Evaluation:

1 After the first week of your project, collect the materials left at your center. Evaluate the effectiveness of your project, and determine if changes are necessary.

2 Which recyclable material is most common in your school? Which material is least used?

3 After one month, make an estimate of the amount of material your center will recycle in a school year. What would have happened to the recycled material if your project had not been started?

Challenge:

1 Make a large poster that illustrates in a graph the amount of material your school is recycling.

2 Research the materials used commonly in your school, such as copier and computer paper. Find out if those materials are available in recycled form. Research the price of new paper and recycled paper.

How Products Are Made

Do you ever think about how things are made? Everyday products are made in factories using different materials and machines. Products such as the pen or pencil you write with, the paper you use, and the bus that may have brought you to school are all made using **processes**. Processes are special machines or operations that help change a raw material into a finished product. Processes can be put into groups that are similar to each other. The major processes used to make products are:

▶ Forming: Molding or shaping materials such as sheetmetal or plastics

▶ Machining: Changing the shape of materials by making chips

▶ Fastening: Hooking materials together with nuts and bolts, welding, or adhesives

▶ Finishing: Painting, varnishing, or plastic coating products or materials

Most of the machines that you might see in a factory can be put into one of the four processes. You learned in Chapter 4 that it is important to measure accurately. Measurement can be done by hand or automatically by machine. As products are made, they are often checked to see if they are the right size or shape. This inspection of parts is called **quality control**. We will learn more about quality control in Chapter 10.

The portable orbital sander can speed up the smoothing process. Abrasive paper (sandpaper) is put on a pad that makes small circles or back-and-forth motions. (Photo by Tom Carney.)

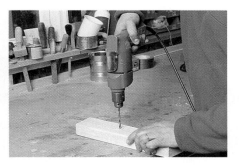

The power hand drill is a very useful tool for drilling holes in woods, metals, or plastics. Other attachments can be added to power drills to do sanding or grinding. (Photo by Tom Carney.)

Show students five different products. Ask them to describe the processes used in their manufacture.

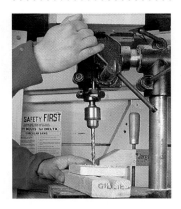

The drill press can drill holes easily and accurately in many materials. You will use a drill press to help make many things as you complete technology activities (Photo by Tom Carney.).

Power sanders help you shape and smooth materials. They can save a lot of time. Disk sanders use a spinning disc with abrasive paper (sandpaper) glued onto its surface. Belt sanders use a long belt of abrasive paper that moves around rollers inside the machine. (Photo by Tom Carney.)

The portable saber saw can be used to cut woods, plastics, or metals if the proper blade is used. The blade moves up and down very rapidly. Be sure to check what is on the other side of the material you are cutting. (Photo by Tom Carney.)

Band saws and scroll saws are used to cut straight lines and curves in different materials. Band saws have a long blade (band) that rotates in one direction. Scroll saws have a blade that moves up and down rapidly. (Photo by Tom Carney.)

A strip heater can be used to melt plastic so that you can bend it into another shape. The strip heater heats a straight line (strip) so you can easily melt and bend sheets of plastic. (Copyright © Pam Benham.)

Manufacturing Processes

You have learned only a few of the many processes that are used to make products. People have been developing new processes and perfecting old processes since the Stone Age. Today the development of processes is especially rapid. Technology has made it possible to use lasers and high-pressure jets of water to cut through tough material, for example. Here are just a few of the hundreds of processes used to manufacture products.

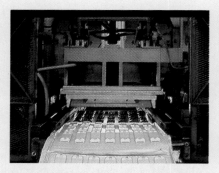

Thermoforming. (Courtesy of Crystal Thermoplastics Inc.)

Turning. (Courtesy of Simmons Machine Tool Corporation.)

Sanding. (Courtesy of Delta International Machinery Corporation.)

Arc Welding. (Courtesy of Miller Electric Manufacturing Company of Appleton, Wisconsin.)

Metal is melted into a liquid in a foundry furnace. The melted metal is poured into molds and allowed to cool into the shape you need. (Courtesy of FMC Gold Company.)

A hot glue gun can fasten two materials together quickly. The gun melts a plastic adhesive (glue) that hardens quickly as it cools (Photo by Tom Carney.).

Sheet metal is often cut with a shear. The shear works like a strong pair of scissors to cut thin sheet metal into useful shapes. The flat sheets are often bent into shape using a break. (Courtesy of Reynolds Metals Company.)

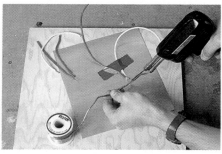

Metals are sometimes fastened together using solder. A solder gun heats metals such as wires so that they can melt solder. Solder is usually made with tin and lead. A chemical called a flux is often used to clean the metals being soldered (Photo by Tom Carney.).

You have already been using processes to do the activities in this book. Now you will learn how to use more machines so that you can safely make more things. The new things you make will help you understand how technology works. Study the following pages carefully so that you feel comfortable using machines. Your teacher will show you how to use the machines in your technology lab. Ask questions about anything that is not clear to you.

THINGS TO SEE AND DO:
Making Products

Introduction:

Not all of the products made today are designed with conservation of resources or recycling in mind. This activity will give you a chance to design a product that can be used to recycle aluminum cans. One of the problems with recycling cans is that they take up a lot of space. By crushing them, many more cans can be transported easily to a recycling center.

TECHNOFACT

Technofact 32

Junked or abandoned cars are a garbage problem. Nationwide, 20,000 cars are junked each day. What can we do about it? A German car manufacturer is trying to produce a car that's 100 percent recyclable by the year 2000. Researchers are disassembling (taking apart) cars to experiment with reusing the materials to make different parts. They are already manufacturing glove compartments from waste products and using fabric scraps to make sound insulation. They plan to use more plastic in their cars, because plastics can easily be reused. Someday your brand new car may be totally built with recycled parts from someone's old Chevy!

Design Brief:

Design and make an aluminum can crusher that will help people save space when recycling. The can crusher must be safe to operate, easy to use, and effective. The cans must be crushed to less than one-half their original volume. The materials used to build the can crusher must also be able to be recycled.

Materials:
- Wood: scrap 2″ × 4″, 2″ × 6″, plywood, etc.
- Steel or aluminum strap: 1″ × 1″ × ⅛″
- Miscellaneous fasteners
- Abrasive paper

Equipment:
- Band or scroll saw
- Belt or disk sander
- Drill press and drill bit set
- Hack saw
- Screwdriver
- Wrenches
- File

Optional: - Computer with CAD software

Procedure:

❶ In this activity, you will build your own can crusher. First you must decide on a design. You should think about how you would like to have your crusher work—by stepping on it or using your hands.

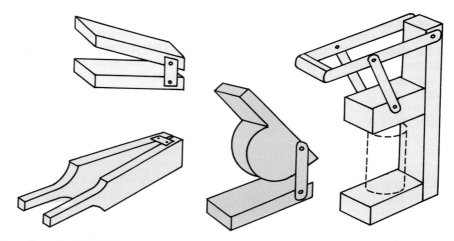

Many ideas are possible for your can crusher design. Here are a few. Be sure to design your crusher with safety in mind.

❷ Your design must consider the safe use of the crusher. You should avoid pinch points, where fingers could be caught during use. Here are just a few possible ideas:

❸ As you put the finishing touches on your design, list the materials that you will need. Your list might look like this:

Quantity	Part Name	Material
2	Linkage	1″ × 1″ × ⅛″ Steel

④ After you have finished your design and list of materials, have your teacher check your plans.

⑤ Follow the safety rules for each machine you use in making your crusher. Remember to ask the teacher for help if you are not sure how to do something.

SAFETY NOTE:
It is important that you wear safety glasses and follow the general safety rules for proper operation of any hand tools or machines.

⑥ Assemble your crusher, and check it for splinters or sharp edges that could cause injury. Use abrasive paper or a file to remove any sharp edges.

Draw the students' attention to the safety notes provided in the text.

Evaluation:

❶ Test your crusher with an empty soda can. Evaluate its performance. Decide how your design could be improved. Discuss the changes with your teacher.

❷ Write a set of instructions for the safe use of your product. Attach the instructions to your crusher before you take it home.

Our environment is too valuable to waste by spoiling it with pollution. (Courtesy of New York State Department of Economic Development.)

Challenge:

❶ Design a machine that could safely crush hundreds of cans per hour.

❷ Research machines that are designed to crush materials such as rock or coal. Make a sketch that illustrates how the machines operate.

❸ Contact the company that recycles aluminum in your community. Ask them what happens to the aluminum cans that you recycle.

Is Our World Disposable?

Are we wasting our resources? Are companies producing products without thinking about the appropriate use of resources? These are hard questions for you to answer, but you need to ask them. We often forget the fact that we live in a world that has limits. Do you ever think about the fact that the air you breathe and the water you drink are recycled? The Earth is a **closed system**. There are no other sources of air or water, for example, than what has always been on earth. It might be unpleasant to think that the air you are breathing now or the water you drank for lunch may have been used by someone else before you. If air and water are polluted, they cannot be used safely again by anyone.

You are surrounded by products that companies claim are "disposable." Think about all the paper and plastic cups, food containers, and packaging materials that your family throws in the garbage each week. What do you think happens to them? These materials often end up being wasted in landfills where they take many years to **decompose** (break down).

When you see a picture of the Earth, do you ever think about the fact that nothing new is being delivered to our planet? Do you see any garbage trucks leaving the Earth? What happens to the things we throw away? Where do we get new materials? (Courtesy of NASA.)

Studying about technology can help you become a better citizen. You will be able to make informed decisions about your future and the future of our world.

TECHNOFACT

Technofact 33

Just think about all the lights left on in U. S. homes and office buildings when there's no one there. That's a huge waste of electricity. New sensor systems can turn lights off when a room is empty and then react to any human movement in the room to turn the lights back on. Researchers predict that these smart lighting systems will save you 20 to 40 percent on your electric lighting bill.

Special Report
Energy, Recycling, & Us

Today, you must use resources carefully and be aware of how they will affect the environment after they are no longer in use. You can do things that will really make a difference.

- ▶ *Recycle Materials:* Aluminum, newspaper, glass, and other materials can be recycled, or used again. Recycling helps in two ways. It saves the energy that would be needed to make new materials, and it keeps the materials from taking up space in landfills. Recycling can also save you money.

- ▶ *Conserve Energy:* Using less electricity means burning less fossil fuel or producing less nuclear waste. If everyone used less electricity, it would help to save you money and help the environment too.

- ▶ *Get Involved:* Be a part of public service organizations in your school, your community, your state, or even at the national level. Help these groups to clean up streets, build parks, fix up old houses, or do other things to conserve resources and improve the environment.

- ▶ *Be Part of the Solution, Not Part of the Problem:* You can make a big difference. Some of the suggestions above might not sound as if they will help solve the problem, but they will. The problems facing us and our environment did not happen overnight. They cannot be solved overnight either. But consider the result of recycling or conservation over your entire lifetime. Even if you recycled only one pound of aluminum each month, you will have recycled nearly 500 pounds by the time you are 40!

- ▶ *Be Technologically Literate:* Studying technology can help you make informed choices about appropriate uses of resources and how they affect the environment. You will also be able to make informed choices when you vote for people who will represent you in government, where many decisions about handling our resources are made.

As you design systems and products in the next chapters you want to be aware of all the resources you can use. Most important, you want to be a good consumer (user) of these resources.

Summary

Before you can make anything, you have to find the appropriate resources to use. People, machines, information, raw materials, energy, money, and time are some of the resources available to you. Choosing a resource depends on its availability, usefulness, and cost. You need to use resources carefully so they are not wasted.

Everyday products you use are made through special processes. The major processes are machining, forming, fastening, and finishing. Using machines requires special safety precautions at all times.

You can make a difference in determining how our resources are used and what happens to them once they are no longer in use. Recycling materials and conserving energy are just two ways you can help save resources and protect the environment.

Challengers:

❶ Research the time it takes materials to decompose. Make a chart using pictures or a graph. Optional: Use a software program and computer to make a graph.

❷ Use a modem to gather information on a related topic from a commercial database.

❸ Interview a business person. Find out how that person is a resource person to you or others.

❹ Research areas of the world where oil is located. Make a map or computer database.

❺ Find out where your electricity comes from and how much it costs per kilowatt hour. Compare the cost with the cost of electricity in another area of the United States and in Japan.

❻ Ask the custodian to help you find out how much waste is collected in one day at your school. How much of it could be recycled?

❼ Contact an environmental protection agency near you. Find out what special problems affect your town and how you can help.

❽ Make safety posters or ads for the machines you use. Optional: Use a video camera, computers, or darkroom processes for special effects.

❾ Select an area somewhere in the United States. Research what resources are available there. Are the main industries there using those resources? Why did those companies choose to be there?

❿ Research the different kinds of indoor pollution such as asbestos, lead, household cleaners, and formaldehyde. How do they affect you and the environment?

Techno Talk
Selecting Raw Materials

Techno Talk
Quality in America

Techno Talk
The Cost of Making Things

Have students brainstorm possible ways they can help improve or protect their environment.

See Teacher's Resource Guide.

Chapter 6

How Do Things Work?

Things to Explore

When you finish this chapter, you will know that:

▶ Complex machines can be divided into systems and subsystems that are easy to understand.

▶ Systems often follow a model—input, process, output, and feedback.

▶ Most machines are made of systems such as electrical, thermal, fluid, mechanical, and chemical.

▶ Troubleshooting can help you diagnose a broken machine.

▶ Systems can be used together. When you combine electrical and mechanical systems, you have an electromechanical system.

▶ Superconductivity can greatly improve the way electrical and electronic products work.

Chapter Opener
How Do Things Work?

TechnoTerms

analogy	flywheel	resistance
chemical	force	semi-conductor
circuit	fulcrum	subsystem
compressor	input	super-conductor
conductor	insulator	symptoms
cylinder	linkage	systems
electrical	maglev	thermal
electrode	mechanical	thermocouple
electro-mechanical	output	trouble-shooting
electronics	petrochemical	
feedback	piston	
fluid	process	
	refining	

(Courtesy of NASA.)

Careers in Technology

Technology can be overwhelming to some people. When you look at something complicated it is easy to think to yourself that you can't understand it or fix it. If you break things down into systems and subsystems, you may be surprised that you can understand them. The Space Shuttle, for example, is one of the most complex machines ever built. If you look at the shuttle systematically, however, you can begin to understand how it works. Here, technicians are installing a thermal shield around the nose landing gear.

Have students make a list of the machines they use every day.

TECHNOFACT

Technofact 34

How about a car that's powered by ice? A Swedish researcher, Bo Nordell, has invented a 260-pound vehicle called the Icy Rider that can go 40 miles per hour. It runs on a pressure tank filled with nine-tenths water and one-tenth oil. When the water freezes it expands and pushes the oil into a small cylinder that contains nitrogen gas. The pressure then drives a hydraulic motor! Imagine the potential for an ice-powered vehicle in cold climates. If you replaced the wheels with runners you could use it like a snowmobile but without the noise and air pollution!

Reinforce the idea of breaking up a complicated machine into simple-to-understand systems.

Techno Teasers
Systems to Make Things Work

Techno Teasers
Answer Segment

What Is a System?

Many people don't know how the machines they commonly use work. Stereos, televisions, automobiles, and bicycles are examples of machines used by millions of people every day. We expect them to work perfectly. If they break, we often have no idea about what could be wrong or what to do about it. Something as simple as a dead battery can often make us frustrated with technology-related devices. Rather than try to learn something about how a machine works, some people say, "Well, I'm just not good with machines."

Sometimes we get frustrated with technology when it doesn't work. Something simple like running out of gas or a dead battery can make some people very angry. (Photo by Tom Carney.)

The fact is, you don't have to be a rocket scientist to understand the basic operation of machines and devices. Even the most complicated machines can be broken down into smaller **systems**. A system is a combination of parts that work together as a whole.

To make it easy for you, there is a general model that fits almost every system. The system model has only four parts: **input**, **process**, **output,** and **feedback**. Inputs are things that go into a system. It's easy to remember because you are putting things *in*. For example, in a computer system, you press keys to put instructions into the computer. The next step is to process the input instructions. In a computer system, the instructions are processed by the CPU. The process is what is done with your inputs. The final results of the process are called the output. Again, it's easy for you to remember this part because it is what comes *out* of the process. In our computer example, the output might be sound or graphics. If the system is working right, the output will be what you wanted from your input. Feedback is information about the output. It can be used to change the output so the result is what you want.

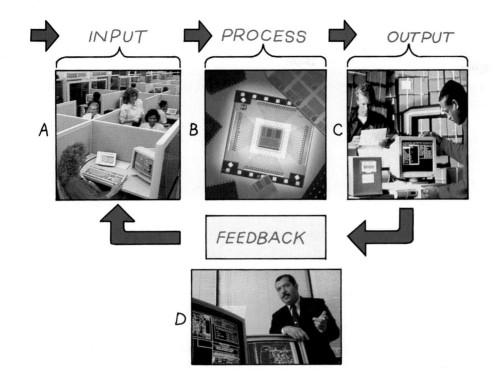

The general model of a system is simple to understand. First, something must be added (input). Then some action is taken (process). The result of the action is the output. Feedback provides information about the results. (Figures A and B Courtesy of Harris Corporation. Figure C Courtesy of United Parcel Service. Figure D Courtesy of Bechtel Corporation.)

The problem-solving strategy introduced in Chapter 1 is really just a system. You've been solving problems **systematically**. Systematically means simply that you are using a system. Even eating at a fast-food restaurant is systematic. For example, first you wait in line. Then you place your order, which is the input. Your order is processed by the workers. The output is your meal. Hopefully, it is what you ordered! If not, you might send it back. That's feedback!

Ask students to list the steps in systematically solving problems. Reinforce the fact that students will need to know how to use these steps throughout the book.

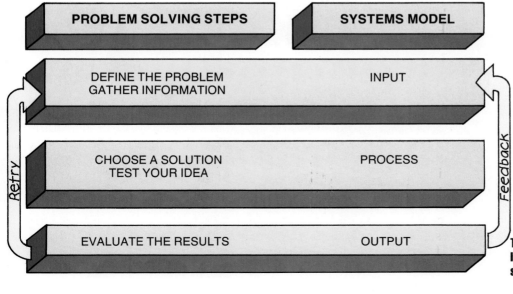

The process of solving a problem is similar to the general system model.

The five systems can easily be found on any car. Can you find examples of the five systems on other machines? (Courtesy of Chrysler Corporation.)

Techno Teasers
What's Wrong with This Car?

Techno Teasers
Answer Segment

Sometimes so many systems are put together, it is hard to figure them out. The problem can often be solved by dividing systems into smaller parts called subsystems. (Courtesy of Cleveland Institute of Electronics.)

Although there are many different kinds of systems, in technology you often use five basic energy systems to make a complete product. The energy systems are **mechanical** systems, **electrical** systems, **fluid** systems, **thermal** systems, and **chemical** systems. The five basic systems can be used independently or in combination to make something work.

Almost any machine has one or more of these systems that make it work. A car, for example, is a complex machine made up of all five systems. Here are some examples of how the five systems work together to make your car run.

- Mechanical: Door latches, fan belts, pulleys, gears
- Fluid: Water pump, shock absorbers, hydraulic brakes
- Electrical: Battery, lights, radio, ignition
- Thermal: Radiator, air conditioner, heater
- Chemical: Fuel, battery fluid, antifreeze

These systems work independently and together to make a car run smoothly and safely. When a car isn't working right, the problem is found by carefully checking each system.

Engineers, designers, technicians, and architects must know how all five systems work alone and how they work together. People who understand technology can break these complicated systems down even further into smaller systems called **subsystems**. Subsystems make it even easier to understand how things work. For example, a bicycle is made mostly of parts in a mechanical system. You can divide that mechanical system into subsystems such as brakes, steering, chains, gears, and so on. If you understand how each bicycle subsystem works, it is easier to find and fix a problem .

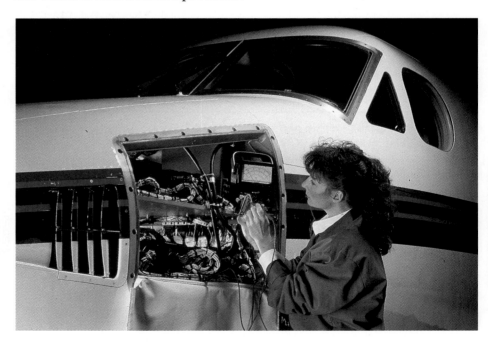

THINGS TO SEE AND DO:
Machine Dissection

Introduction:

You may have dissected a worm or a frog in science class to learn more about its parts. In this activity, you will dissect a machine to learn about its parts. Most machines are a combination of many parts from the following energy systems: mechanical, electrical, fluid, thermal, and chemical. The parts are designed to work together to make a complete system. You will find that even the most complicated machines can be divided into systems or smaller parts called subsystems.

Design Brief:

Dissect a "junk" machine and group the parts into the energy systems listed above. You will need to use hand tools to disassemble (take apart) your "junk" machine. Identify and save all of the parts for possible future use.

Reinforce the fact that the students will be disassembling their piece of "junk" for this activity. Students and parents should be aware that the goal is not to fix the product, but to dissect it and recycle the parts.

Materials:
▶ "Junk" machine
▶ Small bags or paper cups

Equipment:
▶ Various hand tools such as:
▶ screwdrivers
▶ nut drivers
▶ Allen wrenches
▶ adjustable wrench
▶ socket set

Procedure:

❶ In this activity, you will be working individually. Ask your parents, friends, or relatives if they have a "junk" machine that you could have. Please make it clear that you will not be repairing the "junk," and it will not be returned to them. Some possible "junk" items might include: toaster, alarm clock, TV, video game, toy, lawn mower, mixer, drill, blender, VCR, lamp, heater, stereo, record player, exercise machine, or bicycle. Ask your teacher about storing large pieces of "junk" before you bring them to school.

Check the "junk" that students bring in for possible safety hazards. Old lawnmowers, for example, should be checked to be sure oil and gas have been removed.

SAFETY NOTE:
Ask your instructor to inspect your "junk" and warn you of any dangerous parts you should avoid. Specifically, you should be careful with old televisions or computer monitors. They might have an electronic capacitor that can hold a charge even if the machine has been unplugged for many hours. Be careful! Ask your teacher for help.

❷ Plan how you will dissect your machine so that subsystems can be saved for future use. Specifically, you should look for some of the following parts that can be reused: speakers, motors, batteries, pulleys, belts, gears, and fasteners such as nuts and bolts.

This is a good time to introduce unfamiliar tools to students. Demonstrate the proper and safe use of hand tools.

Many of the parts from this activity can be saved for use later.

Encourage students to bring in the instructions from products. Read the instructions to find the troubleshooting sections.

3 Disassemble your machine using the right tools for each part. Ask your teacher for help if you can't loosen or remove a part.

4 Put the pieces into separate bags or cups according to their system. Identify as many of the parts as you can by name. Ask your teacher for help with parts that you can't identify.

5 Save all of your parts for future use. Present your "junk" to the class. Explain how you took the machine apart. Name all of the parts for the class.

Evaluation:

1 Which of the five systems had the most parts in your piece of "junk"?

2 Make a list of each of the parts you found in your "junk".

Do you know what **troubleshooting** is? Trying to find the problem in a system is called troubleshooting. Instructions that come with appliances such as washers, toasters, and microwave ovens often include troubleshooting charts to help you find and fix problems.

Doctors also troubleshoot. When you are sick, doctors investigate your different body systems to locate the problem. Doctors use the term **diagnose** instead of troubleshoot. They look for **symptoms**, or signs, that might give them a clue to possible problems. Part of your job in troubleshooting system problems is to look for symptoms.

Technicians often troubleshoot problems with machines.

TROUBLESHOOTING

PROBLEM	POSSIBLE CAUSE
Computer won't start	Bad plug Blown fuse
Can't save files	Bad disk Wrong command
Screen is too dim	Turn up brightness Clean your glasses

You will use one or more of the basic energy systems when you make a product, so let's find out a little more about each one.

Designing Mechanical Systems

Most machines and products contain at least a few mechanical parts. Levers, springs, nuts and bolts, screws, belts and pulleys are just a few of thousands of mechanical parts that are used on everyday machines. Mechanical parts often use simple machines such as the

Doctors diagnose problems in people. (Courtesy of GLAXO.)

lever, wheel and axle, inclined plane, pulley, wedge, and screw. Can you identify any simple machines that are parts of a bicycle?

Today, you can look in a parts catalog and find the right combination of mechanical parts needed to do a specific job. These parts are often called "off-the-shelf" because they are commonly available. Nuts and bolts, used as mechanical fasteners in many machines, are commonly used off-the-shelf items. Many other mechanical parts are made in **standard sizes** also. That means they are interchangeable and easy to replace.

Show students a catalog of mechanical fasteners.

In designing mechanical systems, you will need to know some information about the subsystems described in this chapter.

Levers and Linkages: People often need to do work that is impossible for human muscles to do alone. Levers help people to multiply their muscle strength by many times. How do levers work? When you push or pull something, you are applying a **force**. The pivot point of a lever is called the **fulcrum**. The load you are trying to move is called the **resistance**. Machines sometimes use levers called **linkages** or **cranks** to apply forces.

Can you identify the use of simple machines on this bicycle? The simple machines include, lever, wheel and axle, inclined plane, pulley, wedge, and screw. (Courtesy of Trek Bicycle Corporation.)

Mechanical fasteners come in many different types and sizes. (Photo by Tom Carney.)

Levers can be used to multiply the force that people can make. This worker is using leverage in the shears to multiply the sqeezing force of his hand. (Photo by Tom Carney.)

THINGS TO SEE AND DO:
Leapin' Links and Levers

Introduction:
Levers have been used throughout history to help move heavy objects. Today we use levers in so many ways that we don't even think about it. The next time you pull a nail with a hammer, crack open a nut with a nutcracker, or tighten a bolt with a pair of pliers, think about levers. In this activity you will make a lever-and-link device that will work like a mechanical hand on the end of a robot arm.

Design Brief:
Build a robot gripper that can be used to pick up a pencil from a desktop. The gripper will be used in the last activity in this chapter. The gripper can use any type of lever or linkage you can think of.

Some of the parts obtained in the machine dissection activity may be used here.

Materials:
▶ Acrylic plastic
▶ Wood
▶ Wood screws
▶ Other?

Equipment:
▶ Drill press or power drill, drill bits
▶ Scroll saw or band saw
▶ Screwdriver
▶ Ruler

Procedure:
❶ Work in groups of four. Each person should brainstorm at least four different ideas for the gripper design. Each member of the group should make four sketches on paper or use graphics or CAD software on a computer. The group will work together to make one gripper. One idea is shown below.

This is one possible idea for a robot gripper. How could you improve this design? (Copyright © Pam Benham.)

❷ Refine your design and have it approved by your teacher. If your idea is original and workable, your teacher will issue a patent to your group. No one else in the class will be able to use your idea without your permission after you receive a patent.

❸ Safely and carefully cut the materials you will need to make your gripper. Have each person in the group make a different part to save time.

❹ Assemble and test your gripper. In another activity, you will attach your gripper to the end of a robot arm.

Evaluation:

❶ As you use tools and machines to make your gripper, make a list of all of the levers and linkages you notice.

❷ Can you think of a lever that is a part of most cars?

❸ What other mechanical parts might help in making a robot gripper?

❹ How can robots pick up small delicate parts such as a watch crystal?

Gears: Gears **transmit** (send) forces. They can be used to change the speed or direction of spinning parts. Gears are made in different shapes depending on how they are to be used. The speed at which a gear turns is measured in **RPM**s, or **revolutions per minute**. Most people think all gears are round, but that's not true. Have you ever seen a gear that isn't round? One example, sometimes found in a car's steering mechanism, is called a **rack and pinion**.

Chain and Belt Drives: When forces have to be transmitted over a longer distance than gears can handle easily, a chain or belt is often used. Look at your bicycle chain. It rides on a toothed wheel called a **sprocket**. The chain transmits the force from your pedals to the back wheel through a chain and sprocket. Belts can do the same things as chains. You may have seen a belt drive on a washing machine or on a cooling fan in a car. Belts use pulleys to grip the sides of the belt just as a chain's holes mesh with the teeth in a sprocket. Belts are lighter and run more quietly than chains, but they can slip more easily.

Power can be transmitted (moved) from one place to another with the use of belts, pulleys, chains, and sprockets. (Copyright © Pam Benham.)

Specialized shapes and designs of mechanical parts can help machines work smoothly and efficiently. (Courtesy of Ford Motor Company.)

Other Mechanical Subsystems: As you have learned, even the most complicated machine can be simplified by looking at the parts individually. An automobile engine uses a mechanical part that changes **rotational** (turning) motion into **reciprocating** (up-and-down) motion. That part is called a **cam**.

Compare the leaf spring with the spring in a retractable ball point pen. How are they similar? How are they different? (Photo by Michael Bombard.)

Special Report
Mechanical to Electrical

Sometimes important machine parts can be very simple. A **flywheel**, for instance, is an important part that helps keep engines in machines such as lawn mowers running. A flywheel is just a metal wheel that is heavy enough to keep spinning because of its own weight.

Machines often store mechanical energy in **springs**. Springs come in many sizes, strengths, and shapes. Take a look at the spring inside your ballpoint pen. Compare it with the spring on a car's suspension system. What do they have in common?

Many modern machines use electrical parts instead of the mechanical parts that were used in older machines. Old typewriters, for example, were mostly all mechanical links and springs. These complicated mechanisms often jammed when two or more keys were struck together. The next generation of typewriters used a combination of mechanical and electrical parts to make typing easier and faster. Machines that contain both electrical and mechanical parts are called **electromechanical** machines.

Today's typewriters are almost entirely *electronic*. The keys you press on the keyboard are actually electric switches!

Designing Electronic Systems

Electronics is a part of technology concerned with the movement of **electrons** (small, negatively charged parts of an atom) through different materials. These materials are called conductors, insulators, semiconductors, and superconductors.

A

B

C

D

Materials can be divided into four categories: conductor, insulator, semiconductor, and superconductor. (Figure A courtesy of Southwire Corporation. Figure B courtesy of Niagara Mohawk Power Corp., Syracuse, NY. Figures C and D courtesy of G.M. Hughes Electronics Corporation.)

▶ *Conductors:* A conductor is a material that lets electrons easily pass from one atom to another. **Electricity** is defined as the flow of electrons through a conductor. Most metals are good conductors. Can you think of other materials that are good conductors?

▶ *Insulators:* Materials that do not allow electrons to flow easily are called insulators. Examples of insulators include plastic, rubber, and glass. Can you think of other materials that are good insulators?

▶ *Semiconductors:* A very important group of materials used in electronics conducts electricity only under certain conditions. These materials are called semiconductors. Silicon is a commonly used semiconductor material. It is used to make microchips or integrated circuits (ICs).

▶ *Superconductors:* Another special group of materials, called superconductors, can conduct electricity perfectly. Even some of the best regular conductors such as copper *resist* (hold back) the flow of electrons a little. Superconductors, on the other hand, have no resistance at all.

Anything that runs on electricity can benefit from superconductivity. Experiments are being conducted to build train systems that would float on a magnetic field instead of rolling on a track. This type of system would allow trains to travel over 300 miles per hour without touching the track. These trains are called **magnetic levitation**, or **maglev**, trains. Can you think of ways superconductivity might change your future?

TECHNOFACT

Technofact 35

Some of the first mechanical machines didn't work quite the way we wanted. Did you know the first vacuum cleaners blew air out instead of pulling it in? They made huge clouds of dust but didn't clean anything! Inventors improved the vacuum so it had suction (pulled air in), but two people were needed to operate it. One person held the hose, while the other turned a crank or pumped a pedal to make the suction. Finally in 1901 Hubert Booth built a vacuum that had a motor. It worked well except for one thing. It weighed several hundred pounds! Portable models like the ones you're used to didn't show up until several years later. Now researchers are working on home robots that will take care of vacuuming for you!

Superconductors are materials that have no resistance to the flow of electricity. Can you think of a way to use superconductors that would make our world better? (Copyright © Pam Benham.)

To be useful to you, conductors, insulators, semiconductors, and even superconductors must be connected in some way. Electronic **components** (parts) are most commonly connected in three basic circuits: series, parallel, and series-parallel. A **circuit** is a complete path for electrons to flow.

Electrical Circuits: When electronic components are connected one after the other, they are in series. **Series circuits** are very common. They are easy to trace because everything is placed in line, one after the other. One disadvantage of series circuits is that if one part in the series fails, the entire circuit will fail. You may have a string of lights that are wired in series, so if one light burns out, the whole string goes out.

Parallel circuits are arranged in a way that allows other parts to continue to work even if one part fails. Parallel circuits are common in all electronic products such as televisions, radios, and stereos.

The third type of circuit is really just putting parts together in a combination of series and parallel circuits. These are called **series-parallel** circuits.

Superconductivity

Inexpensive superconductor demonstration kits are available from science and technology suppliers. Be sure to follow the safety instructions that come with the kits. See Teacher's Resource Guide.

Special Report
Superconductivity—MRI

What is superconductivity? You might think it is something brand new, but scientists have known about superconductivity since 1911. It was discovered by a Dutch physicist, Heike Kamerlingh Onnes. He was researching the effects of extremely cold temperatures on different metals. He discovered that mercury lost all resistance to the flow of electricity when cooled to about 4 **Kelvin** (K). Kelvin is a temperature scale where 0 K equals about −460° Fahrenheit and −273° Celsius, or centigrade. That's cold!

To understand how important this discovery is you need to think about how electricity works. Electricity is the flow of electrons. The electrons flow through materials called conductors, making a current. Copper, aluminum, silver, and gold are excellent conductors. But even the best conductors are not perfect. Some electrical energy is lost because of resistance to the flow of electricity. Before Onnes' discovery, there wasn't any way to eliminate resistance. Superconductors became the new kind of conductor. They let electricity flow with no resistance at all.

Can you imagine the uses for such a discovery? Superconductors could be used to save energy and money in power systems. Because they have no electrical resistance, they don't give off any heat. Generators wound with superconductors instead of copper wire could generate the same amount of electricity with smaller equipment and less work. The electricity could also be distributed through superconductors. This could save up to 20 percent of the electricity used by common systems.

The electronics field has already been using superconductors with some success. Superconductivity is used to make a new kind of integrated circuit (IC). Heat is an enemy of ICs and causes them to fail if the temperature gets too high. In the new integrated circuits, the electric circuits are replaced with superconductors, which produce no heat! The circuits can then be packed closer together.

In medicine, superconductivity is already being used. The **MRI** (magnetic resonance imaging) machine makes detailed images of your body parts such as your heart or brain without having to cut into the skin or put dyes into the blood. MRI machines work by putting you in a powerful magnetic field generated by a superconducting electromagnet.

Many other uses for superconductivity are being developed in transportation and science. Right now, finding a superconductor that will work at room temperature is a problem. Scientists see superconductivity as a very promising, exciting application of technology that will help us save energy and provide better ways to make things work.

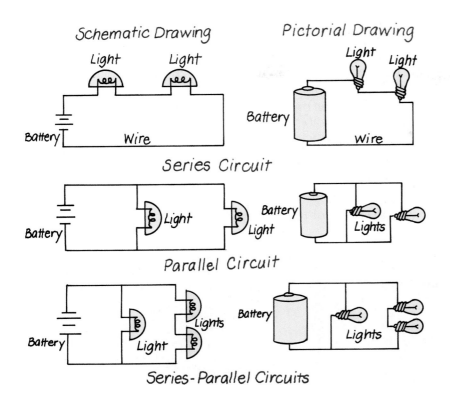

Schematic Drawing　　Pictorial Drawing

Series Circuit

Parallel Circuit

Series-Parallel Circuits

There are only three types of electrical circuits: series, parallel, and series-parallel. It is amazing that all of the complex electronic circuits we use everyday are all made of only these three basic circuits.

The path that the electrons follow is very important. If the electrons don't follow the proper path, the result might be a **short circuit**. Short circuits can be dangerous. They can cause a fire or electrocution. If you are experimenting with electricity, you should use low-voltage batteries to avoid the possibility of a painful or life-threatening electrical shock.

Electronic Components: The parts that are put into circuits are designed to control the flow of electrons. There are many special parts that control electrons. Some you might use to make electrical circuits are **resistors**, **capacitors**, **diodes**, and **transistors**.

▶ **Resistors:** Resistors are probably the most common electronic component. Resistors resist the flow of electricity through them. They come in many sizes and shapes, but the most common types have different-colored stripes to show their resistance.

▶ **Capacitors:** Electronic components that temporarily hold an electrical charge are called capacitors. Capacitors also come in many sizes and shapes . They can be dangerous because even though the power has been turned off, they can hold a charge that could injure someone. Televisions and computer monitors have high-voltage capacitors that can hold a charge for many hours after the power is turned off. Many people believe that if an electrical appliance such as a TV is unplugged, it is safe to work on. That is not true! The capacitors may still be charged. To prevent a possible dangerous electrical shock, you should never work on electrical equipment without the help of a qualified adult.

Challenge students to make a working display of the three types of electrical circuits.

Challenge students to make a display of electronic components. Have them label each one and give a short description of their use.

▶ **Diode**: A diode is like an electronic check valve; it lets electrons flow in only one direction. Diodes come in a variety of sizes, depending on the amount of current in the circuit. Diodes must be marked in some way to show which way the electrons will flow.

▶ **Transistors**: The one electronic component that started a revolution in technology development was the transistor. Suddenly, electronic products were smaller, lighter, more reliable, and less expensive. Transistors are made with a semiconducting material. They have three wires called **leads**. A very small current or voltage applied to one lead can control a large amount of electric current at the other two leads. Transistors are commonly used to switch or to **amplify** (make larger) electrical circuits.

Special Report
Capacitors—A Little Extra Boost

Resistors do just what their name says. They resist the flow of electrons. (Courtesy of Johansen Manufacturing Company.)

Capacitors can hold an electrical charge. (Courtesy of Mallory Capacitor Company.)

Diodes only let electrons flow in one direction. (Courtesy of Motorola, Inc.)

Transistors can be used as switches or as amplifiers in electrical circuits. (Courtesy of Motorola, Inc.)

THINGS TO SEE AND DO:
Motor Motion Magic

Introduction:

A simple battery-operated motor can be used in many ways. Small motors that run on 1½ to 12 DC volts can be found in many products. Portable tape players, toys, and even full-size cars use battery-operated motors. You can probably think of many more examples. In this activity you will learn to reverse a small motor using a switch. You will use a type of switch called a double-pole, double-throw, or **DPDT**. This is one of many different types of switches used to control the flow of electrons.

Students may use some of the parts obtained in the machine dissection activity.

Design Brief:

Make an electric circuit that will reverse the direction of a motor using a DPDT switch.

Materials:

- Small DC motor 1½ –12 volts
- Hookup wire

Equipment:

- Power supply 0–12 volts DC
- Wire strippers

Procedure:

1 Work in pairs. Use a small electric motor from your "junk" dissection activity, or get one from your teacher. You will be connecting your motor in a series circuit with a switch to control the direction of the motor.

2 Electronic circuits are drawn using a set of symbols to represent real parts. The symbol-drawings are called **schematic diagrams,** or simply **schematics.** Connect the circuit according to the following schematic.

Ask students to think of other ways that this type of electrical circuit might be used.

Schematic Drawing

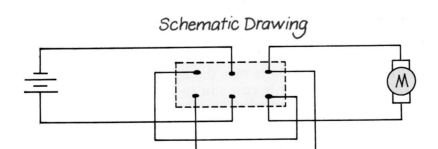

A schematic drawing for wiring a DPDT switch to control a motor.

3 Be sure to set the power supply to zero before you connect it. Have your teacher check your circuit. Turn on the power supply, and check your circuit to see if it will reverse the direction of the motor when you flip the switch.

Pneumatic systems use compressed air to transmit pressure. (Courtesy of Clayton Homes.)

Demonstrate hydraulic force by connecting two different-sized syringes with plastic tubing.

Hydraulic systems use oil to transmit power. (Courtesy of Caterpillar.)

Evaluation:

❶ Did your circuit work the first time you tried it? If not, what was wrong?

❷ How could you use a DPDT switch and motor circuit in a crane or winch?

To get an idea of how electrons flow through a conductor, imagine the flow of water through a hose. In both instances, there is a kind of pressure pushing a given amount of electrons or water through a circuit or hose. This kind of comparison is called an **analogy**. The same sort of analogy can be used to show how fluid systems work.

Designing Fluid Systems

When someone says "fluid," what do you think of first? Most people think of water or some other liquid. The fact is that fluids can be either liquids or gases. Both air and water are examples of fluids.

Fluid systems are one of two types: **hydraulic** or **pneumatic**. Hydraulic systems operate using a liquid, usually oil. Pneumatic systems operate on a gas, usually compressed air. Fluid systems apply pressure on oil or air to do work.

Can you figure out if these are single-or double-acting cylinders? (Courtesy of Caterpillar.)

Use the syringe/hydraulic demonstration with water and with air. Ask students to evaluate the results.

A common component, or part, used in both hydraulic and pneumatic systems is called a **cylinder**. Inside a cylinder is a **piston**. As air or oil is pumped under pressure into the cylinder, it makes the piston move. Cylinders commonly come in two types, **single-acting** and **double-acting**. Single-acting cylinders have pressure applied to the piston in one direction only. Automobile shock absorbers and hydraulic door closers are single-acting cylinders. Double-acting cylinders can have pressure applied to either side of the piston. This type of cylinder can push as well as pull. You may have seen hydraulic cylinders working on a backhoe or a dumptruck. Which kind of cylinder do you think is at work in the backhoe? Which is on a dumptruck?

Fluid systems need a source of power. In hydraulic systems, a **hydraulic pump** is used to force oil under pressure to cylinders or to other hydraulic components. Pneumatic systems use an **air compressor** (a machine that squeezes or compresses air) as a power source.

THINGS TO SEE AND DO:
Fun with Fluid Forces

Techno Teasers
Fluid Systems

Techno Teasers
Answer Segment

Techno Teasers
Answer Segment

Introduction:
You know that a fluid can be a liquid or a gas. Fluid systems that use gases such as compressed air are pneumatic. When a liquid such as oil or water is used, it becomes a hydraulic system. Gases such as air can be compressed into a smaller volume. Liquids cannot easily be compressed, so they can be used to transmit pressure from one place to another through hoses. In this activity, you will build and use a pneumatic and hydraulic system that uses air or water to transmit pressure.

Design Brief:
Build and test a pneumatic and hydraulic system that will make it easier to lift or move objects.

Materials:
- Syringe (no needle) 1–10 cc, 1–50 cc
- Plastic tubing, 24″ long
- Hose clamps
- Water
- Ruler
- Pressure gauge
- Wood, wood screws, screw eyes

Equipment:
- Drill press or power drill, drill bits
- Scroll saw or band saw
- Screwdriver
- C-clamp, 6″
- Spring scale, 0–25 pounds

Syringes without needles are available from technology and science suppliers. See Teacher's Resource Guide. Plastic tubing is available at most hardware stores. Keep the syringes in a safe place as they make great squirt guns!

Review the safe use of equipment used in this activity.

Procedure:

SAFETY NOTE:
You must wear eye protection at all times. Proper eye protection could be safety goggles, safety glasses, or a face shield.

1 Work in groups of four. Each student in the group will cut one of the following pieces of wood, using a band saw or scroll saw.

Quantity	Size	Part Name
1	¾″ × 2½″ × 7″	Base
2	¾″ × ¾″ × 5″	Arms
1	¾″ × 2½″ × 14″	Lever
1	¾″ × 2½″ × 14″	Scale support

2 Drill the lever to hold the syringe in the following order:

a. Measure 7″ from either end, and mark the center of the wood.

b. At the center point, drill a 1″ diameter hole ¼″ deep.

c. Drill a ⅜″ diameter hole all the way through at the center point.

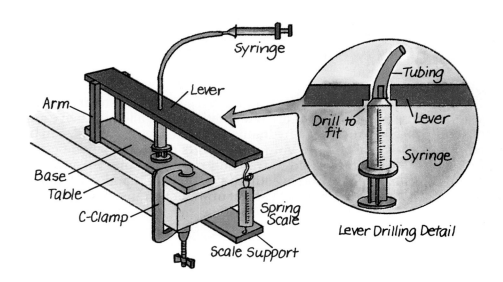

Syringe

Arm

Lever

Base

Table

C-Clamp

Spring Scale

Scale Support

Tubing

Drill to fit

Lever

Syringe

Lever Drilling Detail

Your fluid power test system should look like this.

❸ Using a scroll saw, cut a slot for the tubing as shown. This way, you will be able to remove the syringe easily.

❹ Drill the ends of the two arms for the wood screws. The hole should be slightly larger than the diameter of the screw. This is called a **clearance** hole.

❺ Measure 1″ from the end on the base. Drill a hole that is slightly smaller than the diameter of the screw as shown. This is called a **pilot** hole. It is made smaller so that the threads of the screw will have some wood to hold on to.

❻ Attach the lever to the arm in the same way. Install the screw eyes to hold the spring scale as shown. You are almost ready to do some experiments.

❼ Cut a length of tubing 24″ long to connect the two syringes. Connect the tubing to the syringes using hose clamps. Note: These connections must be very tight to prevent water from leaking out during the experiments.

Evaluation:

❶ Clamp your finished fluid tester to the edge of a table as shown. Connect a spring scale to the screw eyes as shown. You may need to adjust the length of the screw eyes in the arm and scale support, depending on the size of your scale. Adjust the screw eyes so that the scale reads zero.

❷ Pneumatic test: Compress the smaller syringe with only air in the system. What is the reading on the scale?

❸ Hydraulic test: Fill the syringes and tubing with water. Be sure to get all of the air out of the system. What is the reading on the scale this time?

④ Which system, pneumatic or hydraulic, would you use to lift heavy loads? Why?

⑤ Can you think of a situation where pneumatic systems would be better than hydraulic systems?

Save your fluid tester for use in future activities.

Designing Thermal Systems

What do thermostats, thermal underwear, and Thermos bottles have in common? If you said "heat," you're right! Unless something goes wrong, such as your home or school becoming too hot or too cold, you probably don't think too much about thermal systems. Thermal systems also control the temperature of your toaster as well as the temperature of your automobile engine.

THINGS TO SEE AND DO:
Putting Electrons to Work

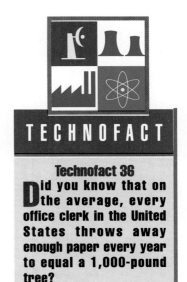

Introduction:

You know that electricity is the flow of electrons through a conductor. You also know that some materials let electrons flow more easily than others. The resistance of materials determines how easily they will let electrons through. In long-distance power lines, resistance is not something we want. However, resistance is not always bad. In this

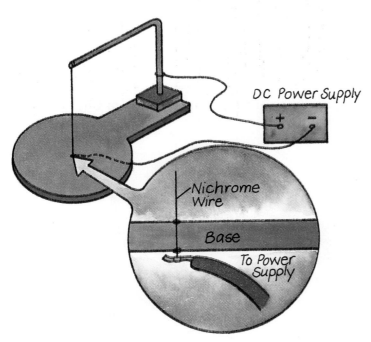

DC Power Supply

Nichrome Wire

Base

To Power Supply

The hot wire cutter can help you test many of your ideas by quickly cutting a Styrofoam model. The cutter is a simple series circuit that uses a nichrome wire to melt the plastic.

activity, you will use resistance to create heat or thermal energy to cut plastic foam.

The hot wire cutter you will make in this activity is an example of a simple series circuit. A power supply will push electrons through a wire to a special type of wire that is made to resist electron flow. The resistance causes the wire to become hot. Electric heaters and toasters use this type of wire, called **nichrome wire** to heat your home or to make toast. The design of the hot wire cutter is basically very simple.

Nichrome wire and DC power supplies are available from technology and science suppliers. Check Teacher's Resource Guide.

Design Brief:

Design and build an electric hot wire cutter that you can use safely and accurately to cut shapes out of the plastic foam.

Materials:

▶ Nichrome wire (28 gauge x 12″ long)
▶ Prefinished particle board, ¾″ × 12″ × 24″
▶ Steel rod, ¼″ diameter × 18″
▶ DC power supply, adjustable 0–12 volts
▶ Hookup wire
▶ Wood glue
▶ Wood block, 1½″ × 3½″ × 3½″
▶ Stick-on plastic "feet"

Equipment:

▶ Drill press or power drill
▶ ¼″ drill bit
▶ Wire cutter or stripper
▶ Hack saw
▶ Band saw and scroll saw
▶ C-clamp, 6″

Procedure:

1 Work in groups of five or six. Design a base for your hot wire cutter. Your design should be sturdy and have a wide area for holding the plastic foam flat. You might design your base using graphics or CAD software on a computer.

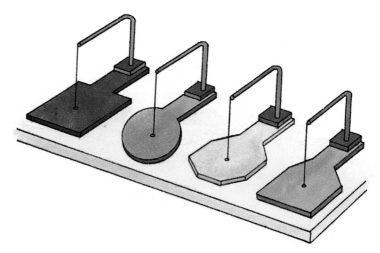

You can design your cutter many different ways. Can you think of other ways to make it?

② Lay out your design on particleboard. Carefully cut the particleboard using a scroll saw for sharp curves and a band saw for straight lines or wide curves.

③ Cut a 2″ × 4″ to the 3½″ length needed , using a hand saw or band saw.

Review the safe use of power equipment.

④ Glue and clamp the wood block to the back of the base to help support the steel rod. Let the glue set. Using a drill press, drill a ¼″ diameter hole in the center of the block and through the base. If you are using a portable power drill, be sure to hold the drill in a vertical position.

⑤ Place an 18″ long piece of ¼″ steel rod in a vise. Carefully bend the rod to almost a 90° angle. You may have trouble bending the rod. Try to think of a safe way to make it easier for you to bend steel. Ask you teacher for help with your idea. When you finish, the nichrome wire will pull the rod into a full 90° angle. The steel rod works like a spring to keep the nichrome wire tight.

⑥ Locate a point in the center of your cutter base for the nichrome wire to pass through. Drill a ¼″ diameter hole.

⑦ Cut and strip the ends of the hookup wire needed to attach your cutter to the power supply. Assemble the cutter so that the nichrome wire is pulled tight enough to bend the steel rod to 90°.

⑧ Finish your cutter by putting on the plastic "feet" to help the cutter sit steady and to protect countertops.

Evaluation:

Testing the Hot Wire Cutter

Caution students about the safe use of the hot wire cutter.

① Ask your teacher to inspect your work. Be sure to start with the power supply turned off and adjusted to 0 volts. Connect the hookup wires to the positive (+) and negative (−) terminals. It doesn't matter which wire is positive or negative.

SAFETY NOTE :
The nichrome wire will become very hot while using your cutter. Do not touch the wire or let it touch your clothing or anything other than the plastic foam you are cutting. Adequate ventilation must be provided to remove the fumes produced when plastic foam is cut. Do not use the hot wire cutter in closed spaces or without teacher supervision.

② Test your hot wire cutter by putting a piece of Styrofoam against the wire. Slowly turn up the voltage until the nichrome wire is hot enough to melt the foam. Try cutting various shapes.

Check the junk supply for an old thermostat. Look for uses of bimetallic strips.

Making Your Hot Wire Cutter Better

❶ Design and test a method to accurately cut the following shapes:

- Long, thin rectangles
- Perfect circles
- Cones
- Other geometric shapes

Sketch your methods on paper. Try to make the shapes above. Turn in your sketches and shapes for a grade.

❷ Find out if your community has a Styrofoam recycling center. Recycle the foam scraps if possible, or think of a way they might be used instead of throwing them away.

Challenge:

Design a hand-held foam cutter. Sketch your design on paper, and discuss it with your teacher. Make and test your hand-held cutter with the help of your teacher.

Devices that control temperature often use a **bimetallic strip**. *Bimetallic* means there are two different metals. As different metals heat up, they expand (get larger). One metal expands more than the other as the temperature increases. This difference in expansion causes the bimetallic strip to bend to one side. As it bends, it can trigger an electric circuit to start or stop heating or cooling equipment.

Another interesting device in the thermal system is a **thermocouple**. A thermocouple is like an electric thermometer. Using a hand thermometer wouldn't work well if you needed to read very high temperatures in dangerous areas such as the core of a nuclear reactor! Thermocouples are connected to gauges in control rooms away from areas affected by extreme heat or cold.

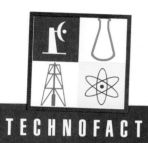

TECHNOFACT

Technofact 37

How do you do laundry in space? Space washing machines can't use too much water and they can't rattle and shake too much or they'll ruin the experiments. One company has been working on a washing machine that doesn't have an agitator and uses only 20 percent of the water most washing machines on Earth use. It uses a special cleaning fluid to save on water. Your clothes should still come out "squeaky clean."

Designing Chemical Systems

Have you ever heard of sodium tallowate, stearic acid, ammonium chloride, or methylchloroisothiazolinone? Read the label on everyday products such as toothpaste, soap, or shampoo and you might find names that sound even stranger. You may be surprised to find that chemicals play an important part in the technology that you take for granted. In fact, without chemical systems, you wouldn't even be able to use a portable stereo.

Chemical reactions are specifically designed to produce electricity in batteries. The batteries give us a portable electrical source. The chemicals used to produce a voltage are called **electrodes** and **electrolytes**.

Small batteries such as those used in a flashlight use a paste of dry chemicals to produce a voltage from a chemical reaction. Larger batteries such as those used to start a car use a liquid. The terms **dry cell** and **wet cell** are sometimes used to describe the two types.

Assign students to make a list of the chemicals used to make normal everyday products.

Batteries are only a small example of how we rely on chemical systems, however. Most of the energy we use every day is a result of some chemical system. Chemical systems are used to produce most of the electrical energy we use at home, in school, and in factories. Many communities count on the chemical energy stored in coal to be changed into heat (thermal) energy and then into electrical energy. Even the gasoline we use in cars or trucks is a product of very specialized chemical systems.

The oil found in the ground is called **crude oil**. Crude oil is not usable in its natural form. It must be changed into useful products by **refining**. Oil refineries are chemical factories that change crude oil into useful products such as lubricating oil, heating oil, gasoline, and kerosene. Special chemicals called **catalysts** are added to the crude oil in the refining process to make the chemical reaction go faster or to change the production of chemicals. Catalysts can be reused many times.

Many other chemicals are refined from crude oil. Did you know that some plastics and medicines come from oil? Products from oil are called **petrochemicals**. Oil is sometimes called **petroleum**. Petrochemicals are chemicals made from petroleum.

Batteries change chemical energy into electrical energy making electricity portable, as is the case with this flashlight. (Courtesy of NASA.)

Challenge a group of students to make a large chart that illustrates how crude oil is refined.

Putting Systems Together

At the beginning of this chapter, you saw examples of how the five energy systems work together and separately in a complex machine like a car. Even simple products, tools, or machines such as an electric drill, a model rocket, or a flashlight are a combination of more than one energy system. Can you name the systems used in each one? Are these systems working independently or together?

Factories called refineries change crude oil into many products we use every day. Plastics, gasoline, jet fuel, and heating oil are just a few of thousands of products we get from oil. These products are called petrochemicals. (Courtesy of American Petroleum Institute.)

TECHNOFACT

Technofact 38

New double-hulled (bottomed) oil tankers are one answer to oil-spill problems. Each new tanker will cost about $50 million. They will also carry about 10 percent less oil than single-hulled tankers. People, especially environmentalists, are excited about the number of sea birds, otters, sea lions, dolphins, and other sea life that will be saved by this simple change in ship technology.

Techno Teasers
Putting a Motor Together

Techno Teasers
Answer Segment

THINGS TO SEE AND DO: Putting Systems Together

Introduction:

You have learned that even large complex machines can be divided into systems and subsystems to make them easier to understand. In this activity, you will have a chance to put systems together to make a machine. The machine you are going to make is a robot! We will call it a robot, but it will really be a device to do a specific job with your help. Real robots can do many different jobs. Real robots use computers to control their actions. In this activity, you will be the computer controller for your robot.

> Students will use the parts gathered in the machine dissection as well as their robot gripper and syringe/hydraulic demonstrators.

Design Brief:

Design, build, test, and refine a robot arm model that can pick up a pencil from the surface of a table. The robot must be able to move the pencil to another part of the table and release it. During the testing, you may touch only the controls of your robot, not the gripper or the arm. The controls must be at least 12″ from the gripper.

Materials:
- Acrylic plastic
- Wood, wood screws
- Other?

Equipment:
- Drill press or power drill, drill bits
- Scroll saw or band saw
- Screwdriver
- Ruler

Procedure:

1 Work in groups of four or five . Your robot can be a combination of any of the following systems you have been studying:

- Mechanical
- Electrical
- Thermal
- Fluid
- Chemical

2 A robot can be a complex combination of systems. In this activity you need to remember how to solve a problem systematically. Do you remember the problem-solving steps? Here they are so that you can solve this challenging problem:

- Identify or define the problem
- Gather ideas or solutions
- Use your best judgment to pick the best solution

- Test your idea
- Evaluate your idea, and refine it until it is the best possible solution to the problem

❸ Brainstorm different ideas of how your robot might operate. Make sketches or use graphic or CAD software on a computer.

❹ Your group should discuss how the robot will operate and what materials will be best. Use any parts from your machine dissection activity that will work.

❺ Make a list of the materials needed, their size, and quantity. This list is called the **bill of materials**. The bill of materials might look like this:

Be sure that all students in each group are participating.

Quantity	Item Name	Description
3	Syringe	50 cc
2	Wood Screw	1½" #8 Flat Head
3 ft	Tubing, plastic	⅛" I.D. (Inside diameter)

A sample bill of materials. Can you think of the best software to use to make this?

❻ Carefully and safely cut the materials to make each part of your robot. Have everyone in the group work on a different part to save time.

❼ Assemble and test your robot arm. Remember that very few new ideas work perfectly the first time. The last step in solving a problem is to evaluate it and refine it as needed.

Evaluation:

❶ List the steps in problem solving. Next to each step, write a description of what you did to solve the robot problem.

❷ Which system do you like to work with the best? Why?

Challenge:

❶ Research ways to use a computer to control a part of your robot.

As you explore technology, you will use the five systems studied in this chapter. They are used in the designing, constructing, and manufacturing of various products.

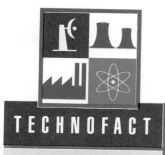

Summary

All machines can be divided into energy systems. Most systems have input, process, output, and feedback. In this chapter you learned five basic energy systems used to make machines work. The systems and some examples of their use are:

- ▶ Mechanical: Levers, gears, pulleys
- ▶ Fluid: Pneumatic and hydraulic cylinders
- ▶ Electrical: Circuits, resistors, ICs, transistors, conductors
- ▶ Thermal: Thermostats, thermocouples
- ▶ Chemical: Batteries, petroleum products

Simple machines might use only one system, while complicated machines use a combination of these systems.

Challengers:

❶ Ask a repair person to explain to the class how to troubleshoot a machine.

❷ Make a chart identifying different kinds of mechanical fasteners by name.

❸ Make a display of electronic components and label them.

❹ Figure out the gear ratios for a bicycle.

❺ Design a flywheel-powered model car.

❻ Find a junk thermostat to take apart. See if you can figure out how it controls heat.

❼ Make an electric question-and-answer matching board using batteries, switches, lightbulb or bell, and wires.

❽ Write to an oil-refining company and ask for information on oil refining and petrochemicals.

❾ Using the hot wire cutter, make a jigsaw puzzle.

❿ Design an electrical maze using wires, batteries, and a buzzer.

See Teacher's Resource Guide.

Techno Talk
Systems of the Future

Chapter 7

Designing Things

Things to Explore

When you finish this chapter, you will know that:

▶ Precision measurement is required to make high-tech equipment.

▶ Most of the world uses the SI metric measuring system. In the United States, we use the English system as a standard.

▶ Ergonomics is the study of how the human body relates to the things around it.

▶ Designers use the measurements of people to determine the appropriate sizes of products.

▶ There are many properties to consider before choosing which material to use for a product.

▶ The materials used to make products should be chosen for their strength as well as for their impact on the environment.

Chapter Opener
Designing Things

TechnoTerms

anthropometric	ferrous	property
biodegradable	greenhouse effect	prosthetic
combustible	hardness	softwood
composite	hardwood	standards
compression strength	landfill	synthetic
control	multimeter	tension
decimal	nanosecond	thermoplastic
ecology	nonferrous	thermoset
ergonomics	Ohm's law	variable
ferrocement	pasteurize	

(Courtesy of United States Department of Energy.)

Careers in Technology

Products are designed with many considerations in mind. Unfortunately, not all products are designed with the environment in mind. A growing group of careers are involved with the effects of technology on our environment. These effects can be very small and difficult to measure. It is important to know the exact condition of the environment now, so that we can compare future changes. This worker is measuring air and water quality in a remote location.

Measuring Things

When you design and make things, you use measurement tools to help you be as accurate as possible. You've used rulers before, but there are many other measurement tools used in technology. Tools such as stopwatches, thermometers, **multimeters** (machines that measure electricity), and meter sticks are some other devices that you will use in designing, building, and testing things.

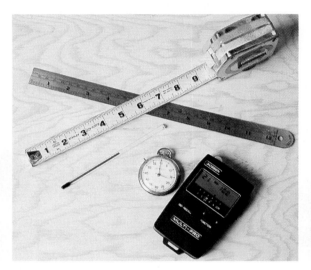

You might only think of rulers as measuring tools. Measuring tools also include things like a stopwatch, thermometer, and multimeter. (Photo by Tom Carney.)

TECHNOFACT

Technofact 40

What are nanoseconds, MIPS, and flops? These are standard measurement terms used in today's computer technology. A nanosecond is one billionth of a second. A nanosecond is a unit of measurement for computer speed. MIPS stands for *m*illions of *i*nstructions *p*er *s*econd. Flops stands for *f*loating point *o*perations *p*er *s*econd. Both MIPS and flops are used as a measurement of how fast supercomputers can process information.

Measuring is nothing new to humans. Even in earlier times, measurement was important to almost everyone, especially to traders and farmers. Beginning measurements were often based on human dimensions. For instance, a foot was the length of the average man's foot. Other measurements, such as an **acre** and a **furlong**, were built around practical activities like plowing. An acre was the amount of land two yoked oxen could plow in one day. A furlong is the distance a horse can pull a plow without stopping for a rest. Can you see some problems with these early systems of measuring? Early measurement systems were not very **precise**, or accurate, as you can tell, but they were easy for people to use.

The field of science especially needed more precise **standards** (exact units used by everyone) of measurement to measure things such as the speed of light, the temperature of liquid nitrogen, and the amount of electricity used by an appliance. This need for more precision in measuring as well as standard ways to measure things produced the measurement systems we use today.

You've probably learned about the metric system or **decimal** (base 10) system of measurement in math. The common base units are meters, liters, and grams. Parts of the metric system make up the **International System of Units** (called **SI**, for the French name, Système International) which is used internationally for trade. SI makes it easier for scientists, engineers, and construction industries all over the world to make materials and parts that are interchangeable.

Measuring Electricity

Measuring electricity is just like measuring anything else. There are standard basic units used throughout the world. The three basic units of measurement used to measure electricity are the volt, the amp, and the ohm.

Have students use a multimeter to test batteries.

▶ **Volt:** Electricity flowing through wires in a circuit is very similar to water flowing through a hose. The pressure that pushes water through a hose is similar to the **voltage** that pushes electrons through a wire. Voltage is a unit of electrical pressure.

▶ **Amp:** The amount of water flowing through a hose might be measured in gallons. In an electrical system, the amount of electrons flowing through a wire is measured in amperes (amps). **Amperage** is the amount of electricity.

▶ **Ohm:** A water hose that has a nozzle at the end will **resist** (hold back) the flow of water. The resistance in electrical systems is measured in ohms. The higher the resistance, the more electrons are held back.

All three units—volts, amps, and ohms—can be measured with a *multimeter*.

Here, we see a multimeter in use. (Courtesy of Cleveland Institute of Electronics.)

The great thing about working with electricity is that there is a simple mathematical relationship between volts, amps, and ohms. This special formula is known as **Ohm's law**.

$$\text{Amps} = \frac{\text{Volts}}{\text{Ohms}}$$

	English Unit	Abbrev.	SI Equivalent	Abbrev.
Distance 1 inch	Inch	In. or "	25.4 Millimeters	mm
	Foot	Ft. or '	305 Millimeters	mm
	Yard	Yd.	.914 meter	m
Area 1 square inch	Square Inch	Sq. In. or In²	645 Square Millimeter	mm²
	Square Foot	Sq. Ft. or Ft.²	.0929 Square Meter	m²
	Square Yard	Sq. Yd. or Yd.²	.836 Square Meter	m²
Volume 1 cubic inch	Cubic Inch	Cu. In. or In.³	16,400 Cubic Millimeter	mm³
	Cubic Foot	Cu. Ft. or Ft.³	.0283 Cubic Meter	m³
	Cubic Yard	Cu. Yd. or Yd.³	.765 Cubic Meter	m³
Mass 1 Lb.	Ounce	Oz.	28.4 Grams	g
	Pound	Lb.	454 Grams	g

We need to use both the English and the metric (SI) system of measurement in technology.

The metric system is used in most parts of the world. The English system, which uses the foot, the pound, and the quart, is still used in the United States today for most of our measurements. At one time, the United States planned to switch totally to the metric system but never did. Today many items show both English and SI measurements. Look at some items at the grocery store to see if both measurements are given.

Sometimes things are too large or too small to make or draw in their actual size. For instance, you couldn't make a full-sized drawing of a

Sometimes it is not possible to draw plans the actual size of a project to be built. Here a set of house plans are drawn to scale so they are useful to both contractors and architects. (Courtesy of Ford Motor Company.)

house because they don't make paper that large! You also would have a difficult time drawing the actual size of an integrated circuit because all the parts are too small. Can you imagine drawing more than 1,000 circuits in a space smaller than the eraser at the end of your pencil? To solve this problem, things are often drawn to **scale**. Drawing to scale means that you make the shape exactly the same as the real object, but the object is actually larger or smaller.

Show a set of house plans to the class. Ask them to explain why the plans are drawn to scale.

THINGS TO SEE AND DO:
Measure Mania

Techno Teasers
Measuring Things

Techno Teasers
Answer Segment

Introduction:

The ability to measure quickly and accurately is an important skill in technology. It is also important to be able to estimate the size of objects without actually measuring them. In this activity, you will first test your ability to use the English and metric (SI) measurement systems. After you are able to measure accurately, you will be challenged to estimate and measure objects quickly and accurately.

A simple measurement quiz will identify students who need a measurement skill refresher.

Design Brief:

Demonstrate your ability to measure accurately and quickly using both the English and metric (SI) measuring systems. Put your measurement skills to use in the measurement game described in part 2.

Materials:

▶ String
▶ Miscellaneous materials for measurement game

Equipment:

▶ Rulers, tape measures
▶ Calculator (optional)

Procedure:

Part 1:

❶ Complete the following measurement test carefully. Write your answers on a separate sheet of paper. You must reduce all fractions and use decimals where needed.

❷ Ask your teacher to check your test. Measurement is a skill of accuracy. If you miss any one of the problems, you must take the test again.

❸ If you miss some problems, your teacher will help you understand your mistakes.

❹ When you are sure you know how to measure in both English and metric (SI) units, go on to the next part of this activity.

Copy this format and fill in the blanks with the help of your teacher.

TECHNOLOGY MEASUREMENT TEST

NAME: ______________________ PERIOD ________

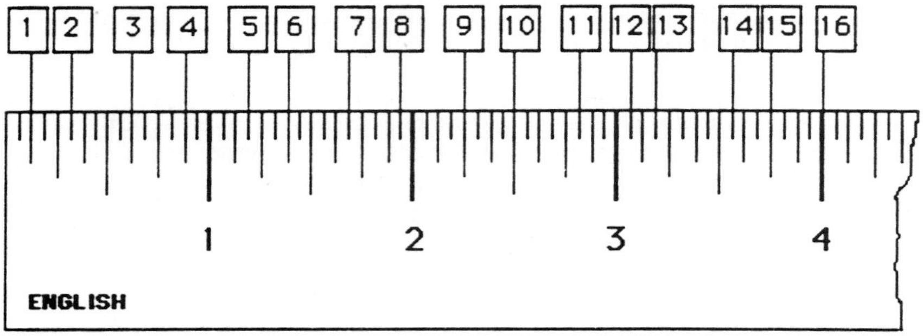
1
2
3
4
5
6
7
8
9
10
11
12
13
14
15
16
1
2
3
4
ENGLISH

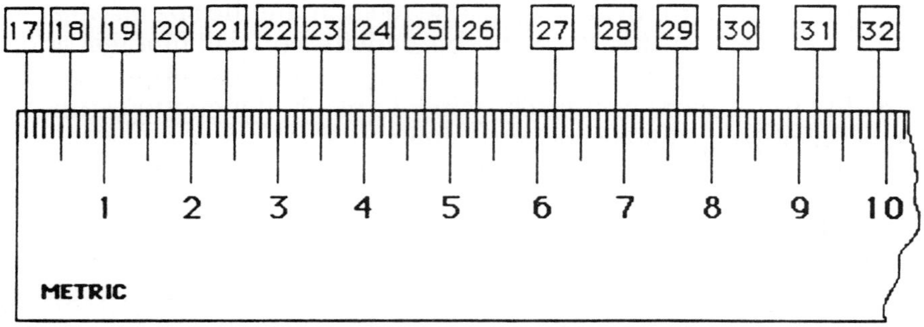
17
18
19
20
21
22
23
24
25
26
27
28
29
30
31
32
1
2
3
4
5
6
7
8
9
10
METRIC

1. ____	9. ____	17. ____	25. ____
2. ____	10. ____	18. ____	26. ____
3. ____	11. ____	19. ____	27. ____
4. ____	12. ____	20. ____	28. ____
5. ____	13. ____	21. ____	29. ____
6. ____	14. ____	22. ____	30. ____
7. ____	15. ____	23. ____	31. ____
8. ____	16. ____	24. ____	32. ____

Part 2:

1 Write your estimate of the size of each object in the following list (use a separate sheet of paper). You must complete both the English and metric (SI) estimates before going on.

Name of Object	Estimate	Actual Size	Difference
Width of sheet of paper			
Diameter of a globe			
Thickness of a pencil			
Height of a desk			
One Meter			
		TOTAL	
Width of classroom door			
Height of this book			
One Inch			
Circumference of globe			
Width of computer disk			
		TOTAL	
		ENGLISH TOTAL	
		GRAND TOTAL	

English brackets the first section; *Metric (SI)* brackets the second section.

2 Using the proper measurement tools, make accurate measurements of each object, and record them on your sheet.

3 Find the difference between your estimate and the actual size by subtraction. You may use a calculator.

4 Make a total of the differences between your estimates and the actual sizes of the objects in both the English and metric (SI) sections. Check with your teacher on how to make an English total.

Evaluation:

1 Which measurement system is easier for you to use? Why?

2 List three occupations that require fast and accurate measurement.

3 How is measurement important in sports such as volleyball, football, basketball, and baseball?

Challenge:

❶ Research what other countries use the English measurement system.

❷ Find out the meaning of the following units: newton, furlong, joule, light-year.

When you are designing and building, you will find it easier if you stick to one measurement system rather than mixing units such as inches and centimeters.

Designing Products for People

Have you ever tried on gloves and found the medium size was too big and the small size was too small? Have you ever wondered how designers decided on what a "medium" size is? Technology can make our lives more comfortable through **ergonomics**. Ergonomics is the study of how the human body relates to things around it. It is also called "human engineering." Places where you live, work, and play are safer, easier for you to use, and more comfortable if they are designed based on the actual human body size.

The actual measurements of your height, width, weight, reach (arms extended), and so on are recorded in books as **anthropometric** information or data. Designers and engineers use the size information to determine the dimensions of products such as clothing, furniture, sporting goods, car interiors, and even spacesuits. When designers made the chair you are sitting in, they used the anthropometric data from many people and found a size that would be comfortable for 90 percent of the people. Five percent of the people will probably be too large and another five percent will be too small for that particular chair. Because there is such a wide range in sizes of people, it is difficult to make one product that everyone would find comfortable. Is your chair really comfortable? Does it fit you?

Ergonomic design of furniture can make the way we live more comfortable. (Courtesy of Steelcase Inc.)

Ergonomic design can even make our leisure time more enjoyable. (Courtesy of Rossignol Ski Company, Inc.)

Not all products are made with ergonomics or anthropometry in mind. Simple everyday things like water faucets and door knobs can sometimes be hard to figure out because they weren't designed with people in mind. Many times they were designed just for looks. In the past, tradition rather than people's needs was also responsible for the way some things were designed. For instance, maybe the reason most automobile engines are located in the front of a car is that the engine replaced the horse. The horse, of course, was in front of the wagon.

You might think that designing products around people is just for appearance and isn't very important. That isn't so. Part of ergonomic design is to make products safe for you to use. The special dashboard and ceiling padding, airbags, seatbelts, harnesses, and other safety features are put in today's automobiles for your safety. Even auto seats are designed to let you sit in a proper driving position.

Ergonomics plays an important part in today's high-tech workplace. Many people spend long hours using keyboards. Studies show that muscles can be damaged by repeating a simple movement over and over again. This can be even a very simple movement like pressing the keys on a keyboard. **Carpal tunnel syndrome** can be a painful result of incorrect ergonomics for people who spend all day at a keyboard. To help prevent this problem, designers have come up with various solutions. These include wrist braces, adjustable chairs, and even specialized keyboards.

NASA put technology and ergonomics to work in designing spacesuits for its astronauts. They used to design a special spacesuit for each astronaut. This approach was very expensive. NASA now designs its new spacesuits, called **extravehicular mobility units** (EMU) in a variety of standard sizes to fit all astronauts.

Ask students to describe products they think are difficult or uncomfortable to use. Ask them how they would change the

Ergonomic design of car seats is important for comfort and safety. (Photo by Michael Bombard.)

TECHNOFACT

Technofact 41

Now there's a sun-safety watch that lets you know when you've had enough sun. The Sundial is about the size of a standard digital watch but it has a computer. All you do is program in your skin type and whether you have put on a sunscreen with a certain SPF (sun protection factor). The watch's computer evaluates all your information and then, like a stopwatch, it counts down to zero. It beeps a warning when you need to head for the shade.

NASA has researched the design of spacesuits since the first space flights. Today, astronauts are able to work in space thanks to very special protective clothes. Here, astronaut Bruce McCandless is working in the cargo bay of the space shuttle. (Courtesy of NASA.)

EVA means Extra Vehicular Activity, a fancy way of saying working in space outside of the Space Shuttle. Astronauts must be able to work in space comfortably and safely. Anthropometric data was used to design a spacesuit that considers the ergonomics of space. (Courtesy of NASA.)

Special Report
Fashion in Spacesuits

THINGS TO SEE AND DO:
Design a Space Helmet

Introduction:

You have learned that anthropometry is the measurement of the size and limits of movements of the human body. Scientists, technologists, designers, architects, and engineers all use anthropometric data to help them design and build products for people to use comfortably. In addition to the clothes we wear or the chairs we sit in, special attention must be given to designing high-tech equipment such as spacesuits. In this activity, you will gather anthropometric data and use it to design a space helmet.

Design Brief:

As a designer, your job is to research and design a new space helmet that can be used easily and comfortably. Your design must consider the following:

▶ Proper size: Your group will determine the best size by taking measurements. This is called anthropometric data.

▶ Appropriate materials: The helmet should be able to withstand impacts (hits). The visor should protect astronauts from the blinding glare of the sun.

▶ Light weight: Even though things are "weightless" in space, the helmet must still be used in training on Earth.

▶ Comfort and safety: Your helmet should be comfortable to work in for many hours. A communication microphone and ventilation should be provided.

Encourage students to work in different groups.

Materials:
▶ Graph paper, pencil

Equipment:
▶ Large calipers
▶ Ruler or tape measure
▶ Computer with spreadsheet and graphics software (optional)

Procedure:

1 Work in groups of four or five. Read through the procedure below, and divide the tasks so that everyone in your group is helping to solve the problem of designing a space helmet. Use a tape measure or calipers and a ruler to take the measurements shown below. These measurements will be your anthropometric data for this activity.

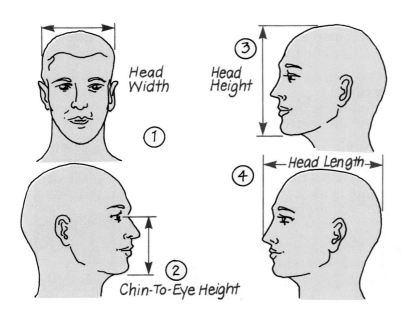

Use calipers or a tape measure to measure the sizes illustrated.

2 Record the measurements for each member in the group on a chart similar to this:

ANTHROPOMETRIC DIMENSIONS					
	1	2	3	4	
Student 1					
Student 2					
Student 3					
Student 4					
Student 5					
Total					
Average					
Maximum					

The data you collect should be put into spreadsheet form.

(You could use a computer and spreadsheet software to record the measurements. Remember, computers do not use fractions very well. You might want to use the metric system to avoid problems.)

3 Using graph paper, choose an appropriate scale to sketch your designs. Make a front view and a side view of your helmet. Sketch a head to scale using the largest dimensions obtained in step 1.

4 Label each part of your space helmet design. Give each part a name, for example, visor, sealing neckring, microphone, air vents, padding, and antenna.

Design Hints:

The following list of materials and their properties may help you choose the right materials for your helmet.

▶ *Fiberglass:* A very strong material called a **composite**. Composite means made up of more than one material. Fiberglass is made of very thin glass strands glued together with a liquid plastic that hardens. The glass fibers go in every direction, giving fiberglass its high strength.

Fiberglass gets its strength from thousands of glass strands glued together so that they are going in all directions. (Courtesy of Bell Helmets.)

▶ *Polycarbonate:* A very clear and strong plastic used to make safety glasses and windshields for snow machines and motorcycles. Polycarbonate is a thermoplastic material that can be bent or shaped with heat.

▶ *Aluminum:* A very strong and lightweight metal that conducts heat and electricity easily. Aluminum can be formed into almost any shape, from very thin foil to thick castings.

▶ *Acrylic:* A thermoplastic material that is easy to cut, bend, or shape with heat. Acrylic plastic is clearer than glass. It expands and contracts with temperature change and can be very brittle in thin sections.

Make a list of materials such as this:

Part Name	Material	Special Properties	Comments
Visor	Polycarbonate	Strong, clear, lightweight	Tinted to reduce glare

Evaluation:

1 Exchange the design your group worked on with another group. Using the design brief as a guide, evaluate the design of another group. Give suggestions as to how the design might be changed.

2 What is your estimate of how much your helmet design would weigh? What do you think it would cost to make? How could the cost and weight be reduced?

3 List two other examples of where anthropometric data are used to design products.

Challenge:

1 Use CAD software on a computer to make a final design of your helmet.

2 With your teacher's help, make a full-size model of your helmet design.

Ergonomic design is also important in designing equipment and products for physically challenged people. Special equipment such as skis, wheelchairs, some artificial body parts, or **prosthetic** devices, furniture, and bicycles are designed to help these people carry out everyday activities.

Look around you. Is the room you're in designed to fit people your size? Can you reach all the shelves? Are the chalkboards at a proper height so you can easily look at them? Can you reach all the materials you need to do your work easily from your desk?

Choosing the Right Material

Not only is it important to design products with people in mind, but it is also important to choose the right material for your product. The materials you choose for a product can make it either useful and long-lasting or dangerous and short-lived. People have been researching new materials and new uses for old materials since the Stone Age. You learned in Chapter 1 that materials were so important that entire periods of history, such as the Bronze Age and the Iron Age, were named after them.

Basically, materials can be divided into two major groups, **synthetic** and **natural**. Synthetic means that people made them, and they cannot be found in nature. The many kinds of plastics are examples of synthetics. Natural materials, such as copper and wood, can be found in nature. Products often are combinations of many kinds of materials. For example, a television has a picture tube made of glass, a cabinet made of plastic or wood, and wires made of copper. A bicycle has rubber tires, steel frames, and a plastic seat.

When you are completing design briefs, you should consider using more than one kind of material in your solution.

Artificial body parts are also designed using anthropometric data and ergonomics. Prosthetics are artificial body parts designed to help physically challenged people. (Courtesy of Hosmer Dorrance Corporation.)

Special Report
Introduction to Materials

Assign a group of students to make a chart that demonstrates natural and synthetic materials.

Classifying Materials

You already know the difference between synthetic (human-made) and natural materials. Did you know that most materials can be further divided into groups? The grouping of materials is based on their properties or their origin. Here is the way to classify some materials:

Woods: There are two types of wood, hardwood and softwood. Sounds simple, but the words *hard* and *soft* have nothing to do with the hardness of the wood. The difference is in the tree that the wood came from. **Hardwoods** come from trees that have broad leaves, for example, walnut and maple. **Softwoods** come from trees that have needles, such as pine and fir.

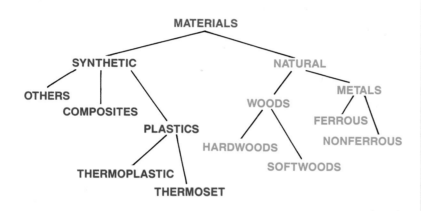

Materials can be classified according to their properties or origin.

Softwood. (Courtesy of United States Department of Agriculture.)

Hardwood. (Courtesy of United States Department of Agriculture.)

Techno Teasers
Answer Segment—Oil

Techno Teasers
Answer Segment—Composites

Classifying Materials (cont.)

Metals: There are two types of metals also, ferrous and nonferrous. *Ferrous* is a Latin word for iron. The difference between ferrous and nonferrous is that **ferrous** metals contain iron and **nonferrous** metals do not. Ferrous metals include iron and the many types of steel. Nonferrous metals include copper, tin, lead, aluminum, gold, and silver.

Nonferrous metals. (Courtesy of Reynolds Metals Company.)

Plastics: There are also two types of plastics: thermoplastic and thermoset. The difference is very simple. **Thermoplastics** can be melted and remelted many times using heat. **Thermosetting** plastics change chemically when they set. They cannot be remelted. Acrylic plastic is an example of thermoplastic. It can be reheated many times to change its shape. Bakelite is a common plastic used for electrical plugs and cooking-pot handles.

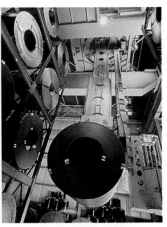

Ferrous metals. (Courtesy of Bethlehem Steel Corporation.)

Thermoset plastic pot handles. (Courtesy of Corning, Inc./ photographer: Frank Petronio)

Thermoplastics. (Courtesy of First Brands.)

Composite materials are made by combining two or more materials. Composites may be organic or synthetic. (Upper figure Courtesy of American Plywood Association. Lower figure Courtesy of Owens-Corning Fiberglass.)

Composite materials: By combining different materials, new and often better properties can be obtained. Composite materials such as fiberglass and carbon graphite or graphite-epoxy are very lightweight and strong. They are used to make high-performance aircraft wings and lightweight sporting goods such as tennis racquets.

Materials are chosen by their characteristics. The characteristics of a material are called its **properties**. Each material has special properties that make it useful for certain products.

THINGS TO SEE AND DO:
Personal Pop Can Press

Introduction:
The proper selection of materials applies not only to products but also to the materials used to package products. The cost of the package must be added to the cost of the product. Inexpensive packaging is sometimes not the best choice for the environment, however. Materials that are strong, lightweight, and able to be recycled are good choices for packaging products. In this activity, you will do a test to see if aluminum is the best choice for soda cans.

Design Brief:
Determine the most appropriate material for packaging liquid products such as beverages.

Materials:
- Empty soda cans
- Paper
- Tape
- Plastic film (overhead transparency film)

Equipment:
- Scissors
- Ruler
- Bathroom scale
- Micrometer, balance (optional)

Procedure:

1. Work in groups of three. With the help of your teacher, cut open an empty aluminum soda can. Use a micrometer to measure the thickness of the aluminum used to make the can.

2. Measure the thickness of different types of paper and plastic film used to make overhead transparencies. Find a piece of paper and plastic that are the same thickness as the aluminum can.

3. Cut the paper and plastic sheets to the same height as a soda can (4⅞″). Roll and tape the materials into a cylinder the same size as a soda can. Your group will need three "cans" to test; one real aluminum can, one made of paper, and one made of plastic film.

4. Carefully weigh each "can," and record its weight.

5. Use two sturdy chairs or desks for support as shown below. Choose one person in your group to be the tester. Each "can" will be tested to see how much weight it can hold. Place one "can" at a time on a bathroom scale for

. . . .
Assist students in reading a micrometer.
. . . .

Balance yourself between two chairs to carefully test the compression strength of a pop can. (Copyright © Pam Benham.)

each test. Carefully watch the scale as the tester places one foot on the top of each "can" and tries to see if it will support all of his or her weight. Note the maximum weight before the "can" was crushed.

SAFETY NOTE:
The tester should be careful to steady his or her weight, using the two chairs for support. Be careful not to twist an ankle when the "can" is crushed.

Assist the tester to prevent possible falls.

6 If you were careful to choose an aluminum can without any dents, you could probably stand on it with all of your weight. Try the test again without the bathroom scale. Be sure the tester is steadying himself or herself on the chairs. This time, with all of the tester's weight on the can, gently and quickly tap the side of the can with a ruler to make a small dent in the can.

Evaluation:

1 What happened when the aluminum can was dented while the tester was standing on it? Can you explain why?

2 Which "can" weighed the most? How does the weight of a package affect the selling price of the product? Why?

3 How could the paper "can" be made waterproof so that it could hold liquids without leaking?

4 Research how long it takes for aluminum, paper, and plastic to decompose in a landfill.

5 Which material (aluminum, coated paper, or plastic) is easiest to recycle?

Challenge:

1 Can you think of other materials besides aluminum, coated paper, and plastic film that might make a good can?

2 Design a test that would test cans laying on their sides rather than standing up.

3 Which properties of aluminum make it a good choice for beverage cans? Which properties make it a bad choice?

Testing Materials: All materials are tested to see how they can best be used to make products. Would you ever build a boat out of cement or glass? Some very large boats are actually made of a cement mixture sprayed over wire mesh, called **ferrocement**. They actually float! Many boats are also made of glass fibers held together with plastic. You learned earlier about fiberglass. It is used to make other products such as skis and crash helmets. Some materials have surprising uses.

There are many different properties of materials that you can test:

▶ **Hardness:** The ability to resist dents.
▶ **Tensile (tension) strength:** The ability of a material to resist stretching or pulling apart.
▶ **Compression strength:** The ability of a material to resist being squashed or smashed.
▶ **Fatigue strength:** Compression and tension sometimes work together as materials are bent back and forth by vibration. This causes them to break.

Special Report
Testing to Select the Best

When some materials are bent or flexed many times, their strength is reduced. The resistance to breaking after many cycles of bending is called fatigue strength. (Photo by Mike Hemberger.)

THINGS TO SEE AND DO:
Have You Ever Felt Fatigue?

Introduction:
When you bend or flex some materials many times they will break. The ability to stand up to many cycles of bending or flexing is called fatigue strength. Fatigue can cause serious accidents if it goes undetected. Airplanes are sometimes damaged by the flexing of metal each time the airplane takes off or lands. Also, the sheetmetal fuselage of an airplane where passengers sit expands and contracts on each flight as the airplane reaches cruising altitude and returns for a landing. Small cracks around rivets caused by fatigue can lead to a disaster if they are not detected. In this activity, you will do a test to determine the fatigue strength of a steel paper clip.

Design Brief:
Design and perform a test method that will determine the fatigue strength of a paper clip. Use math to find the average fatigue strength of at least 10 paper clips.

Materials:
▶ Paper clips
▶ Pencil, paper

Equipment:
▶ Computer with spreadsheet and word processing software (optional)

Procedure:

❶ Work in groups of five or six . Discuss the possible ways a paper clip might be tested by bending it until it breaks. Experiment with a few paper clips to see if your idea will work.

❷ Your group should decide which test method will be used. A Test Procedure should be written out or typed on a computer. Your procedure should be detailed enough so that another group could follow your directions. Your Test Procedure might look like this:

TEST PROCEDURE: PAPER CLIP FATIGUE

Step 1 — Bend a paper clip out to a flat position as shown:

Step 2 — Place the largest side of the paper clip on the side of a table:

Step 3 — Bend the paper clip in the middle to a 90° angle, even with the edge of the table:

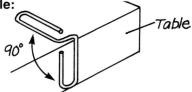

Table

90°

Step 4 — Place the bent paper clip flat on the table and bend it back so that it is the same as in step 1.

Step 5 — Each time the paper clip is bent and straightened will be one cycle. Record the number of bending cycles for three different paper clips in the data table below.

	Bending Cycles
Test Sample # 1	_____
# 2	_____
# 3	_____
Total	_____

Step 6 — Calculate the average number of cycles for the three paper clips tested.

$$\text{Average} = \frac{\text{Total number of cyles}}{\text{Number tested}}$$

Average = _____

This is a sample Fatigue Test Procedure. Can you think of another way to test the paper clip?

❸ Have each person in your group test two or three paper clips until they break. This method of testing is called **destructive testing** because you must break the material to complete the test. Each person should write down the number of bending cycles the paper clip went through before it broke.

❹ Each person in the group should add up the number of bending cycles for each paper clip tested. You will need to find the average number of bends for the paper clips you tested. To find the average, divide the total number of bending cycles for all the paper clips you tested by the number of paper clips you tested. You could use a computer and spreadsheet software to help you find the average.

TECHNOFACT

Technofact 42

Did you know that the term *acid rain* was first used by an English chemist, Robert Angus Smith, in 1852? He wrote a 600-page book about acid rain but hardly anybody read it. Today, acid rain causes damage to water, soil, food crops, buildings, and animals. It is produced by the burning of fossil fuels in power plants, industry sites, and motor vehicles. Sulfur and nitrogen oxides are gases formed when fossil fuels burn. The gases then change to sulfuric and nitric acids when they mix with water. The acids are carried thousands of miles in the atmosphere before they drop to the ground as acid rain.

TECHNOFACT

Technofact 43

NASA engineers are designing a fuel refinery prototype to be used on Mars. The prototype can separate oxygen from carbon dioxide, which makes up 96.5 percent of the Martian atmosphere. The oxygen would be stored as liquid oxygen in refrigerated tanks on the Mars base. The idea is to build a base on Mars where spacecraft can refuel their liquid oxygen supply. Liquid rocket engines use about one part of some combustible (burns easily) material such as liquid hydrogen with 10 parts liquid oxygen. If there were a source of liquid oxygen on Mars, the spacecraft would have to carry only the liquid hydrogen at liftoff from Earth. That would really lighten the 120-ton load of propellant that would otherwise be needed for the trip to and from Mars!

Techno Talk
Acid Rain

Ask a local architect or structural engineer to speak to the class.

5. You should then compare the average number that you determined in your test with the average numbers of others in your group. Find an average number for the entire group by adding all the individual averages together and dividing by the number of people in your group.

Evaluation:

1. Were the average numbers of bending cycles the same for every person in your group? If not, why do you think some people got different numbers?

2. Compare the average determined by your group with the averages of other groups. What is the relationship between the size of the bending angle and the number of bending cycles that can be completed before breaking?

Challenge:

1. Make a bar graph of the average number of bending cycles tested in your group. You can use a computer to help make the graph.

2. Design a test procedure to test other materials for fatigue strength.

Making Models and Prototypes

You know what a model is. Do you remember what a prototype is? A *prototype* is a model of a product being designed for production. Companies can't just start building thousands of products, without being sure that all the defects or problems have been corrected. Prototypes and models are often used by designers, engineers, and architects.

Many problems can be solved without spending a lot of money on the real thing. Here, a model of an oil refinery is being built to see how everything will work together. (Courtesy of American Petroleum Institute.)

When architects design a new skyscraper for a large city, they often build models so that people can visualize the shape of the building. The model lets people see what the building will look like better than a drawing can. Models are usually made to scale. The model of a new building is sometimes placed in a model of an entire city to see how it fits in. In some cases, the entire city model is placed in a wind tunnel to see how the structures will be affected by wind currents. Without planning and testing by using models, expensive mistakes can be made. Some buildings have been built only to find that wind currents were so strong that they sucked the windows out!

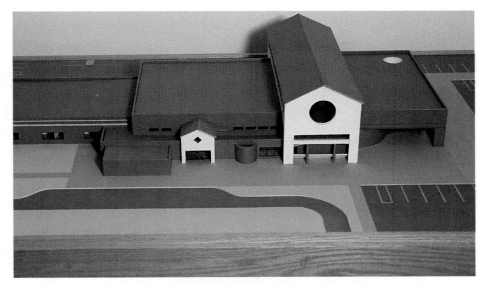

Models of large or small buildings let people visualize or see for themselves how the finished building will look or how it will affect its surroundings.

THINGS TO SEE AND DO:
A Hot Model

Introduction:

You know that models built to test ideas are called prototypes. You also know that time and money can be saved by testing ideas for products or buildings in model form before the real thing is built. In this activity, you will work in groups to design and build a solar cooker that will use the energy of the sun to cook food. This is an important and needed product for developing countries. Solar cookers would prevent people from cutting down trees for fuel. A simple-to-use solar cooker could even save lives in poor countries. Many young children die each year because they do not have clean water to drink. A solar cooker could be used to **pasteurize** (purify) water to kill the harmful bacteria. Your design will be tested to see if it will work.

If constructed carefully, the solar cookers can produce temperatures over 200° F.

Design Brief:

Design, build, and test a solar cooker that will cook food. You may use materials such as the following to make your prototype cooker:

- ▶ Recycled cardboard boxes
- ▶ Newspaper
- ▶ Aluminum foil
- ▶ Clear acrylic or polycarbonate plastic (Plexiglas or Lexan)
- ▶ Flat black spray paint
- ▶ Other materials will require teacher approval

Your solar cooker will be tested to see what temperature can be reached.

Materials:

- ▶ Recycled cardboard boxes
- ▶ Newspaper
- ▶ Aluminum foil
- ▶ Clear acrylic or poly-carbonate plastic

Equipment:

- ▶ Hand and power tools

Procedure:

1 You will be working in groups of four or five. You should consider some of the following facts before starting your design:

- • The cooking area should be insulated so the heat gathered is not lost.
- • More sunlight reflecting into the cooking area will produce higher temperatures.
- • A flat black surface will absorb heat energy rather than reflect it.
- • Clear plastic, like glass, helps to trap heat by the **greenhouse effect**. (The natural buildup of heat trapped by atmospheric gases, mainly carbon dioxide, that let visible light in but keep some of the infrared radiation from leaving the earth's surface.)

2 Design your solar cooker with the facts above in mind. When you have settled on a group design, divide up the work needed to complete the cooker.

Ask the science teachers if they have covered the greenhouse effect.

Evaluation:

When you finish your prototype, test it by following this procedure:

1 Place a one-quart container of water at 72°F inside your cooker. Place a thermometer in the quart of water, and point your cooker toward the sun.

2 After one hour, record the temperature of the water.

3 Continue to test your cooker by moving it so that it points toward the sun each hour for 4 hours.

What are the advantages of using sunlight to cook? What are the disadvantages?

Challenge:

1 Make a graph of the temperatures and times of your solar cooker.

2 How could solar cookers be made more efficient using high-tech materials?

People can look at models and prototypes to decide whether the product will fit their needs before they spend a great deal of time and money building the real thing. Some products, such as airplanes, buildings, and cars, can be modeled through three-dimensional images on a computer. In this way, you can save more time because you don't have to build a real model. You can even *test* some products using the computer **simulations**. A simulation is a way to model a real product. Pilots, for example, can be trained to fly and to handle emergency situations using flight simulators. You can walk through a house simulated by a computer to see if you like the way the rooms are arranged. Some simulations even show you where shade and sunlight would come into your house at different times of the year.

Ask students to list the advantages of using simulators.

Computers can graphically simulate the real world. This is a computer generated image of a new chemical processing plant. Notice the shading and details that make it look like the real thing. (Courtesy of Intergraph Corporation.)

Ecology of a Product

Ecology is the study of how things interact with the environment. Part of designing a product is planning ahead for what will happen to it after it is used. Products might include anything from a newspaper to a jet airplane. Do you ever stop to think what happens to the things you throw away? Most of us don't very often. In fact, we live in a "throwaway" world. When we're finished with a product, we are used to just throwing it away instead of fixing it or **recycling** (reusing) it. Did you know some of these facts?

Special Report
Stereo Lithography

▶ On the average, every man, woman, and child in the United States creates 5 pounds of garbage a day. That's 230 million tons of garbage every year!

▶ Some products that we throw away could easily be replaced with different products that would last longer or that could be reused. For example, every year we throw away the following disposable items:

▶ 2,000,000,000 razors

▶ 16,000,000,000 plastic diapers

▶ 1,600,000,000 ballpoint pens

▶ The cost of throwing away our trash is going up fast because of new regulations to protect the environment and because of lack of space. It can cost $50 per ton to throw away trash. If you multiply the 230,000,000 tons of garbage we throw away each year by $50, we spend $11,500,000,000 just on garbage!

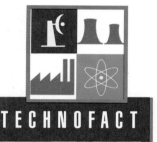

TECHNOFACT

Technofact 44

A frozen pizza plant in a town in Ohio was at first welcomed because it provided jobs for 1,000 people. Less than a year later, however, the plant had produced over 400,000 pounds of "pizza sludge" (tomato paste, flour, cheese, and so on)— waste that couldn't be put through the city sewage system. Environmental specialists wouldn't let the factory bury the sludge either. They were afraid that if the pizza sludge were buried, it would move underground and ooze to other places where it could become a problem.

Do you think we are throwing our world away? What happens to the garbage you throw away? Where does it end up? (Courtesy of Waste Management, Inc.)

Many of the products and materials you throw away end up in **landfills** (garbage dumps) where they are buried or burned, or sometimes the used materials are dumped directly into the oceans. Disposing of materials in these ways can eventually cause air or water pollution. The best thing for you to do is to use, whenever possible, materials that are **biodegradable**. Biodegradable materials break down or decompose naturally like paper and go back into the earth. Other materials that take a long time to decompose, such as aluminum, plastic, and glass, should be recycled. Did you know you should recycle the batteries in flashlights, transistor radios, and other electronic devices? Do you recycle wastepaper at your school?

Does your community have a recycling center? Find out where it is and start using it. If you can't find one, start asking why. (Courtesy of Waste Management, Inc.)

Take a field trip to the local land-fill.

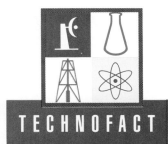

THINGS TO SEE AND DO:
Decomposing Demo

Introduction:
You have heard about the problem of too much garbage and not enough room to throw it away. This is a problem that just can't be thrown away. In this activity, you will take part in a test to see how quickly different materials decompose.

Design Brief:
Bring in a package that can be tested for the length of time it takes to decompose. Some materials claim to be biodegradable. You might bring in different types of packaging such as cardboard, plastic wrap, grocery bags, and paper wrappings.

Materials:
- Packaging materials
- Sample board (¼″ × 36″ × 36″ plywood)
- Tacks, glue

Equipment:
- Measuring tools
- Hand and power tools
- Video camera (optional)

Procedure:
1. This activity will involve the entire class. You will bring in a sample of a packaging material that would normally be thrown in the garbage.

2. Cut your sample into two pieces. One half of your sample will be kept inside. The other half will be exposed to the weather conditions found in your area.

3. Your class will make two test sample boards. The board kept inside is called the **control** sample board. The other board should be made in exactly the same way so that the samples can be identified after the test period. The board placed outside will be called the **variable** board.

4. Each person in your class should mount a test sample on each board using tacks or glue.

TECHNOFACT

Technofact 45
A researcher in California was experimenting with grain wastes and a bacterium that he hoped would dissolve explosive materials. What Larry Rogers discovered was bulletproof wheat! His compound can be used to manufacture products from pasta to lightweight armor.

⑤ Store the control sample board in a safe place inside. Put the variable sample board outside in a place where it will be exposed to the weather. As an optional activity, your class might videotape the start of the experiment and add to the tape each month during the test. In this way, a video record of the degrading process can be shown at the end of the test.

⑥ Compare the variables with the controls after each month. Make a log of the changes in each material.

Evaluation:

❶ What materials break down the fastest in your area?

❷ Which packaging material takes the longest to break down?

❸ Make a list of five products that are commonly purchased in a grocery store. How could packages of each product be changed to help protect the environment?

Challenge:

❶ Write a promotion to encourage people to recycle. These types of short "commercials" are called public service announcements (PSAs). Ask your teacher if your PSA could be read on a local radio station.

❷ Research who makes some of the packaging materials or products you tested. Write to them telling them of the results of your test.

❸ Exchange a test sample board of materials with another school located in an area of the country with a climate different from yours. After one semester, see if the climate makes a difference in how materials degrade.

Summary

As you design and build things, you need to make accurate measurements. Both the English and SI (metric) measurement systems are used today as standard measurement tools.

Things that are too large or too small to draw or make are often made to scale. Prototypes and models are scale representatives of products that help people see what the actual product will look like. Technology helps make your life more comfortable through ergonomics by designing products and places to fit your needs and size.

Choosing the right material for a product is important in determining whether that product will be useful to you and good for the environment. All materials can be divided into two major groups, synthetic and natural. Special properties such as hardness, tensile strength, compression strength, and fatigue strength can be tested before a material is used to make a product.

A sequential videotape of the degrading process can be made to show the stages of decomposition.

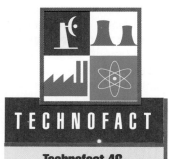

T E C H N O F A C T

Technofact 46

Product testing can give some surprising results. Safety engineers were checking the emergency braking system on a 480-foot mineshaft elevator in England. They raised the elevator to the surface and then sent it into free fall four times before discovering that two miners were trapped inside. The two men bounced up and down for 2 hours in the elevator before the testing stopped. By the way, the braking system passed the test, but the men don't care if they ever ride it again!

Challengers:

1 Research and define five measurement units that you have never used.

2 Find out what measurement standard is used for the exact length of a meter.

3 Collect anthropometric measurements such as height, weight, and reach (arms extended) for your class. Calculate an average for each measurement.

4 Research the composite materials used in making a space shuttle and jet fighters.

5 Compare the properties of steel, aluminum, and titanium. Which metal would be best for a bicycle frame? Explain why.

6 Can you think of a nondestructive test method for a material?

7 Build a scale model of a house you design.

8 Calculate the volume of one soda can or milk carton. Find out how many cans or cartons are consumed (used) in your school each day. Calculate the total volume of space that this garbage takes up in a landfill.

9 Research and report on new forms of biodegradable plastics.

10 Contact the agency that operates your local landfill. Find out their 5-year and 10-year plans for managing the landfill.

See Teacher's Resource Guide.

Chapter 8

What is Automation?

Things to Explore

When you finish this chapter, you will know that:

▶ The automatic control of machines is called automation.

▶ Materials-handling systems bring raw materials to machines and take away finished products.

▶ Robots do some jobs that people find boring or dangerous.

▶ Computers can be used to control the flow of materials and machines.

Chapter Opener
What Is Automation?

TechnoTerms

automation

conveyor

computer-aided manufacturing (CAM)

computer-integrated manufacturing (CIM)

crane

forklift

hoist

industrial robots

manipulator

mass production

materials handling

monorail

optical processing

payload

personnel

photoelectric

pick-and-place maneuver

robot

robotic vision

stepper motor

telemetry

work envelope

(Courtesy of TRW Inc.)

Careers in Technology

Automation is often blamed for taking jobs away from people. Automation technology has helped to create new jobs, too. Would you like to have the job of making all these pistons by hand? The appropriate use of automation can help with heavy or dangerous work. Robots have been made that can perform very delicate and accurate work, but they often lack the judgment needed to deal with unusual events. Factories of the future will probably have more automation and fewer human workers. What does this mean to you when you think about a possible future career?

What about Automation?

Ask students to list examples of automation.

After you have solved the design problems and picked the right resources and materials for your product, you need to have an efficient way to produce and move materials. That's where **automation** comes in. Automation is the automatic control of a process by a machine. Automation is part of almost all areas of modern technology.

Today, automation of the complicated launching process makes it possible to send people into space and bring them back. Robotic arms do welding and other dangerous tasks automatically. Robots explore space and underwater locations in the oceans. Even soft drinks are made by using automated sensors that control the mixture of syrup and water.

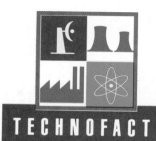

TECHNOFACT

Technofact 47

Seven hundred thousand tons of toxic (poisonous) waste are produced in the United States every day. That's 250 million tons a year, or enough to fill the New Orleans Superdome 15,000 times. Many U.S. manufacturers are trying to reduce the amount of toxic emissions from their factories.

Automation works to make our lives easier and more enjoyable. This conveyor is a part of an automated process used to make spaghetti sauce. Why do you think the workers are wearing hair nets? Why is the technician wearing a hard hat too? (Courtesy of Campbell Soup Company.)

Automation has really affected the modern workplace. Automation does away with jobs that are boring or dangerous. It also creates more need for highly skilled people to control, repair, and program machines. Automation can speed production in many instances and can also save time and money.

There are also problems with automation. Automation tends to take jobs that require less skill away from workers. Sometimes these workers cannot be retrained to do the higher-skilled jobs. In addition, automated machines are expensive to buy and maintain.

The high-speed production of food products is an example of where automation has helped to make the high-quality products we are used to. Would you like to work in a potato-chip factory? Would you like to do the same job all day, every day, every week, for years? (Courtesy of Borden, Inc./David Joel)

It would be hard for us to live without automation now. Think about the automatic washer and dryer your family uses. Do they go through the cycles automatically, or do you have to **manually** (by hand) run through the cycles? What about the copy machine or the bell system used in your school? Can you think of other examples of automation that affect you?

Before factories as you know them existed, people made products in their homes. Different households made different products and had different skills or crafts. In craft manufacturing, each item was produced by one person. It often took a very long time to complete one item. The next step was the beginning of factories. Instead of individuals making products, groups of craftspeople worked in factories making products faster. To make the factory more efficient, a system of making things was developed by Eli Whitney in 1798. Henry Ford perfected the system we know as **mass production**. Mass production uses assembly lines, where products move past a worker who does a specific job.

Ask students to research Henry Ford and Eli Whitney.

TECHNOFACT

Technofact 48
Our world today creates more waste in one month than people used to make in one year!

Early products were made one at a time in people's homes. Today, assembly line mass production is a common process used to make products quickly. Here, workers on an assembly line in England are producing automobiles. (Courtesy of Ingersoll Rand.)

Techno Teasers
Handmade Violins

Techno Teasers
Answer Segment

THINGS TO SEE AND DO:
Mass Production versus Craft Production

Introduction:
You have learned about the early days of manufacturing when people made products one at a time in their homes.   Today, we rely on mass production to make products from cars to cookies. In fact, in this activity, you will manufacture cookies yourself!

Design Brief:
Determine the difference in efficiency between craft production and mass production. You will be assigned a specific job in either the craft production or mass production factory. Your job is to work as fast and as accurately as possible.

Materials:
- Vanilla wafer cookies (plain)
- Assorted flavors of frosting (squeeze tubes)
- M&M candies (plain)
- Napkins or paper towels
- Plastic gloves

Equipment:
- Measuring tools

Procedure:
❶ Divide the class into two groups. Flip a coin to see which group will be the craftspeople and which will be mass-production workers.

❷ The class should decide on a design for the finished cookies.

❸ The supplies of cookies, candy, and frosting should be divided equally among each group.   Each of the craftspeople will need a supply of each supply item.

❹ The teacher will accept or reject the cookies from each group after the manufacturing period has ended. (Only the perfect cookies will be eaten at the end of class.)

❺ Each person in the mass-production group will need a specific job. Your teacher will assign a job to each person in the assembly line. The mass-production workers should sit at a long table or rearrange their desks so that they can easily pass the cookies to the next worker.

❻ All workers in both groups should wear plastic gloves to keep the cookies clean. When each group is ready, your teacher will tell you to start production. The craftspeople will each make complete cookies. Each mass-production worker will do a specific job in the production process and pass the cookie down the assembly line until it is finished.

Here a robot is being used with a conveyor belt to make cookies. Have you ever seen a robot work in a factory? What are the advantages and disadvantages of using robots? (Courtesy of Campbell Soup Company.)

7 After a set time, your teacher will ask both groups to stop. At that time, you must stop where you are. Do not finish the cookie you were working on.

8 Each group should gather their completed cookies. All of the completed cookies will be inspected by your teacher.

Evaluation:

1 Which group produced the most cookies?

2 Which group had the most rejected cookies?

3 What do you think the advantages and disadvantages of craft production are?

4 What do you think the advantages and disadvantages of mass production are?

5 Did the production of cookies turn out the way you expected? Explain.

Challenge students to make a chart showing the difference in the production of products.

Challenges:

1 Research to find out what products are custom made by craftspeople today.

2 Try the production again with different products. Do you think you will get the same results?

3 Try the production process again, but this time switch the groups. Do you think you will get the same results?

4 Research to find out how mass production in Japan is different from that in the United States.

5 List five ways that the mass production of cookies could have been improved.

For machines to work automatically, they must have parts and materials come to them. People keep trying to perfect the mass-production process. Automation has helped improve the efficiency of mass production in today's factories.

Ask students to brainstorm various methods of moving and storing materials and products.

Special Report
Moving Music

Moving Materials

The Industrial Revolution, which you learned about in Chapter 1, marked the beginning of a need for efficient **materials handling** (moving and storing materials). Steam power and machines were put to work to help workers do their jobs better and faster. Automation and assembly lines became part of the factory setup.

Today, various kinds of equipment are used in materials handling. Materials handling systems usually include these basic types of equipment.

▶ *Conveyors:* These are used to move materials and parts from one place to another. They always travel over a fixed path. There are many different kinds of conveyors. Examples are roller conveyors, skate-wheel conveyors, and belt conveyors. Have you used the people-mover walkways at an airport, for instance? These conveyors make it easier for people to get from one place to another. In automated assembly lines, parts that are being worked on are moved by conveyor from one workstation to another. Even supermarkets have conveyors at checkout counters to make it easier and faster for you and the clerk.

A

B

Conveyors are used to handle materials in factories or wherever fast, efficient movement of items is needed. Can you tell which of these conveyors use a belt or rollers? How could these products be moved without conveyors? What would happen to the speed of production if people had to carry all of these products from one place to another? (Figure A Courtesy of TRW, Inc. Figure B Courtesy of Wollard Airport Equipment Company.)

- ▶ *Trucks:* Materials-handling trucks are different from the trucks you normally see on the highways. An example is a **forklift**. Forklifts have forks on the front and are used for moving materials from one place to another.

- ▶ *Containers:* Large parts are moved onto **pallets** or special platforms. The pallets can then be stacked in racks. Small parts are moved in containers such as trays or tubs.

- ▶ *Hoists, Cranes, and Monorails:* Hoists are used for lifting heavy loads. Cranes are really hoists that move in a limited area. For example, an overhead crane lifts large items and then travels along an overhead rail to another spot close by. A monorail is a hoist that moves on one overhead rail.

B

Many methods are used to handle materials. Here, products are being moved using a forklift, crane, truck container, and pallet. The size, shape, and weight of products determine which materials-handling system would be best. (Figure A Courtesy of CSX Corporation. Figure B Courtesy of Texas Instruments.)

A

- ▶ *Automated Storage and Retrieval System (AS/RS):* This system used in manufacturing has a computer-controlled crane. It travels between the pallet racks and automatically loads and unloads the racks. There is even a miniload AS/RS that handles small items that would fit in a drawer or tray.

- ▶ *Automatic Guided Vehicle System (AGVS):* In this materials-handling system, a computer controls the movement of several driverless carts called **AGV**s. These carts follow wire paths that are built into the floor. The computer keeps track of the location of all the carts so they don't run into each other as they move materials around a factory. What do you think would happen if one cart got off track?

Challenge a group of students to make a display of materials handling devices.

As computers become more powerful, they can be used to control automated materials-handling systems as well as machines. These automated lifts automatically deliver parts to the proper place in a warehouse. (Courtesy of SI Handling Systems Inc.)

THINGS TO SEE AND DO:
Can You Handle It?

Introduction:

You have learned that efficient production methods require the movement of parts and materials to the machines and workers on an assembly line. Materials handling is an important part of designing factories that can make high-quality products quickly. In this activity, you will work in groups to make a set of conveyor belts that can be used in your classroom to simulate an assembly line.

Design Brief:

Build and test a conveyor belt system that will transport parts from one point to another in your classroom. The conveyor system must be safe to operate and have adjustable speed so that it can meet the needs of a future mass-production simulation activity (coming in Chapter 9).

Materials:

- Curtain pleating tape
- ⅛" tempered hardboard
- ¼"–20 threaded rod
- ¼"–20 nuts
- ⅛" acrylic sheet (Plexiglas)
- 1½" PVC pipe
- ¼" fender washers
- Masking tape
- Wood glue, screws

Equipment:

- Scroll or band saw
- Disk or belt sander
- Hacksaw
- Adjustable wrench
- Power hand drill
- Sewing machine (optional)

Materials gathered in the machine dissection activity may be used here also.

Procedure:

1. This activity will require you to work in large groups. There are many parts that need to fit together to make your conveyor belt. Your class will need two or three conveyors for the mass-production activity in Chapter 9. Your group should be made up of one-half or one-third of the class. Everyone will help make the conveyor belt.

❷ Your group or your teacher might decide to change the design of your conveyor belt system to meet the needs of your classroom. The plans below will help you get a start on your finished design. You may decide to use some of the materials you saved from the machine dissection activity in Chapter 6. Gearhead motors, rollers, screws, and so on might come in handy. Note that the length of the conveyor is up to you and your teacher. The width of the conveyor is determined by the material you use for the belt. The minimum size should be 4 inches.

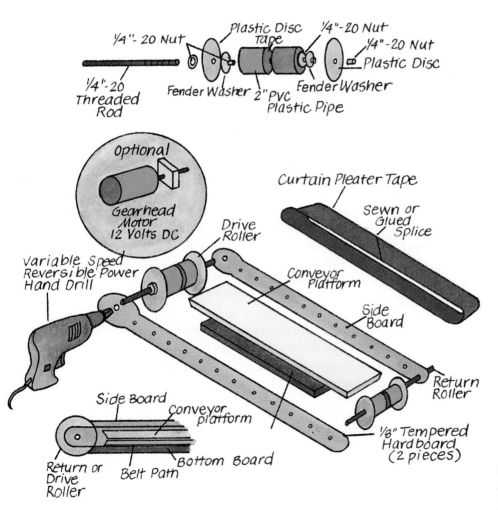

Here is a simple way to make a conveyor belt simulator. Your conveyor belt system design can be changed to meet your needs.

❸ The design that you choose will be used by all the groups so that the conveyors will match. Divide the following tasks among your group. Some jobs might require two students.

Review the proper use of equipment and tools.

SAFETY NOTE:
The following activities require all students to wear eye protection at all times.

- With a hacksaw, cut the ¼"–20 threaded rod to length for the drive roller and the return roller. (The lengths will depend on the width of your conveyor belt.)

- Cut the 1 ½" PVC pipe to length with a hacksaw. Wrap and tape a piece of paper around the place where you are cutting to guide the saw for a square cut.

- Wrap the center of each PVC pipe with two layers of masking tape. This will make the center of the pipe larger in diameter than the ends. This helps the belt track properly without going off to one side.

- Cut four acrylic (Plexiglas) disks 3 inches in diameter using the scroll saw.

SAFETY NOTE:
Ask your teacher to help you cut the disks on the scroll saw. Remember to keep your fingers away from the front of the blade.

- Assemble the drive and return rollers as shown in the drawing. The fender washers are used inside the PVC pipe to keep the threaded rod centered. You might want to put a drop of glue on the threads near the nuts to keep them from coming loose.

- Cut the conveyor platform and bottom board to length and width on the band saw. Ask your teacher to help you cut the ends at a 45 degree angle. The length of the boards depends on your design.

- Cut and sand the ⅛" tempered hardboard to make the sides of the conveyor. Round the ends to match the plastic disks on the rollers.

- Temporarily assemble the rollers, sides, conveyor platform, and bottom board as illustrated. Check to see if the axles are parallel to each other. This is important for the belt to track properly. You might want to file the return roller holes into slots to make it adjustable.

- Use a piece of string or a tape measure to find the proper length of the belt. Add 4 inches for the pleating tape to overlap at the splice. Cut the curtain pleating tape to the proper length.

- Sew or glue the splice together at the proper distance for your conveyor. Be sure the edges of the pleating tape remain straight.

- Glue and screw the completed conveyor system together.

Ask the home economics department to sew the conveyor belt together.

Evaluation:

❶ Attach the power hand drill to the drive roller. (Your teacher might have a special gearhead motor for your conveyor instead of a drill.) Slowly start the drill or motor. Watch carefully to see if your conveyor belt stays on track. Make adjustments as needed.

❷ Determine the speed of your conveyor belt system. The speed of a conveyor is often measured in parts per minute. You can easily measure the speed by doing the following:

How to Determine Parts Per Minute Speed:

- Set the the drill or motor to a constant speed.
- Have one or two students feed the input end of your conveyor with small parts such as washers or nuts. Space the products (washers or nuts) evenly and close together. Count the number of products as they start coming off the end of the conveyor for one minute.

What was the number of parts per minute for your conveyor system?

❸ List two other possible materials that could be used for the conveyor belt.

❹ Think of a way to slow down a high-speed motor so that it could power a conveyor belt system. Make a sketch of your idea.

Ask students how they would reverse the direction of the conveyor belt electrically.

Challenges:

❶ The speed of conveyors is also measured in feet per second. Calculate the speed of your conveyor system by doing the following:

How to Determine Feet Per Second Speed:

- Set the drill or motor to a constant speed.
- Put a mark or piece of tape on the edge of the conveyor belt. This mark will be used to measure how fast the belt is moving.
- Have someone with a watch call out every second. As the timing mark comes into view, use a pencil to mark the side of the conveyor every second. Make at least two or three marks. Turn off the conveyor.
- Measure the distance between each mark and the next. Find an average distance by measuring the total distance and dividing by the number of spaces measured.

$$\text{average inches per second} = \frac{\text{total distance}}{\text{number of spaces}}$$

- Calculate the speed in feet per second.

$$\text{feet per second} = \frac{\text{average inches per second}}{12}$$

- What was the speed of your conveyor in feet per second?

❷ Can you think of another way to measure the speed or capacity of a conveyor belt system? Explain.

❸ What would determine the speed of a gravity-fed roller conveyor system?

❹ How can conveyors be used to move parts around corners? Make a sketch of your idea.

❺ Design and build an easy method for adjusting the tension and tracking of the conveyor belt on the rollers.

Robots are also sometimes used in materials handling but they have many other uses in modern technology.

Techno Teasers
Robots Everywhere!

Techno Teasers
Answer Segment

Ask students to make a sketch of what they think a robot looks like.

Robots

Robots, robots everywhere! Robots are entertaining you in the movies. They are doing jobs for you that you don't want to do or that are too dangerous for you. Robots are one of most exciting inventions in technology. Today's **robots** are highly advanced computer-controlled machines. That means robots are programmed to do special jobs.

What can today's robots do? How would you like a robot that could do all your homework? In order for a robot to do your homework, you would have to program the robot with the correct answers. In other words, you would have to do your homework first and then program the robot to do it! That would be true for every different assignment. Doesn't sound like much help to you, does it? Robots cannot think for themselves yet. People still have to program robots to do special tasks.

Industrial robots are controlled by computers to do many different jobs. This robot arm has been programmed to weld metal parts. The robot can move very accurately making perfect welds every time. Can you think of other jobs robots could do in a factory? (Courtesy cf Motoman, Inc.)

The word *robot* comes from a play by a Czech writer named Karel Capek. Robot means "to work" in Czech. But the idea of robots was around long before Capek used it in the 1920s. In early times a robot might have been simply any machine that could do work without a person running it all the time. By that definition, many devices such as clock radios, clothes washers, microwave ovens, and mechanical toys are robots. But a true robot is controlled by a computer and can be reprogrammed to do different jobs.

Make a bulletin board illustrating early robot designs. See Teacher's Resource Guide.

Some of the appliances we use at home work automatically. A clothes washer, for instance, does a great job of automatically washing and rinsing our clothes and spinning out most of the water. However, a washer is not a robot. It could not be programmed to wash dishes or to walk the dog. Robots are reprogrammable and are able to do many jobs. (Courtesy of Maytag Corporation.)

TECHNOFACT

Technofact 49

How about a robot guide dog for the blind? A robot guide dog will be easier to train than a real guide dog. It will have a synthetic voicebox that talks instead of barks. That way it can tell the person about potholes, bus stops, red lights, or where elevator buttons are. To be able to do this, the robot has an "outdoor data base." Its computer is capable of recognizing sounds from the outside environment as well as those from inside the house. The robot dog will be about the size of a German shepherd and will have one paw that can pick things up

Computer control will eventually give us robots that have full mobility (movement), vision, hearing, speech, and the ability to make decisions. You'll be able to tell your robot what to do, and because it is programmed to recognize your voice and perform special tasks, it will do that job for you. The further development of artificial intelligence (AI) will make it possible for computer-controlled robots to "think" what step to do next. As computers get better and better, so will robots!

How robots look depends on the job they do. Robotic arms are used in manufacturing cars, for instance. These are called **industrial robots**. The robotic arm can be programmed to pick up a door and place it in a certain spot. Another robotic arm fastens it while a third robotic arm spray paints or spot welds certain pieces on the automobile assembly line. The Space Shuttle uses a robotic arm or **manipulator** called the **remote manipulator system (RMS)** to move the **payload** (satellite) from the cargo bay when the shuttle is in space.

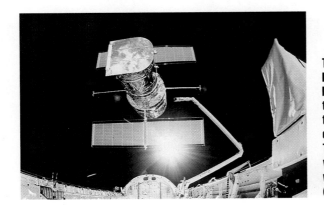

The Space Shuttle uses a robot-like arm called the Remote Manipulator System (RMS). Here the arm is being used to move the Hubble Space Telescope out of the cargo bay of the shuttle. The arm does not have to be very powerful because of the weightless condition in orbit. (Courtesy of NASA.)

The *Viking* landers that NASA sent to Mars moved around on the surface like huge bugs, picking up soil and doing experiments. Other robots handle dangerous materials such as radioactive wastes. Can you think of other dangerous jobs robots can do that you wouldn't want to do?

Industrial robots can run on different power systems. Some move with electric motors. Some use special electric motors that turn a little bit at a time. These are called **stepper motors**. Pneumatic robots run on compressed air. Hydraulic robots use oil pumped under pressure to make the robot operate. The type of energy system used depends on the size of the parts being moved or lifted.

Robots can be equipped with special sensors to help them do jobs. Some robots can hold fragile parts with just the right pressure so they do not drop them or squeeze them too tightly. Other robotic light sensors guide robots along pathways. In some cases, television camera eyes assist robots. These robots can tell the difference between parts with different shapes.

Have you heard that robots are taking jobs away from people? In many instances, robots do jobs that people don't want to do. These jobs are either too boring or too dangerous for humans. Sometimes robots can do a better job because they can be programmed to do the exact same thing over and over and still be **precise** (accurate) to one one-thousandth of an inch. What you need to remember is that people still need to program the robot to do the job. The best person to program or train a robot to weld is a welder. Often people replaced by robots can be retrained to become robot programmers.

Visit a factory that uses robots or show a videotape.

Industrial robots come in many sizes and shapes. Some of the robots used to make products are powered by electric stepper motors. Hydraulic or pneumatic robots are also used in factories. (Courtesy of Motoman, Inc.)

Robots do take some jobs away from people. They also open new jobs that require more training and education. Here, students are learning about the end effector of an industrial robot. (Courtesy of Ford Motor Company.)

What's an End Effector?

The **end effector** of a robot is whatever device is at the working end of a robot arm. End effectors must be designed specifically for the material they are going to handle. We often forget that our hands and fingers are used to grip thousands of objects every day. Your fingers can pick up small pins and needles as well as catch a football or grasp a tennis racket. It is not as easy for only one robot gripper to work as robot "hands" to pick up very small and large objects. The problem of designing a universal gripper that will work for every object gets even trickier with other materials. If the robot has to pick up liquids, hot parts, radioactive materials, or glass sheets, for example, the design must be changed.

Instead of trying to make a universal end effector, designers have created some robot arms that can change hands for different jobs. For example, one robot arm can be programmed to weld two parts together, put the welding gripper down, and grab a grinder or other tool. Where mass-production assembly lines are used, robots are most often dedicated to doing only one job such as spray painting or welding.

Specialized end effectors are made for robots to work with specific materials. An electromagnet might be used to attract ferrous metals such as steel. The electromagnet can be turned on and off by the robot's controlling computer. This method would not work for materials that are not attracted to magnets, however. Aluminum, copper, glass, plastics, or wood, for example require a different kind of end effector. To solve this problem, engineers use suction cups to pick up the materials. The controlling computer can start or stop a vacuum pump to pick up parts such as car windshields.

THINGS TO SEE AND DO:
Programming a Robot

Introduction:

People have been trying to have machines do work since the invention of machines. The idea of people telling machines what to do is now possible thanks to automation and robotics. In this activity, you will first control a robot arm simulator to move objects. When you understand the controls of the robot, you will learn to program it to work for you automatically.

Design Brief:

Design a flowchart that will make a step-by-step list of movements the robot will make. Determine the work envelope of the robot. Program a robot arm simulator to perform a pick-and-place maneuver.

Educational programmable robotic arms are available through technology and science suppliers. See Teacher's Resource Guide.

Materials:

▶ Conveyor belt system (see "Can You Handle It?" activity)

Equipment:

▶ Programmable robot arm
▶ Wood or plastic blocks (sized for the robot to grasp)
▶ Graph paper
▶ Tape measure
▶ Protractor

Procedure:

1 Work in groups of three or four. To program a robot, you must understand each of its movements. Robots are like humans in that they can only move their arm or wrist a certain distance before it hurts. The maximum distance that each part of a robot arm can move is called its **work envelope**.

The work envelope of a robot is how far it can reach in all directions. This multiple-exposure picture shows the work envelope of the robot arm. (Courtesy of Motoman, Inc.)

❷ Ask your teacher to show you how to safely operate a robot arm simulator, or use the robot arm you made in Chapter 6. Each of the joints of a robot arm has a name such as base, shoulder, or wrist. Make a list of the names of each joint in the robot arm you are using. Next to each name, write the maximum movement possible. Your list might look like this:

- Bases, 180°
- Shoulder, 120°
- Elbow, 110°
- Wrist rotation, 360°
- Pitch, 180°
- Gripper, 0″ to 3″

❸ Your next job is to determine the work envelope of your robot. Use graph paper to make a scaled sketch of the top and side views of your robot arm. Choose a scale that will let you draw the robot and its maximum movements. Your sketch might look like this:

Ask the language arts department to help the technology students to write specifications for their robot.

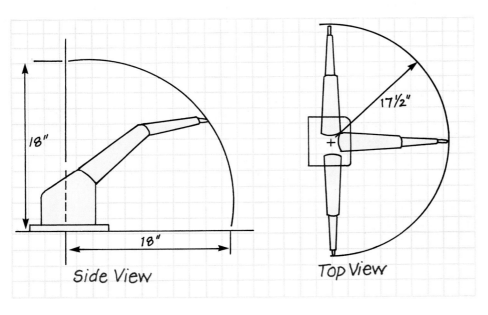

Side View Top View

Sketch the side and top views of your robot arm to scale. Show the maximum reach for each of the arm positions. This is called the work envelope of the robot arm.

❹ Use a tape measure to find the maximum reach for the robot. Make a line on your sketch to show the work envelope for your robot arm. Show the work envelope on both the top and side views.

❺ Now that you know the limits of your robot, it's time to put it to work. From your sketch, determine the best place to put your conveyor system. You will be using the robot arm to off-load (take parts off) your conveyor belt.

The type of program you will make is called a pick-and-place maneuver. This name makes sense because your robot will *pick* a part and *place* it somewhere else.

If a programmable robot is unavailable, use an inexpensive robot simulator. See Teacher's Resource Guide.

6 Set up the conveyor and robot according to your plan. Ask your teacher to show you how to program the robot arm. Each robot has a slightly different way of putting steps into memory. Your robot may be programmable directly, or it might require the use of a computer. Robots can sometimes be programmed with a remote keypad called a **teach pendant**.

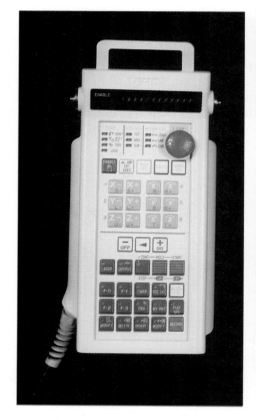 

Some robots have a remote keyboard that is used to control and program the robot arm. The photo on the left is a teach pendant, the remote keyboard that is used to program an industrial robot. (Courtesy of Motoman, Inc.)

7 Make a flowchart of each step you are programming into the robot memory. A **flowchart** is a list of all the steps in a process. Engineers often make flowcharts so that they can plan the production of a product. Flowcharts are also helpful for **editing** (changing) programs in a robot or computer. Computer programmers often make a flowchart before writing a program. This way, they can plan the most efficient way of doing it.

CAD/CAM is a combination of designing and manufacturing parts or products using a computer. Here an engineer is using a computer to design a new car. You might notice that she is "drawing" on the computer rather than typing on it. This device is called a graphics tablet. The movement of the special pen will be shown on the computer screen. (Courtesy of Ford Motor Company.)

8️⃣ When you have learned to program the robot arm, adjust your conveyor to a slow speed.

SAFETY NOTE:
Always be careful with any electrical equipment. Keep liquids away, and place extension cords so that people will not trip. Most educational robots operate on low voltage. However, you should always ask your teacher before changing any electrical connection. Do not attach or unplug equipment from computers while they are turned on. Computer circuits are easily damaged.

Have one student in your group place **products** (wood or plastic blocks) on the conveyor belt at an even rate. Try to program the robot to grab the parts at the end of the conveyor and drop them into a box.

9️⃣ You might find it hard to place the parts on the conveyor belt in the exact spot the robot expects to find them. One solution might be to design and build an attachment for the conveyor that will direct the products to the center of the belt. You could also have a stop at the end of the conveyor that will give the robot time to grab each product.

Ask students to apply the system model (input, process, output, feedback) to this activity.

Evaluation:

1️⃣ What was the hardest part of this activity? How could it have been made easier?

2️⃣ What do you think would happen if the products on the conveyor were all different sizes?

Ask students to express what they think robots of the future will be like.

❸ Make a sketch of an attachment for your conveyor that would line up products so that the robot could easily pick them up.

❹ What would be the advantage of having one computer control the robot and the conveyor belt?

Challenges:

❶ Design an electrical system that will start and stop the conveyor belt so that products don't pile up at the end of the belt.

❷ Make a sketch of how a laser beam going across the conveyor belt might be used to control the system.

❸ Visit a factory that uses conveyors to move parts or products. Make a sketch of how parts move through the factory. Can you think of a way that robots could be used in the factory?

❹ Design a people-mover conveyor system that could help students move faster in your school.

❺ Research the safety hazards of industrial robots. Why is it important for workers to know the working envelope of a robot? What precautions are taken in factories to prevent people from being injured by robots?

Robots can do jobs that make life easier for you. Today, robots are used to help physically challenged people in hospitals. Robots are being used in homes for entertainment as well as for doing specialized jobs such as walking the dog. In factories, robots are helping to produce goods more efficiently and **cost-effectively** (saving money). In one case, a Japanese factory uses *only* robot workers! As technology advances, you will find robots being used in many new ways.

Working with Technology

Conveyor systems and robots are only a part of the way modern factories operate. Computers are not only used to design products, as you learned earlier, but they also are used to control machines to make products. Using computers for this purpose is called **computer-aided manufacturing** (**CAM**).

Another technology used today is called **CAD/CAM**. It joins CAD (computer-aided design) with CAM so that you can design a part on the computer. Then the design information, for instance, the shape or size of some part, is sent directly to a machine tool. The machine tool makes the part. An advantage of CAD/CAM is that you can design a part on the computer and test it using the computer before you actually produce the real part. Just think of the time and money you could save using CAD/CAM.

Computer-integrated manufacturing (**CIM**) puts manufacturing, design, and business systems together. In CIM, the computers control machines, store design information and data on materials and parts, and schedule how materials will be purchased and shipped. The computers even make inventory reports, do the accounting, and produce **financial** (money) reports. If you needed to get a total picture of what was happening in your factory, you could just check the computer screen. CIM makes it easier for management **personnel** (people) to locate and solve problems in different parts of the process.

Computers can be used to control the entire operation of a factory. They are used to design the products, control the machines, and keep track of inventory (the number of parts or products in storage). This is called CIM—Computer Integrated Manufacturing. (Courtesy of Xerox Corporation.)

Earlier, you learned that some robots use television camera eyes to help them "see." This is called **robotic vision**. Vision systems help robots perform tasks such as installing automobile windows or making sure packages are lined up in the right order to be labeled.

CAD/CAM simulators are available through technology suppliers.

THINGS TO SEE AND DO: *Telemetry Technology*

Introduction:
You probably are asking yourself what telemetry technology is. By now, you should have a pretty good idea of what technology is. **Telemetry** is the process of sending and receiving information from a distance. For example, space satellites send us information about the Earth and other planets. We send signals from Earth to satellites to control cameras, computers, or even to turn on rocket motors to change the direction of the satellite. The sending and receiving of all that information is called telemetry. In this activity, you will use telemetry and robotic vision to control a robot from a remote distance.

Design Brief:
Control a robot from a remote location using telemetry. Your job will be to simulate a robot space probe going to Mars to bring back a sample of Martian soil to Earth.

Be careful not to use materials that may damage delicate equipment.

Inexpensive walkie-talkies are available from electronics and science suppliers.

Materials:

- Tray of "Martian soil" (unpopped popcorn kernels)
- Cardboard, tape, scissors
- Empty plastic bowl or jar (to hold soil sample)

Equipment:

- Robot
- Video camera and VCR or camcorder
- Video monitor
- Tripod or ceiling camera mount
- Walkie-talkies or intercom
- Scale or balance (30 g)

Procedure:

1 In this activity, you will work with a partner to communicate telemetry instructions. The first task for you and your partner will be to design, build, and test a robot end effector. Your end effector will be attached to the end of a robot arm. It must be able to scoop up a 30-gram sample of Martian soil (unpopped popcorn kernels).

2 Your end effector will be made of cardboard. It should have a place for the robot to grip it, and it should be able to scoop up at least 30 grams of Martian soil at one time. You should design the end effector so that it can be folded into shape. Your design might look like this:

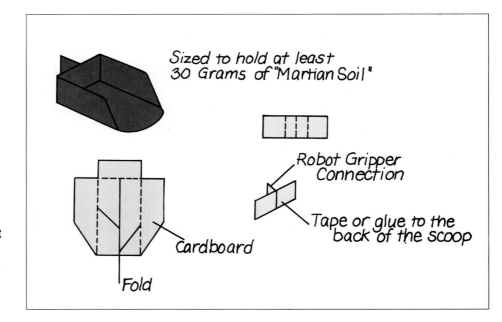

Sized to hold at least 30 Grams of "Martian Soil"

Robot Gripper Connection

Tape or glue to the back of the scoop

Cardboard

Fold

Your robot scoop end effector must be designed to hold at least 30 grams of "Martian soil". The robot gripper connection should be designed to work with the robot you will be using. Can you think of a better design?

3 When you have finished the scoop end effector, the next step is to understand the operation of the telemetry system. In this activity, one person will operate the robot controls without watching the robot. The robot operator will receive instructions from his or her partner using walkie-talkies. The Earth-based partner will be in a remote part of the classroom or in another room watching the movements of the robot on a video monitor. The setup sounds complicated, but it is simple.

Adjustments to this setup may be made depending on availability of equipment.

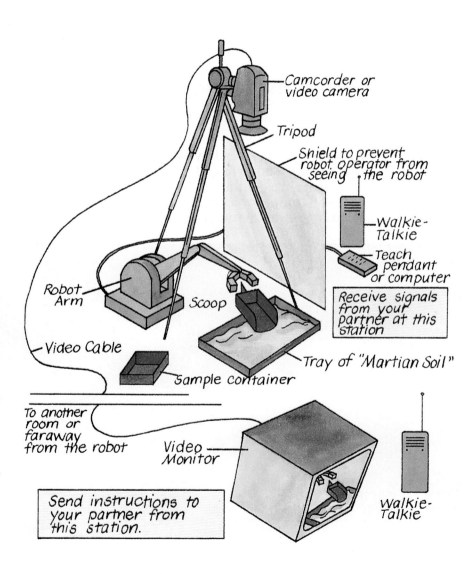

Camcorder or video camera

Tripod

Shield to prevent robot operator from seeing the robot

Walkie-Talkie

Teach pendant or computer

Receive signals from your partner at this station

Robot Arm

Scoop

Video Cable

Tray of "Martian Soil"

Sample container

To another room or faraway from the robot

Video Monitor

Walkie-Talkie

Send instructions to your partner from this station.

The setup of equipment for the telemetry activity is really very simple. You and your partner should practice using the equipment and communicating on the walkie-talkies before starting your experiment.

Assist students with camera movement and tripod adjustments.

④ You and your partner must be able to communicate instructions clearly and quickly. In this activity, the robot operator is taking the place of a telemetry computer controller on a Martian probe. Signals from Earth are being simulated by your partner watching the video monitor that shows the movement of the robot arm. You should practice sending and receiving instructions with your partner before collecting the Martian soil sample.

⑤ Run a wire from the video camera to a monitor in another room or far away from the robot operator in your technology classroom. The camera should be directly over the robot. Adjust the camera so it will include the work envelope of the robot arm.

⑥ Place a cardboard wall around the robot so the robot operator cannot see it. Set up the video camera on a tripod, and adjust the camera to point down toward the top of the robot.

⑦ When everything is ready, your teacher will time you to see how long it takes to gather at least 30 grams of soil and place it in a container. The reason for trying to gather the soil sample quickly is to simulate the conservation of battery power on the Mars probe.

Should we send people or robots to explore Mars? What is your opinion? Can you think of any famous explorers who found something new? Can you think of any robots that have explored new lands? (Courtesy of NASA.)

Evaluation:

① How long did it take you and your partner to gather 30 grams of Martian soil from the tray and place it in the sample container? Which team in your class took the least amount of time?

❷ If a robot had a video camera on its arm, would it be able to see the parts on a conveyor and go to their location? Explain.

❸ How could your end effector scoop be designed so that it could work in dusty, sandy, or rocky soil?

❹ Why would we send robots to Mars instead of people?

❺ Can you think of other experiments that could be done with robot vision and telemetry?

Have each student express their opinion regarding manned spaced missions versus robotic space missions.

Challenges:

❶ You probably noticed that it was sometimes hard to tell where the robot end effector was located. Your eyes give you **stereoscopic vision**. That means that you can tell how far away objects are. Looking through the lens of the video camera is like looking through one eye. Training robots to see things is difficult. Robotic vision can't tell how far away objects are. The ability to judge distances is called **depth of field**. Can you think of a way to change the camera setup so that you would have a greater depth of field?

❷ What do you think about sending robots to other planets instead of sending astronauts? Do you think all space exploration should be made by humans or by robots? Explain.

❸ Sketch a Martian lander that would not only have a robot arm, but would also be able to travel over the Martian soil on wheels or legs.

❹ Research the idea of using small robots shaped like insects to explore planets. What do you think a telemetry-controlled robot should look like? Make a sketch, or draw your design on a computer.

❺ One of the reasons scientists want soil samples from Mars is to test for signs of living organisms. Do you think they will find life on Mars? Do you think there might be life in any other part of the universe? Explain.

The simplest form of machine vision uses **photoelectric** switches or sensors. You have seen one type, called the **beam-break** system, at the supermarket checkout counter. They are designed to stop the conveyor belt automatically when a product reaches the checker. The beam-break system makes robots useful in assembly and production lines, where the working environment is kept as simple and orderly as possible. An example would be a **pick-and-place maneuver** (programmed movement), where the robot picks up the same-sized object each time and places each object in a particular place.

In the real world, robots must be able to deal with problems such as one object blocking the robot's view of another part. Robots that can

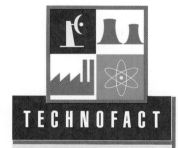

T E C H N O F A C T

Technofact 51

Credit cards are part of banking automation. The idea of the credit card began a long time ago in 1887 when an American writer, Edward Bellamy, made up the term *credit card*. He thought that by the year 2000, no one would need cash anymore. We've come pretty close to the cashless world that Bellamy predicted. The average American now carries seven credit cards. One man, Walter Cavanagh, has more than 1173 valid (still usable) credit cards. He keeps them in a specially made wallet that is 250 feet long and weighs 35 pounds!

identify a part from a lot of parts have to be very complex. Space robots of the future may use **optical processing** systems (gathering and comparing thousands of images in a few seconds) to recognize other robots, astronauts, or the earth.

Robots with vision do jobs that are often too detailed for humans to carry out. A good use for robots with vision is inspecting hundreds of microscopic connections on semiconductor chips. Can you imagine doing that kind of close-up work over and over without making any mistakes?

The Factory of the Future

In the future, you may find more robots replacing people in industry. In fact, many factories will be run totally by robots. As computers improve, so will robots. They will eventually be able to assemble products as well as make the parts. This means that people will have to be retrained to find other jobs.

People like you will have shorter working hours. Some people predict that the work week, which now averages around 40 hours per week, will be only half that in less than 50 years. What will you do with all your free time?

Manufacturing will be important in space, too. Factories on space bases will operate automatically. They will use natural materials from the space environment and will use solar energy for power. Space robots will perform well-defined tasks in some cases, but, with the improvement of artificial intelligence, robots will be more "intelligent." They will be capable of making logic decisions and with robotic vision, will actually recognize different objects.

TECHNOFACT

Technofact 52
More manufacturing operations in the future will be done in very clean and controlled environments. That's because even small specks of dust or a human hair could totally ruin a part being manufactured.

Today's robots are faster and much more capable than those of just a few years ago. Assembly lines like this one have more and more robots doing the work each year. The future factory may not require people at all. The entire manufacturing process will be handled by computers and robots. (Courtesy of Ford Motor Company.)

THINGS TO SEE AND DO:
Science Fiction or Fact?

Introduction:
You probably have seen science fiction movies that show robots as lovable little friends, big monsters, or evil insectlike machines. It is sometimes strange the way science fiction seems eventually to come true with advancements in technology. No one knows for certain what the future holds. We can make predictions based on current developments, however. In this chapter, you have learned about some of the things robots can and cannot do. In this activity, you will try to predict the future of robots and make your prediction come alive in a science fiction video production.

Design Brief:
Produce a science fiction video program that expresses your thoughts about the future of robots. Your video will be shown to elementary technology students. It should be be exciting, humorous, and educational. The video production should be between 5 and 15 minutes long.

Encourage imagination and creativity without letting students lose sight of the technology theme in their videos.

Materials:
▶ Cardboard boxes, Styrofoam packaging, and other inexpensive or recycled materials for costumes or props
▶ Scissors, tape, paints, brushes

Equipment:
▶ Video camera and VCR or camcorder
▶ Sound effects tapes or CDs (optional)
▶ Computer with word-processing software (optional)

Procedure:

1 In this activity, you will work in groups of five or six. Elect a producer for your video production from within your group. The producer will be in charge of organizing and planning the production of your video. The producer should take notes as your group discusses the ideas for your video.

Work with the drama teacher to help with this activity.

2 Your group will need to brainstorm possible plots for your science fiction video. Try to avoid using the same plot of a movie that you have seen. Think of some of the possible advancements in computers and robots. Use these new technologies in your story. Some things you might think about include:

• Voice recognition: robots you can talk to

• Artificial intelligence: robots that can learn

• Virtual reality: computer graphics that make people feel as if they are in another place

• Superconducting batteries

Challenge a student to make a chart of Isaac Asimov's robot laws.

Be sure that the student "robot" can move freely in case of an emergency.

Robots from science fiction stories sometimes give us a look into the future. What do you think the robot of the future will look like? Will they look like humans or machines? (Courtesy of Lucasfilms, Inc.)

- Android: a robotlike machine that is programmed by people to do work
- Cyborg: an advanced android that can learn on its own
- Fiber optics: sending information as light pulses through thin glass fibers
- Other technologies?

Decide what the general theme, or plot, of your video will be. Start a script that will be easy for elementary students to understand and enjoy. Don't forget that the purpose of the video is to teach about the future of technology. Students of all ages enjoy props and costumes. You need to think of a way to make your video more interesting than just people talking with each other.

4 The future of robots is easy to predict because almost anything can happen. Science fiction writer Isaac Asimov thought the following laws should be used to keep the future of robots under control:

1. A robot must not injure a human being or, through inaction on its part, allow a human being to come to harm.

2. A robot must always obey orders given to it by a human being, except where these conflict with Law 1.

3. A robot must protect its own existence, but only as long as this does not conflict with Law 1 or Law 2.

Maybe you could use these "robot laws" as a part of your video story.

5 Design a robot costume that your group can make. You might want to design it around one of the members of your group. Ideas for the costume might come from pictures in science fiction books or in comic books. The robot prop will be controlled with remote levers or hydraulic syringes. Use what you have learned about the electrical, fluid, mechanical, and thermal systems. Ask your teacher before using other equipment in the preparations for production. Maybe you could have the art class help in the design and construction of your robot costume. If you have time, design and make a futuristic background for your video. You could project an image onto a cardboard backdrop using an overhead or opaque projector.

6 Rehearse your production so that everyone knows what they are going to do.

7 Produce a finished videotape of your story. Share the story with an elementary classroom. Maybe the elementary teacher will let your group talk to the younger students and explain more about robots and how exciting the future can be.

Evaluation:

1 What could have been done to make your video production better?

2 How did the elementary students react?

3 Do you think robots will be used to do housework like vacuuming in your future? What else do you think robots will be used for?

4 How could you have improved your performance in the video? Explain.

Challenges:

1 Make a list of science fiction books or movies you know. Which parts of the books or movies were real to you? Which seemed fake? Explain why.

2 Look for books written by science fiction authors such as Isaac Asimov, Jules Verne, H. G. Wells, and Gene Roddenberry in your school or community library. Read a science fiction book related to future technology.

3 Research the topic of special effects photography. Work with your teacher to try to simulate some special effects using the equipment in your technology class.

4 Ask your teacher if one of your school's computers could be used to put titles or other graphic images onto your video. With your teacher's help, experiment with the equipment available to you.

Character generators are also available for video titling.

No one knows what the future of robots will be. It is certain that their abilities and intelligence will continue to increase. Science fiction stories sometimes make robots into evil mechanical villains. The fact is that robots are used today to help people rather than to hurt them. Today, robots are used by people who are paralyzed and unable to move their hands or arms. Robots are even used to help surgeons perform delicate brain operations. The ability of a robot to be directed to a precise location in the human brain has made brain surgery much safer. If robots are being used like this today, what do you think the future holds?

Summary

Automation is the automatic control of a process by a machine. Today, automation is part of almost every technology. For machines to work automatically, they must have parts and materials come to them. Materials-handling systems might use conveyors, trucks, containers, hoists, cranes, or computer-controlled cranes.

Robots do many of the jobs that people find too boring or too dangerous to do. They sometimes have vision systems or special sensors that help them "hear" or "see." Robots used in manufacturing are called industrial robots. Robots are becoming better as computers become better.

Computer-aided manufacturing (CAM) uses computers to control machines that make products. In computer-integrated manufacturing (CIM), computers tie manufacturing, design, and business functions together. Another technology, called CAD/CAM joins computer-aided design (CAD) with CAM. With this system you can design a part on the computer and then send the information to a machine tool that makes the part.

Factories of the future will have more robot workers and will be more highly automated. Workers will have shorter work weeks and more leisure time. In space, manufacturing may be totally automated and geared to using materials found in space.

The automated factories of the future may give us more leisure time, and technology can improve our standard of living. As automation makes our jobs easier, we will have more leisure time to exercise our bodies to keep fit. (Courtesy of NIKE Inc.)

Challengers:

❶ Research the working conditions for people, including children, in U.S. factories around 1900. How have conditions changed? How has technology played a part in the changes?

❷ Have your teacher show a science fiction movie that has robots in it. Write your opinion on whether you think robots will be used that way in the future.

❸ Write to a company such as Chrysler, Ford, or General Motors for information on how they use robots in the manufacture of cars.

❹ Design a personal robot that would do a special job that you don't like to do.

❺ Read a science fiction book. Compare how automation and robotics are used in the book with what is happening right now in your world.

❻ Write your own science fiction story. Use word-processing software on the computer.

❼ Build a robot out of recycled materials gathered in the machine dissection activity. Use an energy system—electrical, thermal, pneumatic, or hydraulic—to provide movement for your robot.

❽ List the automated machines in your school or home. Are any of these robots?

❾ Write instructions for a robot to do a certain job in the classroom. Have another student act as the robot. Read the directions, and have the student follow them. Evaluate whether your instructions are accurate enough.

❿ Describe and design a factory that you think could run efficiently with only robot workers.

See Teacher's Resource Guide.

Techno Talk
Computers Make Everything?

Chapter 9

How Does Business Work?

Things to Explore

When you finish this chapter, you will know that:

▶ Businesses are organized in different ways to manage production of products.

▶ Corporations are owned by stockholders who expect to receive a dividend for their investment.

▶ Employees negotiate with managers for wages, working conditions, and benefits.

▶ Consumer surveys give businesses feedback about product ideas.

▶ Jigs and fixtures help workers make interchangeable parts more accurately.

▶ Quality-assurance departments inspect parts and finished products for defects.

▶ Products are often sold through a system of distributors, wholesalers, or retail outlets.

TechnoTerms

administration	marketing
capital	negotiate
consumer	on line
corporation	partnership
defect	personnel
dividend	proprietorship
distribution	quality control (QC)
downtime	research and development (R&D)
interchangeable	statistical process control
just-in-time manufacturing (JIT)	stockholders
liquidation	survey
manage	

Careers in Technology

Every day millions of people go to work to manufacture the products we depend on. Businesses are started to make a profit. Large companies need managers to make decisions that keep the company going. Employees are hired to do a specific job in manufacturing. Here, an assembler and a managing engineer are working together to make changes in the design of a 240-ton mining truck. No matter what product is being manufactured, the quality of the finished product is in the hands of the employees.

Ask students to list as many names of companies as they can.

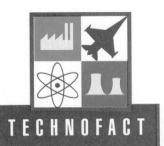

TECHNOFACT

Technofact 53

Some companies do some interesting advertising to get their product recognized. One automobile dealer held a "Kiss the Car" contest. The person who could kiss a car for the longest time would win the car. The winner lasted two and a half days before the others puckered out.

Partners can form a type of company called a partnership. There are different ways to start a company or corporation. Governments regulate the formation of companies and corporations. (Courtesy of Deere & Co.)

What Is a Company?

You've probably seen ads in magazines and on television sponsored by names such as Xerox, IBM, General Motors, and Exxon. These are huge companies that hire thousands of workers to produce goods or services for you, the consumer. Thousands of companies make products or provide services for you. A company is an organized group of people doing business.

A company is an organized group of people in business to make a profit. We often forget about the factories and companies behind the products we use everyday. Here, technicians are inspecting the quality of bars of soap in a factory. (Courtesy of Colgate Palmolive Company.)

In the last chapter you learned that craftspeople used to make products in their homes. These were not companies but individuals. Production of products was increased by organizing people in factories. Factories became parts of companies that produced products. In this chapter you will investigate the business end of how factories and companies operate today.

Every company is in business to sell products and to make a profit. Remember that one resource needed to make things is money, or **capital**. In order to get a company started, someone has to **invest** (put in) money to start the company. Then, once the company is organized, you have to have some way to run it. There are different ways to **manage** (run) a company.

▶ **Proprietorship:** A proprietorship is a business owned by just one person. It is the easiest type of business to form because you as the owner have complete control over everything. Besides that, you get to keep all the profits.

▶ **Partnership:** A business owned by two or more people is a partnership. It is also easy to form. The partners share the profits. An advantage to being in a partnership is that you can share the workload and responsibilities.

▶ **Corporation:** A corporation is a company organized and owned by stockholders. A stockholder is anyone who buys a share in a company. Would you like to own some stock in a company? It's not hard to do. Have you ever heard the stock market report on the news, or have you seen it in the newspaper? Stockholders purchase stocks in companies hoping that the value of their stock will increase. If the stock's value does increase you might sell your shares for a profit. While you own the stock, you will also receive what's called a **dividend**. A dividend is a payment you get as part owner of the company.

People buy stock in a company in the hope that its value will go up. The stockholder can then sell the stock and make a profit. This is a stock exchange where thousands of stocks are bought and sold every working day. (Courtesy of New York State Department of Economic Development.)

You can see that forming proprietorships and partnerships is fairly easy. Corporations are more complicated. Sometimes corporations start out as proprietorships or partnerships. They have a very structured organization because there are so many people involved.

If you wanted to start a corporation along with some other people, you would first have to make out an **application** (special form) that must be approved by a government **agency** (department). Once it is approved, you can sell stock or get a loan to raise the money needed to start and run the business. This money is called **capital**. After the capital is raised, you need to fill the positions in the corporate structure.

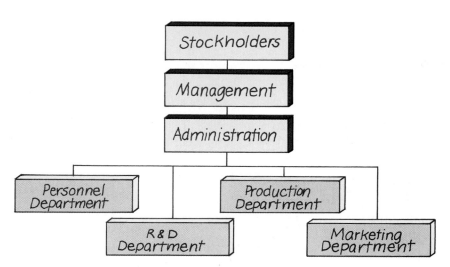

This is a diagram of the corporate structure.

TECHNOFACT

Technofact 54

Some interesting business facts:

• A new business has a 50-50 chance of being successful.

• Only 4.5 percent of top corporation executives are female.

Send for the annual reports from various companies. Have the students identify management positions within the companies.

TECHNOFACT

Technofact 55

Would you like to check out your own groceries? According to David Humble, who invented security tags for clothes, his first system for self-checkout in supermarkets is already up and running. The automated checkout machine (ACM) lets you scan your own purchases. Once the item is scanned, it goes down a conveyor belt with built-in security. The price and description of each item are displayed on a special touch screen, which keeps a running total of your purchases. You can always push the "HELP" button if you get confused. People are still needed to bag your groceries. The best part is that the checkout lines move faster!

The corporate structure includes the following:

▶ **Stockholders:** People who have bought shares in the company. Stockholders hold an **annual** (yearly) meeting. At the annual meeting, they elect the board of directors for the next year.

▶ **Board of Directors:** Board members are elected. They set company policies and determine the main company goals. They report how the company is doing to the stockholders. The board also hires a company manager or president to run the company.

▶ **Management:** These people run the company. They have to be good leaders and hard workers to make the company successful. Management people have to pick the products the company will make, decide how to raise money to buy or rent buildings, where to get raw materials, and how much to pay workers.

▶ **Administration:** The company president is the administrator in charge of the company. The people in administration carry out the decisions made by the management department. The company president makes sure things are done right. Depending on the size of the corporation, the president might have one or more vice presidents who are responsible for certain parts of the company. In your school, you might have a vice principal who has special duties to help the principal. One very important job of the company administration is to keep records. You can see how computers have been a big help in modern businesses because they can easily organize information, calculate numbers, and retrieve information.

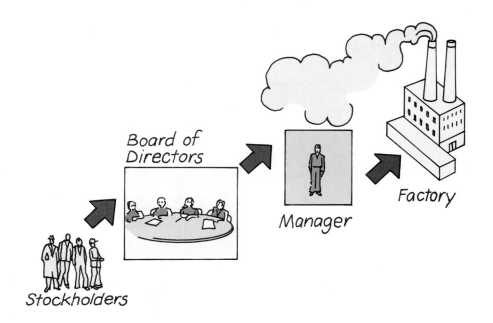

Many decisions have to be made in the operation of a company. The stockholders elect a board of directors who hire a manager or president to run the company.

There are many other workers in a company, too. As you learn what each department does, think about what jobs you would like to do or would feel most qualified to do. These are some of the main departments in most companies.

Human Resources: In some companies, this department is called Personnel. This department makes sure that the people who have the skills needed for certain jobs are hired. If extra training is needed, the Personnel department makes sure workers get the training. The Human Resources department also makes sure you are rewarded if you are a good worker. Usually you are advanced to a position with higher pay. They may also be in charge of safety and other working conditions. Any disagreements about working conditions, salaries, or **fringe benefits** (extras) between workers and management have to be worked out. The personnel department may help **negotiate** agreements between workers and management.

Ask the students to list examples of what they think are "fringe benefits."

Managers negotiate with workers to work out a solution that is agreeable to both sides. Working conditions, salaries, and fringe benefits are commonly negotiated. (Photo by Paul E. Meyers.)

Research and Development Department (R&D): The **research and development** department improves existing products or designs new products that people really want. They look at existing products and try to predict new products that they think people will want to buy. Can you name a product that you would like to see developed? Right now many R&D departments in the automobile industry are trying to develop cars that use alternate energy sources instead of gasoline. Companies depend on R&D departments to find efficient ways to make products so that they can save money and make a profit.

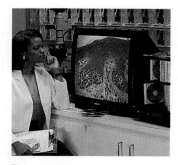

B

Research and development departments try to perfect existing products and design new products. Technology can be applied to help solve problems related to energy use, medicine, and many other fields. (Figure A Courtesy of Corning Incorporated/ Photo by Frank Petronio. Figure B Courtesy of Eastman Kodak Company.)

A

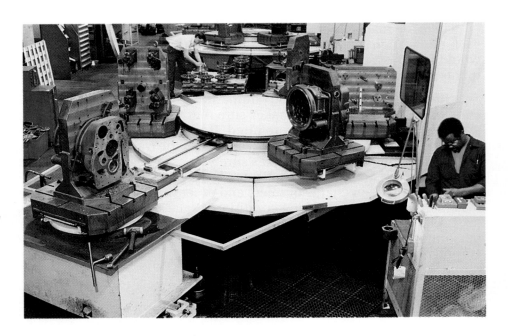

The production department actually makes the product. If machines break or parts become unavailable, the production line comes to a halt. Downtime for companies can be very costly. When it happens, many people must work quickly to get the production line back on line. (Courtesy of Cincinnati Milacron.)

Marketing departments promote the product and conduct consumer surveys. The results of surveys provide feedback for the company so that managers can make decisions about the future of the product. (Courtesy of Colgate Palmolive Company.)

Production Department: The production department is in charge of actually making the products for the company. The production workers must turn materials into parts for products and then assemble the products. The production department usually plans a production system to make sure each job is done as accurately, quickly, and safely as possible. A mistake in production caused by poor planning can cost a company a lot of money or slow production down. The production department has to plan ahead so there is no **downtime**, or stop in production. Let's say your company produces computers and you run out of memory chips. That will slow down the entire production line, because the people working on the outside case and the people in the testing **division** (department) will not be able to continue until the computer boards are **on line** (back in production) again. Running out of parts is an expensive mistake for a company because you lose money when products are not being produced. In addition, you will probably have to pay the workers while they are waiting to go on line again.

Marketing Department: The job of the marketing department is to sell the company products. Sometimes they conduct **consumer surveys** to find out what people want, how much they are willing to pay for a product, and who would probably buy that product. The marketing department must also have a **marketing plan** (strategy) for **promoting** (advertising) and selling the product. Lots of companies hire outside advertising firms to promote products. Sometimes they will send out catalogs and brochures even before the products come off the assembly line. From the number of orders the company gets, the marketing department can **estimate** (predict) how many of each product they should make. Larger companies will also have sales representatives who cover a certain **territory** (area). Their job is to contact customers and show them the company products. One salesperson might have a certain city or several states as a territory.

THINGS TO SEE AND DO:
Starting a Company

Techno Teasers
Which Department Is Which?

Techno Teasers
Answer Segment

Introduction:
Companies get started so that people can make money. A product or a service is made available for sale to consumers. If the product is successful, the company may expand production. Companies often sell stock to raise money to buy equipment or build factories. In the activities in this chapter, your class will organize into a company. You will sell stock, design products and advertisements, produce the product, package it, sell it, and make a profit (we hope). As in any business, there is no guarantee that there will be money left over after all the bills are paid.

Design Brief:
The goal of all of the activities in this chapter is to organize your class into a company to design, produce, and sell a product. Each student will be a worker in the company. You will also be able to buy stock in the company. In this activity, you will design a prototype and conduct a consumer survey.

Students will be able to role play the operation of a company in this activity and the activities that follow.

Materials:
▶ Materials needed will depend on prototype designs

Equipment:
▶ Computer with graphics and spreadsheet software
▶ Machines and tools as needed to make prototypes

Procedure:
Designing a Product:

The first thing your company will need is a product. Your idea should be something your class can produce in your technology lab. The product should cost less than $5. It must be designed with safety in mind. Sharp edges, electrical shock hazards, or pinch points must be avoided.

You also should consider your market. The **market** is the customers who will buy your product. If your product is going to be sold in school, you might think of the students as your market.

❶ Work with your teacher to brainstorm product ideas. Make a list of possible markets and products. Your list might look like this:

Market	Elementary School	Middle School	Parents	Teachers
Product	Pencil holder	Clipboard	Clipboard	Clipboard ideas
	Locker shelf	Locker shelf	Desk tray	Desk tray
	Bookmark	Book holder	Can holder	Key chain
	Toy	Game	Game	Game
	Puzzle	CD or tape holder	CD or tape holder	CD or tape holder
	Picture	School logo	School logo	School logo

Students should be encouraged to design their own product.

❷ Vote on each of the products on your list. Find the five products your class thinks have the best chance of selling. Work as a class to refine the ideas and make plans for building them.

❸ Divide the work of building a prototype for each of the product ideas among the class. Remember that prototypes are models of product ideas. Your prototype does not need to be perfect. There will be time later to improve the design. You will need to get some market information. That is, you have to find out about what the people you hope will buy your product would like. The next step is very important.

SAFETY NOTE:
Remember, you must have safety instruction and a demonstration on machines before you use them. Ask your teacher for help if you have forgotten how to use a machine safely. Don't forget to wear eye protection.

❹ Now that you have some prototypes to compare, you need to conduct a consumer survey. A **consumer** is the buyer of the product. A **survey** is a series of questions that will give you information. The consumer survey will give you information called **feedback**. All five of the products should be included on the form. The results of the survey will help you decide which of the prototypes will become the product of your company. Your survey might look like this:

> If unfamiliar tools or equipment are needed, be sure to demonstrate their safe use to the entire class.

CONSUMER SURVEY

1. **Are you tired of a cluttered desk?**
   ☐ Yes   ☐ No
2. **Would you buy a plastic letter holder?**
   ☐ Yes   ☐ No
3. **Do you think the letter holder would make a good gift?**
   ☐ Yes   ☐ No
4. **What color would you think is best for your letter holder?**
   ☐ Clear   ☐ Red   ☐ Blue   ☐ Yellow   ☐ Other _____
5. **Can you think of other ways to use the letter holder?**
   _____

If you return this survey, you will be given a coupon for 20% off of your letter holder purchase. Thank you for your cooperation.

Your survey questions should be written to get reactions of possible customers. Surveys should be easy to fill out. If they are too long or hard to complete, people won't want to spend the time to do them. You might include a coupon or other method of thanking people for taking the time to fill out your survey.

❺ Give your survey to students, parents, or teachers—your market—and to other groups as well. The purpose of the survey is to get an idea of who will buy your product. The people you ask to complete your survey should be picked at **random** (by chance). If you ask only your friends, for example, you might not get an **unbiased** (fair) answer.

❻ When you evaluate the results of the surveys, you are **compiling data**. The data can be presented to the class in the form of a graph. It is important to evaluate the results without letting your feelings interfere with your judgment. If your favorite product idea is not selected, don't worry, there is plenty still to do in the next activity.

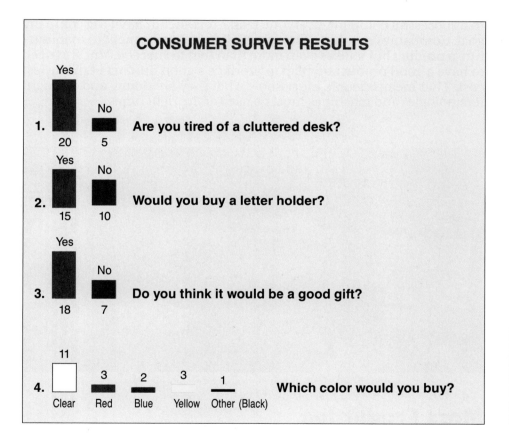

CONSUMER SURVEY RESULTS

1. Yes 20 / No 5 — **Are you tired of a cluttered desk?**

2. Yes 15 / No 10 — **Would you buy a letter holder?**

3. Yes 18 / No 7 — **Do you think it would be a good gift?**

4. Clear 11 / Red 3 / Blue 2 / Yellow 3 / Other (Black) 1 — **Which color would you buy?**

Data can be compiled and illustrated using graphs. Most people can understand data better by looking at a graph than looking at numbers. Graphs are often used in business presentations to make a point easily.

Evaluation:

❶ Make a list of the prototypes, from most popular to least popular.

❷ Have you ever taken part in a survey?

❸ Could your survey have been done by phone? What questions would you change?

❹ Why do companies make prototypes of products?

Challenges:

1. Put the consumer survey results on a spreadsheet.

2. Design a consumer survey that people could fill out easily. What types of questions do you think most people hate to fill out? Why?

3. Design another prototype where cost is not a consideration. Let your imagination work to come up with an idea that you think could be manufactured and sold.

4. Why do people sometimes offer food samples in grocery stores? What other techniques can you think of to get people to try a new product?

As you can see, there are many different kinds of jobs in a company. In a successful company, everyone works together to meet the company goal. Companies may have the right people and equipment to manufacture a product but it takes even more than that to succeed. You also need to have a good production plan to produce a good product at the lowest cost. That means people must know what they are doing, and the right technologies and machines must be used at the right times.

When proper planning and automation technology are used, the cost of production can be kept down, and the rate of production can be kept high. As you know, computers play an important part in any efficient production process. (Courtesy of Kim Steele Photography, New York.)

Mass Production

Mass production enables companies to produce large quantities of parts and products within a short time. To do this, each worker on a factory assembly line is assigned only one job. Each person does the same job over and over. Mass production works only if all the same parts of components are **standardized** (the same size and shape). The parts must be **interchangeable**. Any one of several hundred pieces of one part must fit with any one of several hundred pieces of a joining part. Today's automobiles are made mostly with standardized, or interchangeable, parts. If one part, like a headlight or a door handle breaks, you can buy another one just like it from an automobile dealer.

A

B

Many of the products you use are made by mass production. Standardized (interchangeable) parts are used to make repairs easier and to keep costs low. As a general rule, parts made in large quantities cost less than parts made one at a time. (Figure A Courtesy of Honda North America Inc. Figure B Courtesy of PPG Industries.)

Through mass production and the assembly line, a greater number of products can be made in a given amount of time. Mass production also makes products less expensive to produce and therefore less expensive to buy. Did you know that your tennis shoes are most likely mass produced by machine? One machine cuts the material, another punches holes for laces, another glues the soles, and so on until the final product looks like a tennis shoe to you.

Where do companies put materials and purchased parts until they are needed on the production line? Many companies have to order large quantities (amounts) of materials ahead of time. They then must pay for storage space in a warehouse. The company must also hire people (or buy robots) to move the materials to the production line. Once the finished product is made, it often spends time in a warehouse, waiting to be shipped. One way to cut down on **inventory** (things in storage) and costs is to use a computer to schedule deliveries just in time. **Just-in-time manufacturing (JIT)** is a method that many companies are turning to because it **eliminates** (gets rid of) the need for storage space and extra people to manage the inventory. All the materials and ordered parts get to the factory just in time to be used in production. When the product is finished, it is immediately shipped (sent) to the customer. Just-in-time manufacturing enables companies to cut back inventory as much as possible.

You can see that good management and teamwork are important to just-in-time manufacturing, or **synchronized production**. If one part is not there on time, then the rest of the product must wait. To be successful, production workers, managers, suppliers, and workers who transport materials must work closely together. Computers play a large part in **linking** (hooking) all those people together so that they all know what they are supposed to do and when they are supposed to do it.

Recall the cookie manufacturing activity. Discuss the results.

Mass production makes it possible to make quality products in a variety of styles and sizes. Everyday products such as shoes are made by the thousands in factories that use assembly line, mass-production technology. (Courtesy of NIKE Inc.)

When products are finished, they must be stored in a warehouse before they are shipped to the customers. Storing products can take up a large space. Products in storage are not making any profit for the company either. Just-in-time inventory attempts to predict the demand for products and to avoid having thousands of products in storage. Here, large rolls of paper used to print newspapers are being stored. (Courtesy of Scott Paper Company.)

Techno Teasers
A Picture of Productivity

Techno Teasers
Answer Segment

THINGS TO SEE AND DO: *Organizing a Company*

Introduction:

In the last activity, your class decided on a product that can be manufactured in your student company. It's time to organize your company and get down to the business of doing business. In this activity, you will sell stock, organize into departments, and start producing your product.

Design Brief:

Organize a marketing, production, bookkeeping, and personnel department. Design a stock certificate for your company. Sell stock in your company to raise capital. Apply and interview for a job in the company. Learn your job and start work.

Materials:

▶ Materials needed to produce your product

Equipment:

▶ Equipment needed to produce your product

▶ Computer with graphics, database, word processing, and spreadsheet software

Procedure:

Every student in the class will have a role in this student company.

❶ Your company must be organized into departments so that everyone can help make it a success. The company will need a dependable student to be president. The president will organize and supervise all of the company's operations. If your class is large or your product very complex, you might decide to elect a vice president to help the president. Your teacher will take nominations for the president, and everyone in the class will vote on the best person for the job. Your new president will then put the names of the major departments on the board. You may volunteer for any of the departments.

❷ Each of the groups will need a leader who can help make sure that all the duties of the department are finished. Your leader will be a department vice president, such as vice president in charge of production. The vice president of each department will report to the company president for assignments.

❸ Each department must complete specific jobs before production can start. Here are the assignments:

- All departments: Decide on a name for your product and a name for your company.

- Bookkeeping: Design a company stock certificate. The certificate must have the following information: company name, stockholder's name, date, value, receipt, and number. It might look like the example.

Avoid naming the company or the product after an individual.

EGGSTRA SPECIAL CORPORATION

STOCK CERTIFICATE

This certifies that _____ is the owner of _____ shares of stock in the Eggstra Special Corporation. This stock is non-transferable and can be redeemed for $0.25 per share at any time, or held until liquidation.

President _____ V.P. _____

Bookkeeper _____ Stock Number _____

Stock certificates can be designed any way you like. They should have a place for signatures and a stock number so that you can keep track of who owns them. Design your certificate so that it would be hard for someone to counterfeit (copy).

Sell stock ($0.25 per share) to people in your company or outside of your company. Keep track of the stockholders on a spreadsheet. Be sure to caution people that there is no guarantee of profit. Give the money and receipts to your teacher to help pay the bills for materials used to make your product.

This activity may require funds that will guarantee that students will break even in case of product failure. Check with school administrators first.

- Production: Refine the prototype so that it is ready to be mass produced. Make a final drawing of the product that shows **dimensions** (sizes). Make a list of all of the parts of the product, their material, part name, and size. Make a flow chart of the production process.

There should be a place on the flowchart for every operation that must be done. Your teacher will help you design and make special attachments for machines called **jigs** and **fixtures**. A jig holds an object and guides the tool during work. A fixture keeps the object in the proper place while it's being made. The jigs and fixtures will make your product easier to mass produce.

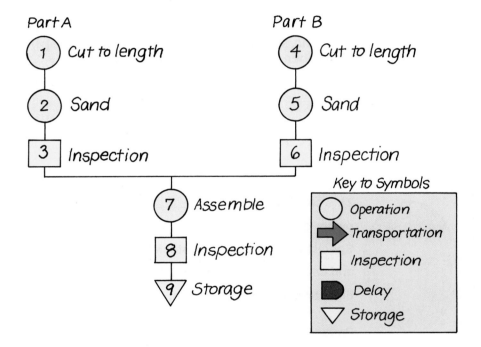

Flowcharts help you list the steps of production in order. This chart should list every job in the manufacturing of your product. Your product may have many branches for each part. All of the branches come together for final assembly at the end of the flowchart.

- Marketing: An advertising campaign will help improve sales of your product. If people don't know about the product, they can't possibly buy it. Your group should consider as many ways as possible to make your product more visible. Some ideas include posters, flyers, announcements, demonstrations, video or audio commercials, and school newspaper ads.

- Personnel: Your department will place all the workers (everyone in the class including the president) in the jobs for which they are best qualified. First, design and make copies of an application form. Then get a list of the jobs from the production department. Post a list of positions available. Have each student (including those in the personnel department) complete an application for a job. The Vice President of Personnel and the company president will interview the job applicants and hire people for the jobs for which they are qualified.

4 When each department has completed its assignment,

BETTER WAY MFG. CO.

APPLICATION FOR EMPLOYMENT

Prospective employees will receive consideration without discrimination because of race, religion, color, sex, age, national origin or physical handicap.

Last Name	First	Middle	Date

Street Address			Home Phone ( )-

City, State, Zip			Business Phone ( )-

Are you over the age of 18? ☐ Yes ☐ No	If not, do you have valid working papers? ☐ Yes ☐ No	Position Desired:

Do you have any impairments — physical, mental or medical — which would interfere with your ability to perform the job for which you have applied?	☐ Yes If yes, please explain. ☐ No	Date Available for Work

Have you ever been convicted of a crime? ☐ Yes ☐ No	Charge, Date & Disposition:	Salary Expected

Are you legally eligible for employment in the U.S.? (Proof of citizenship or immigration status may be required upon employment.) ☐ Yes ☐ No		Have you ever been bonded? ☐ Yes ☐ No

How did you learn of our organization?	If the job requires, are you available to work overtime? ☐ Yes ☐ No

Indicate skills you possess:

Typing Speed _____ wpm   Steno Speed _____ wpm

Computers (specify):   Software (specify):

List any professional memberships which you feel would be an asset for the position applied:

Military

Served in U.S. Armed Forces ☐ Yes ☐ No	Date Inducted:	Date Discharged:

Rank at Discharge	Briefly describe your duties:

You will have to fill out an application for any job you have in a business. Applications give the employer information about your experience and background. It is important that you fill out applications clearly and accurately.

your company is ready for production to start. Each worker (everyone in the class) must know his or her specific job. Your teacher will help the president to teach you your specific job in the assembly line. If possible, set up the conveyor belt systems you made in Chapter 8. The conveyors should be placed so that the movement of materials follows the flowchart made by the production department.

SAFETY NOTE:
Safety on the job is important to millions of workers. Always keep safety in mind. If a worker is hurt on the job, it costs money and slows production. Be careful!

5 As products are completed, they should be placed in storage for the next activity.

Evaluation:

1 What was your specific job in the assembly line? What did you like or dislike about your job?

2 How could your job have been changed to make the production of products more efficient?

3 Could a robot replace you on the assembly line? How would you feel if you were fired and your job was taken over by a robot? Explain.

4 How could you avoid losing your job because of automation or robots?

Challenges:

1 Design your place on the assembly line with ergonomics in mind. How far do you have to reach for parts or tools? How would you feel after doing the job all day?

2 How could the conveyor systems be changed to make them better? Make a sketch of your ideas.

3 Visit a local factory that makes products using the assembly line process. Ask some of the workers what they like or dislike about their jobs.

4 Use computer graphics software to draw a floor plan (view looking from above) of the assembly line. How could you rearrange the equipment or conveyors to make the flow of materials more efficient?

What will business be like in the year 2000? Some business specialists say that the following trends will continue.

▶ Product life cycles are becoming shorter. That means products are made that will not last long.

▶ People want more new products that can be developed in a

shorter time.
- ▶ Managers and executives need more information faster to help them make decisions more quickly.

Automation and mass-production technologies are important in making more products faster. But there's another part of manufacturing that you care about, and that's the quality of the product.

Total Quality Improvement

Have you ever bought a new shirt or some other piece of clothing only to find out once you got it home that there was already a rip in the sleeve? Or maybe the zipper on your coat never worked right? Maybe you bought a compact disk (CD) or tape only to find a **defect** (something wrong) in it that ruins the way it sounds? Then you know one reason why we want quality control of products.

The **quality** of a product is how well it is made. **Quality assurance** means that a product is produced according to specific plans. Another name for quality assurance is **quality control (QC)**. Companies want to make high-quality products. They want to make each product work the way it's supposed to so that people will be satisfied with it. To do this, companies often set a quality standard before they make the product.

The quality control department checks all **phases** (stages) of production. In manufacturing, parts are checked for strength, and to meet standards set up in the production line. Inspectors examine parts, materials, and processes. Inspections often are made during three key times in the production cycle.

The quality of a product is very important for the consumer (buyer) and the company. Defective products can ruin the reputation of a company quickly. This gas turbine blade is used to power a jet. If it were defective, people might not buy products from the company in the future. (Courtesy of Pratt & Whitney.)

The inspection of parts for quality is an important part of production. Have you ever purchased something and been unhappy because it didn't work? Quality control inspectors try to make sure the product is made properly. (Courtesy of Carver Corporation.)

Special Report
S.P.C. at B.F. Goodrich

▶ **Delivery of materials:** Materials are inspected as they arrive. If the materials don't meet the standards, they are rejected and sent back.

▶ **Work in process (WIP):** Inspectors check that work is being done in the right way and that the right parts are being used.

▶ **Finished product:** This is the final inspection, where everything should work and look right!

Sometimes many products are made in large quantities. It is not possible to check each product, so inspectors use a procedure called **acceptance sampling**. Acceptance sampling means that you select a few **samples** (typical products) and inspect them to see if they meet the standards. If the samples pass inspection, then the whole batch is approved. If the samples don't pass inspection, the other parts or products are rejected. Can you see how hard it would be to check every M&M candy as it passes through on the conveyor belt? Do you see now how occasionally you might find one without the letter M?

To check materials, parts, and products, inspectors use some special tools. Some are measuring devices such as **gauges**. They simply compare a part with the gauge, but they don't do any actual measuring. Some devices emit sound waves or x-rays. Even laser beams are used to make very precise (exact) measurements. New quality control techniques make it possible for you to check inside a product without damaging the product. You used to have to take an engine apart to look at the insides. Today you can use a **boroscope** to look inside the engine without taking it apart. A boroscope is an optical inspection device that goes inside a machine or even your body so that specialists can look around!

Some special quality control techniques are used in manufacturing. **Statistical process control (SPC)** is used to make sure that a process is being done right. If the process is correct, then the product doesn't need to be inspected. Computers are very helpful here in keeping track of each machine. For example, suppose you have a certain machine that automatically fills empty boxes with crackers. The box label says that there are 12 ounces of crackers in the box. But not all boxes are filled exactly the same because the machine is set to fill each box within a certain limit. In this case, the machine might fill the box with 11.5 ounces or 12 .5 ounces. If that is within the limit set for the machine, then the product passes inspection. A control chart on a computer keeps track of each box so if the machine goes over or under the limit, the workers can stop and adjust the machine.

Burn in is another special quality assurance test. It is done mostly on electronic products like computers. Usually if a computer is going to fail, it does so in the first couple of hours of operation. Manufacturers run every computer for a couple of hours as the burn-in test. Computers that don't pass the burn-in test are repaired if possible.

Remember, manufacturers want you to like their products. They want to be sure their products do what they're supposed to do.

Computer parts sometimes fail within a short period of time after they are first used. Many companies put computer circuits through a burn-in period to be sure the products shipped are going to work properly. Circuit boards that do not make it through the burn in are further inspected to find out exactly which part failed and why. (Courtesy of Hewlett Packard Company.)

THINGS TO SEE AND DO:
Selling Quality Products

Introduction:
Now that you have some products made, it's time to try to sell them. It is important that every product sold be of the highest quality possible. Selling badly made products is not fair to your customers and it will hurt the sales of more products. In this activity, you will design ways to test the quality of your products to ensure customer satisfaction.

One of the problems with the production of your product may be that not all the parts of your product are the same size. There is no way to make every part exactly the same size. It depends on how accurate your measurement tools are. Fortunately, most parts do not require extreme accuracy. The accuracy of a part adds to its cost to produce. Most machine parts can be a few thousandths of an inch off and still work. Wood or plastic parts may work just fine if they are as much as ¹⁄₁₆″ off the desired size.

Design Brief:
Design and perfect a method of quality control that will help to make your products better. Work as a production team to suggest ways to improve the production of your product.

Materials:
- Wood or acrylic (Plexiglas)

Equipment:
- Machines and tools as needed

Procedure:
1. The accuracy of each part you make directly affects the quality of your product. Manufacturers of products like cars often buy parts made by other companies. All the parts must fit together perfectly, or the car will not work properly or last very long. The reputation of a company can be hurt by just a few bad products. Your task is to invent a way to ensure the quality of each of the parts you make. Work in groups of four or five to analyze the jobs in the assembly line. Look for ways to improve quality.

2 Every part of a product made today has a drawing that shows exactly how to make it. The drawings show the size of the part in exact dimensions. Because most parts do not need to be perfect, a range of sizes is often given, for example, 1.5″ ± 0.005″. This range of acceptable sizes is called the **tolerance**. The range in the example can be 0.005″ over or under the 1.5″ and still be just fine. Your team should decide on the acceptable tolerance for each of your product's dimensions.

3 As they are being produced, parts can be checked quickly to see if they fall within the tolerance limits. This quick check requires a measurement tool called a **go-no-go gauge**. The go-no-go gauge is a simple tool that does not require workers to take time to read a measurement tool. It can be made to fit the tolerance you have assigned to your parts. Your group should design and make a go-no-go gauge for the parts of your product. Your gauge might look like this:

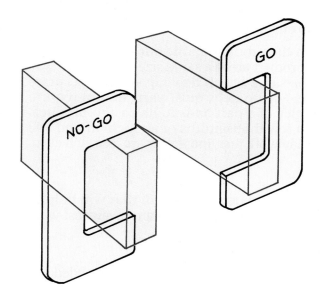

A go-no-go gauge is an easy and fast way to check parts to see if they are within tolerance. After many uses, the gauge must be measured to be sure it has not worn or stretched.

4 To use the gauge in the assembly line, all you have to do is put the part next to the gauge. If the part is smaller than the "no-go" side or larger than the "go" side, it is not acceptable.

5 Your product may have parts that are difficult to measure with a go-no-go gauge. Your group should then design and build a testing device that will test your parts for accuracy.

Discuss the difficulty of making products to an exact size.

❻ The quality of a product is so important that many companies have a quality control department. This department is responsible for checking the accuracy of the product being made and the quality of the parts used to make the product. If they find **defective** (bad) parts, they must stop production and have the problem fixed quickly. Otherwise, time and materials are wasted. Your president and all the vice presidents of the departments should decide if some of the workers on the assembly line could be put into a new department called quality control.

Evaluation:

❶ Have you ever purchased a product and found that it was defective? What was the problem? What did you do about it? How did you feel about the company that made the product?

❷ List five products that you feel are made with quality in mind.

❸ What happens to the quality of products when workers are tired, bored, or mad at their boss?

❹ How could the quality of your product be improved?

Challenges:

❶ Visit a factory, and ask how the quality of the product is measured.

❷ How do you think modern automation and robotics in factories have changed product quality? Do you think a car built by robots is better than one built by people? Explain.

❸ Can you think of a way to use robots or automation to inspect the quality of products? Explain.

❹ What quality control methods are you used to dealing with in school? Why are people and products always being tested? Can you think of a better way to ensure quality?

Packaging Products

How do you feel when you order some item from a catalog and by the time you get it, it looks like a truck ran over it? That's one reason packaging is important. One of the main goals of packaging is to get a product from one place to another without a lot of damage.

There are many other reasons for packaging products. Sometimes products even need more than one package. Think of a bandage or a piece of chewing gum. Each one is individually wrapped or packaged inside a larger package. Packaging protects the product from damage

Ask students to devise a quality control system that would guarantee every student has a quality education.

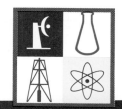

TECHNOFACT

Technofact 57

How would you solve this packaging problem? An ice cream store in Boston wanted a microwaveable hot fudge sundae. It needed to be packaged in such a way that the hot fudge would melt but the ice cream would stay cold. At first, researchers tried using different materials in the ice cream to keep it from melting so fast. But the new product tasted awful! Finally, they turned to packaging. After a year of work, the team found the perfect plan. The ice cream is placed in a completely foil-lined cup so the microwaves can't sneak in, and the fudge is suspended above the ice cream in a plastic dome. When microwaved, the fudge melts and drizzles over the ice cream. Yum!

Ask students to list reasons why products are packaged.

while it is on the shelf at a store. Packaging also can help prevent the product from being stolen by shoplifters. Small parts are often packaged in larger containers to make them easier to keep track of in a store. If your product is a liquid, then the packaging has to contain the liquid safely. Packaging also is a way of making your product more attractive and appealing to customers.

Company logos are symbols that people can relate to and identify easily. Can you identify these company logos? What kind of logo would be a good design for your company? (Apple and the Apple logo are registered trademarks of Apple Computer, Inc. Used with permission. NBC logo Courtesy of National Broadcasting Company, Inc.

For years, the garbage created by discarded packaging was almost always thrown away. Now packaging is created so that it can be recycled. Packages are important for protecting products, advertising, and preventing shoplifting. Can you help design packages that are environmentally safe and still useful? (Courtesy of Rick Siciliano/Price Chopper Supermarkets.)

Packaging is done at the end of the assembly process. Most companies buy packages from a package manufacturer. The package might already be printed with the company label, or logo. You often buy a product just because you recognize a special logo like Nike shoes, Apple Computer, Coca Cola, or Pepsi. Do you know what these company logos look like? Companies want to build an image that you recognize and can remember easily.

THINGS TO SEE AND DO:
Pack It Up and Sell It!

Introduction:
Now your company is well on its way to making a profit. Your products are being made with quality in mind, and your company is ready to roll. The next step is to package and sell your products. In this activity, you will design a package for your product.

Design Brief:
Design a package for your product that will include the specifications (descriptions) listed below. Your package idea should:

- Protect the product from being damaged
- Provide information about the cost of the product
- Include instructions on the safe use of the product
- Help prevent theft by shoplifting
- Be attractive and eye-catching
- Be environmentally safe and biodegradable

Materials:
- Packaging materials

Equipment:
- (Optional) packaging equipment such as:
  - heat gun
  - plastic film sealer
- Silk screen (optional)
- Computer with graphics software

Procedure:
1. The design of your package might take the form of a box, card, bag, label, or other shape. You should work in groups of four or five to brainstorm ideas for possible package designs. Each group will make a sample package for your company's product. Choose the idea that you think best meets the requirements listed in the design brief.

2. Sketch your design on paper, or make a **preliminary** (first draft) computer drawing. Your design depends on the product and the type of package you have picked.

You might consider some of these designs:

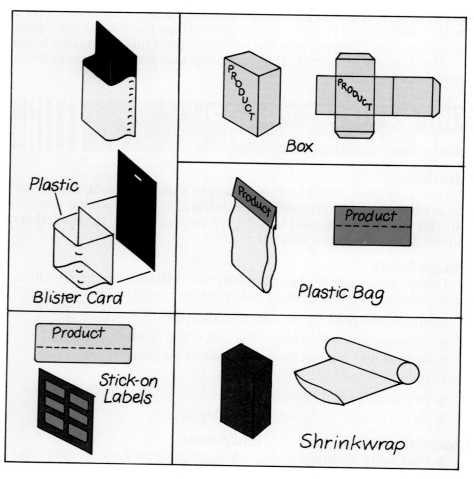

Your package design depends on the size and shape of your product. The cost of the package must be added to the cost of the product. Does it make sense to spend a lot of money on something that will be thrown in the garbage? How can you help to solve this problem?

❸ The package you design must be produced in the same quantities as your product. Keep your design simple so that most of the company's effort can be directed toward making a quality product. Your group may decide to print graphics or instructions on your package. One way to reproduce the graphics is to silk screen the package. Your teacher will help you decide the best method.

❹ Most packages are thrown away after they are opened. You have a responsibility to design a package that will either decompose in a landfill, be recycled, or be reused. Is your package design environmentally safe? You should check the progress of the experiment you started in Chapter 7. Which materials seem to be decomposing the fastest? Can you use any of these materials in your package?

❺ Some packages use plastic foam to protect the product during shipping. The hot wire cutter you made in

Be sure to use adequate ventilation when cutting plastics on the hot wire cutter.

Chapter 6 might be put to use if your product requires this kind of packaging. Find out if your community recycling center accepts Styrofoam.

6 Each group should present its package idea to the class. Evaluate each of the packages on the basis of the design brief requirements. If possible, use the best idea for packaging your product.

Evaluation:

1 List five reasons for packaging a product. Can you think of other reasons?

2 When you walk through a grocery store aisle, which products catch your eye? Can you explain why? What role does the package play?

3 What products have you purchased that seemed to be "overpackaged"?

4 List five products that are impossible to package.

Challenges:

1 Evaluate five packages of products you have at home, using the design brief specifications.

2 List five products that have packages that are heavier than the product.

3 Choose a product that you think is packaged without thinking about the environment. Redesign the package. Make a sketch of your design, and send it with a letter to the company that made the product.

4 What do you think packages will look like in the future? How have products such as milk and soda containers changed?

Marketing and Advertising Products

Before a company begins to sell a product, the marketing department makes a special plan called a marketing plan or strategy. There are all kinds of ways to market products. The plan might include a **sales forecast,** or prediction of how many sales the company will make. If the market research is not accurate, the company could lose a lot of money. Can you imagine making thousands of a product but selling only a few hundred?

Sometimes companies will do test marketing. To do this, they produce just a small number of products and sell them in maybe just one city. If sales are good in the test market, then the company expects the product to sell in other places, too.

Advertising tells people about your product. The main goal of advertising is to convince customers that they need your product. Many companies have a public relations division. Its job is to make the prod-

Challenge students to bring in misleading advertisements for products.

uct name familiar to the public. Have you ever bought something just because an ad caught your attention? Companies count on that happening. Lots of companies try different ways of advertising such as television commercials, billboards, and ads in magazines, newspapers, and on the radio. Sometimes a company will sponsor a contest or give away a free prize to get people to try their product for the first time.

Brand names like McDonald's, Disney, Honda, and Hershey also make it easier to sell products. People already know the name, so if something new comes out from that company you are quick to notice it.

Actually selling and getting the products to the buyers are also part of the marketing strategy. If a company can't sell its products, then it will not be successful. There are three main ways to sell your product.

Make a display of company brandnames.

Advertising, company logos, and brand names go together to help make a product's marketing plan. If the plan works, a product can be instantly recognized all over the world. (Courtesy of The Coca-Cola Company.)

▶ *Direct Sales:* A manufacturer sells its product directly to the customer. Usually a company will have salespeople who make a **commission** (certain percentage) on the amount they sell. The more the salespeople sell, the more money they make. The commission is an **incentive** (encouragement) for them to sell more products. If an area is large and the company does not have enough salespeople, they might hire a sales representative to sell their product.

▶ *Wholesale Sales:* Wholesalers are people or companies that buy large quantities of products from manufacturers. Then they sell the products to other businesses in large quantities.

▶ *Retail Sales:* Retailers buy products either from wholesalers or directly from the manufacturer. Then they sell them to you. If you've been to a shopping mall, you've seen many retail stores from discount stores to large department stores.

Products commonly end up on the shelves of retail stores like this one. Predictions of sales are important so that stores do not run out of popular products or have many products that do not sell very fast. (Courtesy of Tandy Corporation.)

In addition to selling their products, companies must have ways to get their goods to the buyers. This is called **distribution.** Sometimes the distribution path for a product can be short, as in direct sales. Sometimes products must be temporarily stored in warehouses until the right time for distribution.

At some point, transportation is involved in distributing the products. Depending on the product or the need, different kinds of transporation such as air freight, trucks, trains, or ships might be used. Most of the "18-wheelers" you see on the highways are carrying products to wholesalers or retail stores. Getting the product distributed on time is very important to the success of your company!

THINGS TO SEE AND DO:
Making a Profit (or not)

Introduction:
The activities in this chapter have given you a good look at how companies work. We shouldn't lose sight of the fact that companies are started for the purpose of making money. In this chapter, you will finish the production of your product, advertise and sell the products you have made, and calculate the profit your company made.

Design Brief:
Design and make advertisements that will help sell your product. Sell as many products as possible. Collect money, and give it to the bookkeeping department. Calculate the break-even point for your company. Finish your production activity by calculating the profit per share and distributing the profit.

Materials:
▶ Posterboard, markers

Equipment:
▶ Silk screen equipment (optional)
▶ Computer with spreadsheet and graphics software
▶ Calculator

Procedure:

Check with school administration for rules regarding handling of student funds.

❶ Finish the production of your product. Your class and your teacher may decide to extend or shorten the production run of your product, depending on the demand for the product. Sale of the products can begin whenever there is a willing customer. Advertising your product will help. Work in groups of four or five to design and make some form of advertising.

❷ The advertising department will help each group decide on the type of ad to make. They have already made preliminary plans for advertising your product. Each group must produce an ad.

❸ There is an old joke in business that says there are three keys to success: location, location, and location. This joke makes a point. If your ads (or the place where people can buy your product) are not in a highly visible, busy place, you won't sell many products. Consider where to put the ads for the greatest impact. Decide where to sell your product. Will you sell it in the school store, in homerooms, or in the cafeteria?

Ask the math department for help in teaching students how to calculate their break-even point.

❹ Your company president and the bookkeeping department should work together to determine the point at which your company is making a profit. This point is called the **break-even point**. To calculate the break-even point, you need to make a graph. Your teacher will give you the cost of materials for your product. Your graph might look like this:

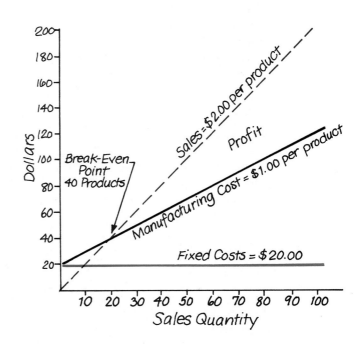

A break-even chart will graphically show you when your company is making a profit. As sales of products increase, you can predict the profit.

⑤ After all the products are sold or everyone in the class has tried to sell the product, it is time for your company to come to an end. In business, closing out a company is called **liquidation**. The profit that your company makes will be divided among the shareholders. Here is the formula for calculating the profit per share.

$$\text{Profit per share} \quad = \quad \frac{\text{Total profit}}{\text{Total number of shares sold}}$$

To calculate the profit you have made in this activity, multiply the number of shares you bought by the profit per share. Even if you didn't make millions of dollars, you now know a lot about how companies operate.

Ask students to bring in advertisements of liquidation sales.

Evaluation:

❶ What was the profit per share for your company?

❷ What part of the student company did you like the best? Why?

❸ Look at the ads for products in a magazine. What ads attract your attention? Why?

❹ What was the biggest expense in your company? How could this expense be reduced?

Challenges:

❶ With your teacher's permission, ask some local stores if they would let you sell your products in their store. Negotiate (make a deal) an agreement with the store owner to divide the profits, or come to an agreement that you both think is fair.

❷ If you had your own company in the future, what do you think your product would be? What would you name your company?

❸ How could a spreadsheet be used to calculate the break-even point?

❹ How could companies use spreadsheets to make "what-if" simulations of product sales or production costs? Set up a spreadsheet simulation for a fictional company. Experiment with a "what-if" simulation that could help company executives make important decisions about the future of the company.

Spreadsheet software makes it possible to try different numbers to see what effect they will have on a company's budget. This is called a "what-if" simulation. It is often possible to predict the result of changes in the way a company does business before they happen. (Courtesy of Texaco, Inc.)

Doing Business in the Future

Most companies are starting to use more automation to produce more goods faster and to improve the product quality. They must do both in order to compete with other manufacturers all over the world. Maintaining large inventories is costly, so many companies are moving toward just-in-time methods that use the computer to schedule deliver-

ies of parts just in time for use. Teamwork among all the people involved in making the product is important to its success.

The success of a business in the future will depend a lot on how it is managed. The R&D department will continue to be important in bringing in new ideas and new ways to produce products. Companies will need well-planned marketing strategies and ways to control company money so that the company can continue to make a profit and grow.

Computers, as we further develop their uses, will be doing many more jobs in all business areas, from administration to production to marketing. Who knows? You might even do most of your shopping at home by computer in the future. Even today, you can order a car through a computerized marketing system. You can examine the car options and then check on the car as it moves through the manufacturing process. Businesses will have to produce quality products quickly and efficiently in order to be successful.

Summary

Every company is in business to sell products and make a profit. A company may be managed as a proprietorship, a partnership, or a corporation. Corporations have many different departments that handle special jobs such as administration, R&D, marketing, production, and personnel.

Businesses today mostly use mass production. Mass production produces many products quickly. Just-in-time manufacturing is a special system that companies use to keep their inventories low.

Each department in the company plays a role in getting a product made and marketed. The quality of a product is important for a company's reputation. Quality control departments inspect parts using gauges and measuring tools. Packaging departments make packages that protect products from shipping damage, help prevent shoplifting, and attract customers. Marketing departments conduct surveys and test market products. Advertising tells people about products using television, radio, billboard, newspaper and magazine ads. Distribution departments get products to you through direct, wholesale, and retail outlets.

All the departments must work together in order to make a quality product efficiently that you will buy. That's how a business stays in business.

Challengers:

In this chapter, the challenge questions were included at the end of each activity.

Techno Talk
Plan to Make Plans

Chapter 10

Building Things

Things to Explore

When you finish this chapter, you will know that:

▶ The construction industry started when people began to build their own shelters.

▶ New materials change the way structures are designed and built.

▶ Scale drawings are made so that people can visualize how to build a structure.

▶ Static loads and dynamic loads affect the structural design of a building.

▶ Building materials must be used to resist forces such as tension, compression, and shear.

▶ Community planning can make better use of resources and make a better place to live.

▶ Buildings should be designed with energy conservation in mind.

TechnoTerms

compression
construction
dead load
dynamic load
elevation drawing
floor plan
live load
load
orthographic
 projection

pictorial
 drawing
shear
shelter
site plan
solar energy
static load
structural
 drawings
tension

(Courtesy of Owens Corning Fiberglass Company.)

Careers in Technology

One of the largest industries in the world, the construction industry, deals with building structures such as houses, bridges, schools, and factories. The people who design and build these structures often take great pride in their work. Some structures, such as bridges and skyscrapers, are even considered works of art. Construction can involve designing and building many things, from a backyard doghouse to a tunnel under the English Channel. Can you see yourself doing the work of an architect, building contractor, structural engineer, or carpenter?

Techno Teasers
Introduction to Construction

Techno Teasers
Answer Segment

What Is Construction?

Look around you at all the houses, office buildings, schools, roads, bridges, factories, and dams that people have built, or **constructed**. Can you name some other **structures**? How much do you know about building structures?

You've learned about the production of products from cars to crayons. Now you will learn about the production of larger products such as houses and bridges. This area of production is called **construction**.

A

D

Construction is the part of the production industry that makes houses, large buildings, bridges, dams, and many other structures. Can you think of other examples of construction projects? (Figure A Courtesy of Northern Homes. Figure B Courtesy of Greater Boston Convention & Visitors' Bureau. Figure C Courtesy of San Francisco Convention & Visitors' Bureau. Figure D Courtesy of Bechtel Corporation.)

B

C

The construction industry uses materials, money, land, and technology to produce the buildings and other structures you're used to seeing. Construction is a large and complex industry that affects you in many ways.

Most people think of construction as building houses. Construction also gives us structures like highways, airports, tunnels, to name a few. Sometimes structures are important for their design or historical meaning like the Washington Monument or the Lincoln Memorial. Most of the time, we design our buildings and structures to meet a special need.

In the beginning, people needed a **shelter** (place out of the cold and rain). Early shelters were caves or structures made mostly of animal skins, twigs, branches, mud, or anything natural that people could find. As people started to settle in one place, they needed more permanent dwellings. They needed shelters that wouldn't blow over in a strong wind and that could withstand different kinds of seasonal weather.

Though no one knows the name of the world's first structural engineer, humans learned to build some amazing structures in early history even without machinery and iron tools. Have you seen pictures of the pyramids? The Great Pyramid at Gizeh, Egypt, was built in 2600 B.C. from huge limestone blocks cut with copper chisels and saws. The people who built it had no mortar or cement, so the entire pyramid is held together only by the weight of the stones. Later, the Romans built huge arenas, bridges, dams, aqueducts (waterways), and temples to serve their needs. You can still visit many of them today.

By the 19th century, new materials changed the ways structures could be built. High-quality steel, developed by Sir Henry Bessemer in 1854, made it possible to build structures that were stronger, more reliable, and lasted longer. Then new concretes that could be poured and would set under water were developed. Concrete combined with steel made reinforced concrete that had the best properties of both materials.

Ask students to brainstorm examples of structures that are made by the construction industry. Ask them if any of their friends or relatives work in jobs related to the construction industry.

Have students research early shelter designs. Ask them to make sketches of different types of shelters used by different cultures around the world. Make a bulletin board display of their work.

An extra credit assignment might be to research the construction methods used to build the pyramids. Have students make a model of what they find in their research.

Challenge students to build an arch using Styrofoam blocks and design a complete structure using this construction technique.

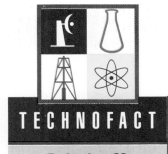

TECHNOFACT

Technofact 60
The Great Wall of China is the longest structure ever built. It stretches 4,000 miles across the northern plain of China. It can be spotted from the moon without a telescope. That's 244,000 miles away!

People need shelter from the effects of weather such as rain, wind, and sun. Shelter can take many forms. Even today, some cultures use building materials and methods that date back thousands of years. (Courtesy of UNHCR.)

Today, the most common building material in the world is concrete. Adding steel reinforcing bars (rebar) to the concrete, it can be made strong enough to withstand tension (stretching forces) and compression (squeezing forces). Concrete is also fireproof, making it an ideal building material for many buildings and other structures such as this bridge. (Courtesy of Portland Cement Association.)

Today, plastics, fiberglass, and improved metals enable architects and structural engineers to build even better and stronger structures. Better sources of power and new technologies also have helped the construction industry move ahead.

Techno Teasers
Roots of Construction

Techno Teasers
Answer Segment

Design

Before any actual building begins on a site, you have to plan your design. If you were an architect or a structural engineer for a project, you would meet with your client (the person you are building for) to discuss the client's needs. Some things you as the designer would need to know are:

- What the building will be used for
- How much space is needed to build the structure
- How much money the client wants to spend on this project
- Where the **site** (building location) is
- Soil, water, and other conditions at the site
- Any special community building codes that must be followed

After discussing your client's specific needs, you would start with some **preliminary** (first) sketches. From these you would make more detailed plans until you and your client were both satisfied. Then you would make accurate, detailed scale drawings called **floor plans** using either CAD (computer-aided design) software or traditional drafting tools. Often, house plans and other small-buildings are drawn to the ¼-inch scale. In this scale, each foot of the actual building is represented by ¼ inch on the plans.

Ask students why it is important to draw plans to scale and why they are not drawn actual size. Students must be aware of the importance of measuring to scale.

Many kinds of working drawings are needed to build a house. You need the site plans, plan views, elevation drawings, section views, and maybe other special information. **Orthographic-projection** drawings are used to show three views of the project in three dimensions. You can be looking down on top of the structure, looking directly at the front of the structure, or looking at the sides of the structure.

See Teacher's Resource Guide for hints on teaching students to visualize in three dimensions.

▶ **Site Plan:** A site plan is drawn to show the property boundaries and the exact location of the structure. Surveyors are sent to the site to accurately locate boundaries and to stake out the building site.

▶ **Plan View:** This is a view looking down on the house without the roof. You can see the location of the walls, all door and window openings, and the size of each one. You would have a separate plan for each floor of the building, such as a second floor or a basement.

▶ **Elevation Drawing:** These are views looking at the building from the front, the sides, or the back.

▶ **Section Views:** These are views looking into the building through a part that has been removed.

▶ **Special Details:** These drawings are usually done to a large scale and have more information about a particular part of the project.

All of the construction activities needed to make a building are controlled by plans. Here, engineers are working with a contractor to make sure the plans are being carried out in the actual construction project. What do you think would happen if there weren't any plans for a construction job? (Courtesy of Toyota Motor Corporation.)

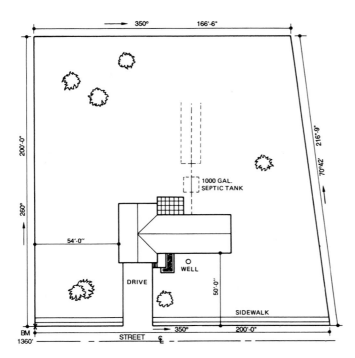

Site plans are made from surveyor's measurements. Site plans show the boundaries of the property, structure location, and other features of the land such as hills, ponds, rivers, and trees. (Photo Courtesy of Paul E. Meyers. Art from Huth, *Construction Technology, 2nd Edition,* copyright 1989 by Delmar Publishers Inc.)

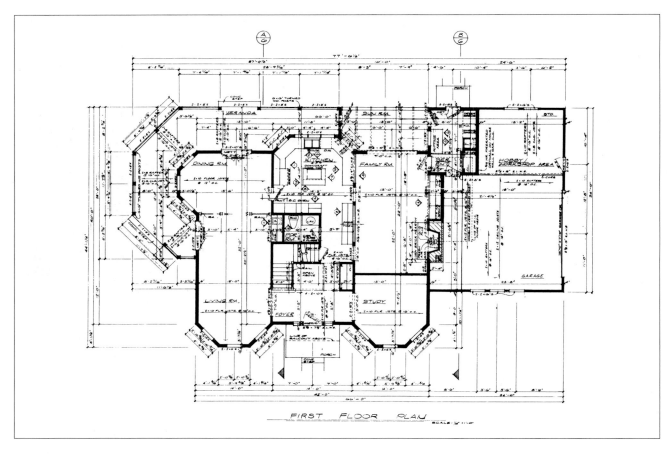

A. Plan View

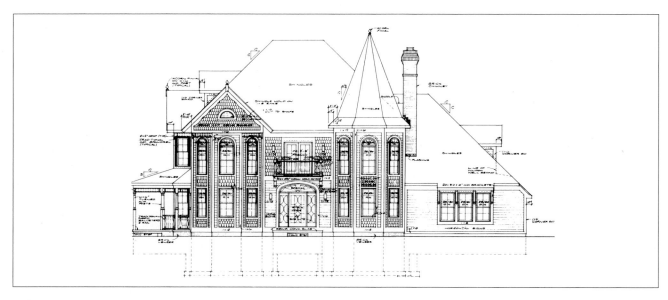

B. Elevation

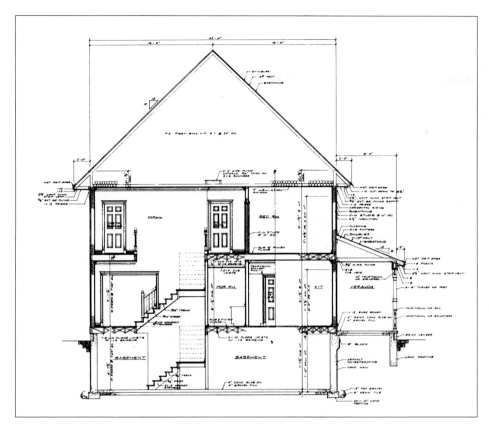

The plans for building a structure include a plan view (floor
plan), elevation (side view),
section (cut-away), and detail
(enlarged) drawings. Contractors and construction workers
build the structure according to
the plans. Costly mistakes can
be made if the plans are not
accurate. (Figures A,B, and C
Courtesy of Home Planners, Inc.)

C. Section

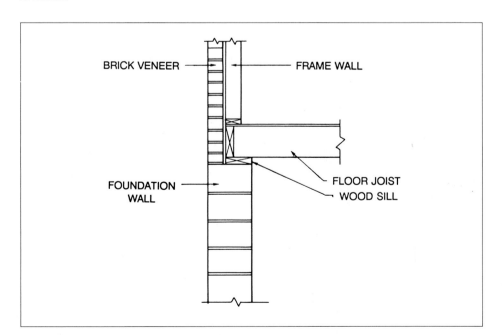

BRICK VENEER — FRAME WALL

FOUNDATION
WALL

FLOOR JOIST
WOOD SILL

Challenge students to make a
bulletin board display of all of the
different drawings that make up
a set of building plans.

**D. Detail (from Huth, *Construction Technology, 2nd* Edition, copyright 1989 by Delmar
Publishers Inc.)**

A **pictorial drawing** also shows a structure in three dimensions, but it shows only one view. Sometimes you might use it to show a client what a structure will look like.

ENTRY COURT

MASTER PLAN FOR **UNION STATION**
ONEIDA COUNTY
UTICA, NEW YORK

DATE: MARCH 2, 19
PROJECT #: 44988

Einhorn
Yaffee
Prescott

Pictorial drawings are used to give clients a "picture" of what the finished structure will look like. An impressive drawing that pleases the client can help an architect get the job. How do you think computers could help in this part of the design process? (Courtesy of Einhorn Yaffee Prescott Architecture & Engineering, P.C./Lance Ferson)

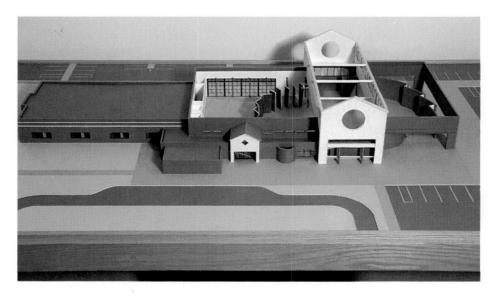

In addition to a drawing that looks like a picture of a structure, a three-dimensional model can make plans come alive. An architectural model with a removable roof lets you actually see inside a proposed building.

You might decide to build a scale model. Models make it easier to show a client what you have designed. Models also make it easier and less expensive to test some engineering problems such as wind resistance and strength before the actual structure is built. Can you imagine building a tall skyscraper only to find out that the wind is so strong at the top that all the windows pop out? That's happened before!

Ask a local architect to show plans, building models, and model building techniques.

THINGS TO SEE AND DO:
Designing Your Dream House

Special Report
An Architect's Point of View

Introduction:

House designs are often made by the owner of the house. Many people think they can design their own house without much thought. There are many things that must be considered in the design of a house, however. The cost of the house can be reduced and more livable space can be built if you plan the floor plan of your house carefully. In this activity, you will design a dream house that you might build some day.

Design Brief:

Design and draw the floor plan for a house that you would like to build some day. Your design must be practical and efficient. Your design must include the following:

- Three bedrooms
- Two bathrooms
- Kitchen and dining area
- Two-car garage
- Utility room and laundry area
- Living room or den
- Closet space

Students should be encouraged to keep their designs practical while expressing their ideas. Some students tend to get carried away with designs that include bowling alleys and swimming pools.

Materials:

- Graph paper
- Model materials (optional)

Equipment:

- Computer with CAD software (optional)
- Drafting tools

Procedure:

1. This is a chance for you to express your creativity in designing a house. Every student will design a house and explain his or her design to the class.

2. Make some preliminary sketches on graph paper. It is easiest to use ¼" graph paper and use a scale of ¼" = 1'. Remember, your first idea may not be the best idea you can think of. Don't get trapped into an idea that is a dead end.

3. Architects and contractors use a set of symbols to draw floor plans. A wall, for example, is not drawn as a thin line. Walls are really 4½" to 12" thick, depending on the construction materials used. The drawing should be made to show the actual thickness of walls. Some of the common symbols you might need to use in your design are shown here.

Some students get discouraged easily when their first idea doesn't work. Students should be encouraged to make many changes and adaptations to their overall plan. Tracing paper placed over a student's design can make it easy to make suggestions. This is a good time to practice sketching skills.

Challenge students to make a bulletin board display of architectural symbols and magazine clippings that illustrate the symbols.

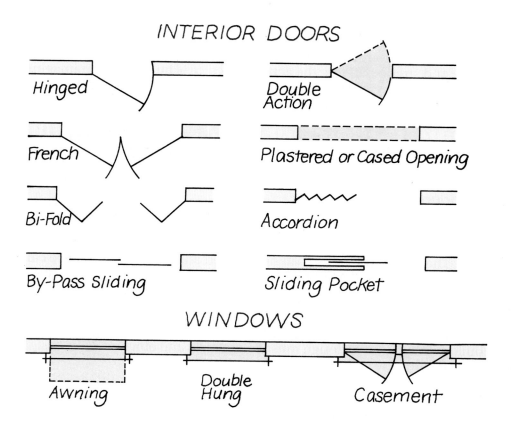

INTERIOR DOORS

Hinged Double Action French Plastered or Cased Opening Bi-Fold Accordion By-Pass Sliding Sliding Pocket

WINDOWS

Awning Double Hung Casement

Symbols are used to represent parts of a structure. Here are some of the standard symbols used by architects, drafters, and building contractors.

Ask an architect to illustrate the use of computer imaging to "walk through" a house design. If local architects are unable to visit, show a videotape of computer 3D software.

4 Here are some general hints to help you design a house that will be livable and efficient. These are guidelines, not hard-and-fast rules.

- Imagine yourself walking through the house you've designed. Can you get to the rooms easily without bothering people in other rooms? Are the bedrooms located near noisy areas such as the kitchen or living room?

- Remember that safety is an important part of your design. There must be at least two outside doors in case of fire. The windows should be large enough for people to exit through in an emergency.

- Too many hallways are usually a sign of wasted space. The cost of a house is usually calculated by the square foot. You pay as much per square foot for a hallway that you use just to walk through as you do for living space.

- If your design will include a second story or a basement, you will need a floor plan for each level. Be sure the stairs for each level are located carefully.

- Stairways should be made so that furniture can be

moved upstairs or downstairs easily.

⑤ When you have refined your preliminary sketch, it is time to make a finished drawing. This is another place where a computer can make your job easier. If you use a computer to make your floor plan, and decide to change the plan, it is very easy to do on a computer.

⑥ Complete your floor plan, and present your design to the class.

Evaluation:

❶ How many square feet (area) are in your design? Include all living spaces and hallways in your calculations.

❷ Ask your teacher to find out the average cost per square foot for building a house in your area. Calculate the cost of building your house.

Cost = Cost per square foot × Area in square feet

❸ How could the building cost of a house be reduced?

❹ Could your floor plan be used for a duplex (two houses built together) or a row house (many houses attached together)?

Work with the math teachers to help students calculate the area of the house plan and cost per square foot.

Challenges:

❶ Design a solar-heated house. Provide for large windows that face south. Make the inside area of the house from a material that can heat up during the day and radiate the heat at night.

❷ Design an underground house that uses the earth for insulation. What kinds of special materials or construction techniques would an underground house need?

❸ Design an alternative energy system for your house. Could your area use a hydroelectric generator (electricity produced from a stream or river)? Could you use wind energy to generate electricity? Can you think of any other practical alternative energy source?

❹ Design a double-wall super insulated house for very hot or cold climates. How long do you think it would take to pay back the cost of the extra construction materials with the money you save in energy costs?

❺ Build a scale model from your floor plan.

Ask the local library or architects for back issues of architectural magazines or catalogs.

Structural Design

Now that you've made the design plans, you need **structural drawings.** These drawings show the structural parts of the house or whatever you're building. Information that should be included in the structural drawings are the kind and sizes of materials that will be used, where

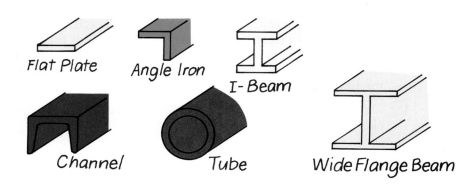

Many different shapes of structural steel are used to build structures such as skyscrapers. Can you identify these famous buildings? (Photo Courtesy of Portland Cement Association.)

parts go, and how the parts will be fastened. For example, if you were constructing a large building, you might need different kinds of glass for windows. The structural drawings would show where reflective glass should go or where you are supposed to put some wired glass for security reasons.

All structures, no matter what their shape or function, are in a game of tug-of-war, with nature's forces on one side and the strength of the structure's design and materials on the other. Structural engineers have to plan structures that have enough strength to stand a sudden gust of wind, an extreme increase in temperature, an earthquake, the pull of gravity, and even the wearing effects of water. The San Francisco earthquake in 1989 caused skyscrapers to move around, bridges to buckle and collapse, and buildings to flatten. Structural engineers were amazed that most of the area's structures made it through the quake.

TECHNOFACT

Technofact 62

Did you know that the only human-made material that comes close to matching the strength of a spider's silk is steel? If you've ever watched a spider spin its web you've actually watched it build a suspension bridge complete with silk cables and anchored supports. The silk on the inside of the web is more elastic than the silk on the outside so that it can cope with dynamic forces such as strong winds!

Earthquakes shake structures very violently. Scientists and engineers are experimenting with different building techniques and materials to prevent injuries to people in case of a natural disaster such as an earthquake, tornado, hurricane, flood, or fire. Here are houses that collapsed in the 1989 San Francisco earthquake. (Courtesy of The Bettman Archive.)

Ask students to research the San Francisco earthquake and make a display or model.

Do you know what forces work on structures? These forces are called **loads**. The structural engineer's first job is to figure out which loads will act on a structure and the strongest they might be in an unusual situation. **Static loads** are loads that are unchanging or slowly changing. Static loads are broken down into two groups, dead loads and live loads.

- **Dead Loads:** These include the entire weight of the structure itself, the beams, floors, walls, insulation materials, columns, ceilings of a building, or the deck of a bridge, etc.

- **Live Loads:** These are forces that a structure supports through its use under normal weather conditions. Live loads would be other weights like people, furniture, equipment, or stored materials. Cars and trucks that pass over a bridge or even the weight of snow, rain, or ice on a structure are live loads.

- **Dynamic loads** are loads that change rapidly, like a sudden gust of powerful wind or an earthquake. Dynamic loads can be very dangerous because they happen so quickly and they produce forces greater than normal. Sonic booms, vibrations from heavy machinery, and even vibrations from people walking along a floor are dynamic loads.

- **Wind loads** are extremely important, especially since we started building tall buildings. That's because the taller the structure, the more it is affected by wind. In very tall buildings up to 10 percent of the structural weight goes into **wind bracing** (resisting the wind).

Special Report
Subsystems

TECHNOFACT

Technofact 63

Robotics and computers will really change construction. New architectural software and advanced graphics allow people to "walk through" a structure long before any construction starts. Researchers use virtual reality software to design, build, change, and "test" a structure under different static and dynamic loads. They can conduct all their tests on the computer in minutes. Construction robots are already used to apply plaster and to spray on fireproofing. Architects and engineers want robots that will use artificial intelligence to make decisions and ones that can move around a construction site easily. That day is not far away!

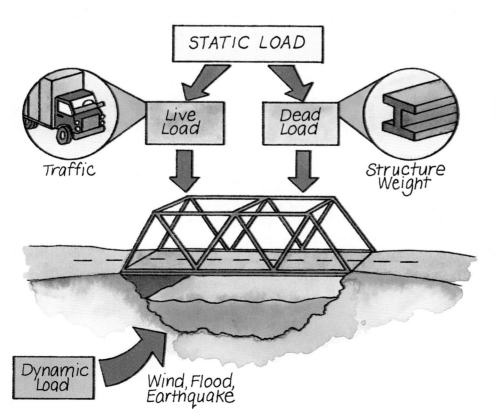

STATIC LOAD

Live Load

Dead Load

Traffic

Structure Weight

Dynamic Load

Wind, Flood, Earthquake

Structures must be designed to stand up to static and dynamic forces. Can you think of other dynamic forces that might act on a structure?

Tension and Compression

Ask students to brainstorm lists of live loads, dead loads, and dynamic loads.

The two most important forces that act on all parts of a structure are tension and compression. Tension is a pulling force. Compression is a squeezing force. The structural parts of buildings, bridges, and even car frames are designed to resist the forces of tension and compression. Sometimes the two forces act together on the same structural part. A beam in the floor, for example, is being stretched (tension) on the bottom and squeezed (compressed) on the top. There is a thin area near the center of the beam that is neutral.

Tension and compression also act on the structure of your body. The bones in your skeleton must resist these forces too. Can you think of how the bones in your body must resist tension or compression when you walk, run, throw a ball, or sit?

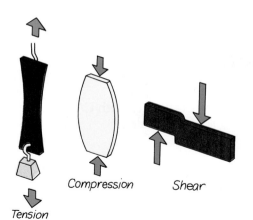

Compression   Shear

Tension

Many different forces act on the structure of a building. Tension is a pulling force. Compression is a squeezing force, and shearing is a tearing or cutting force. Concrete is strong under compression, but weak under tension. Steel bars are strong under tension, but bend under compression. How could concrete and steel be mixed together to make a very strong structure?

Other stresses from compression, tension, and shear affect structures too. Under **compression** an object is pushed together. A standing column is always under compression. Steel and concrete are the best materials for withstanding compression. That's why they are used so much in building. **Tension** is the opposite stress from compression. It is a pulling force that stretches materials. Tension is the main force at work in the steel cables of a suspension bridge or the inflated domes of some stadiums. **Shear** is a sliding force. The blades of scissors cut through paper by forcing it in opposite directions as the blades slide by each other. This is an example of shear stress.

When you're building anything, you need to remember how important the loads and other stresses on the structure are. It is important to plan for these before you start the actual building process.

Have students make a bulletin board display showing examples of tension, compression, and shear.

THINGS TO SEE AND DO: *What Holds Buildings Up?*

Introduction:
The basic structure of a building is often covered with other building materials so that people can't see it. But it is the most important part of the building. Without it, the building would not stand up. The way buildings are built has changed as technology has grown. The structure of a bridge is usually exposed so that you can see exactly how it was built. Bridges probably started when early people used a downed tree to cross a stream. Today, automobiles and trains commonly cross rivers and even deep canyons. In this activity, you will build two types of bridges and compare their strengths.

Techno Teasers
Bridges that Hold

Techno Teasers
Answer Segment

Design Brief:
Design and build a model of a suspension bridge and one of a truss bridge. Your bridges will be made so that they will span a distance of 4 feet. Similar materials will be used for the two bridges.

Have students research bridge designs in the library.

Materials:
- Plywood
- String
- Hardboard
- Wood glue

Equipment:
- Scroll or band saw
- Scissors
- Power hand drill
- Hot glue gun
- Computer with CAD software (optional)

Procedure:

1 Divide the class into two groups. One group will design and make a suspension bridge. The other group will design and build a truss bridge. Your designs might look like these:

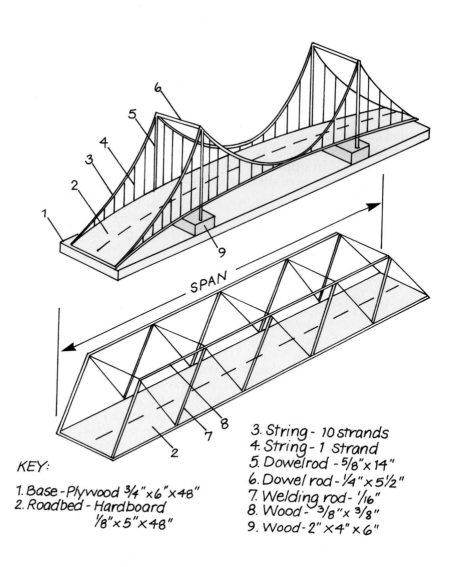

SPAN

KEY:

1. Base - Plywood ¾"x6"x48"
2. Roadbed - Hardboard
   ⅛"x5"x48"

3. String - 10 strands
4. String - 1 strand
5. Dowel rod - ⅝"x14"
6. Dowel rod - ¼"x5½"
7. Welding rod - ¹⁄₁₆"
8. Wood - ⅜"x⅜"
9. Wood - 2"x4"x6"

Your bridge design might look like one of these. Your group might research other types of bridges or modify these drawings.

2 Elect a job contractor from your group. The contractor will be responsible for organizing the process of building your bridge. This is a big job. Your group may also elect an assistant contractor to help make the contractor's job easier.

A touch of realism can be added by assigning a value to each of the materials and coming in with competitive bids.

3 Most large construction jobs have a definite date set for their completion. Sometimes, the contract between the builder and the contractor will have a penalty clause. This means that if the project is not finished on time, the contractor must pay a penalty. Your penalty might be a lower grade for this activity.

4 Design your bridge, and make plans for how it will be built. You might use a computer with CAD software to help. When both groups have completed their design, the class should meet to discuss the completion date for both bridges. Your group should negotiate with your teacher to decide on the due date and a penalty clause.

5 Gather the materials that your group will need. Large construction jobs require a large storage area for construction materials. Your contractor will assign a specific job for everyone in the group to complete.

SAFETY NOTE:
Follow the general safety rules for working in the technology lab area. Remember to wear eye protection. Take your time, and work safely with machines and materials. Even though you are trying to work quickly, your first consideration is safety. This is true in this activity and on a real job too.

This would be a good time to give a safety quiz to students to reinforce safety rules.

6 Complete the construction of your bridge. If your due date has passed, you might try to renegotiate the penalty clause so that it won't hurt your grade so much.

Evaluation:

1 Put the two bridges near each other so that their ends are supported on a block of wood about 3 inches off the floor. Start testing the bridges by placing weights (books or bricks, for example) in the middle of the roadbed. Be careful not to let the weights fall on you when the bridge fails. How much weight did your bridge support?

2 What are the advantages and disadvantages of suspension bridges?

3 What are the advantages and disadvantages of truss bridges?

4 Where is the longest bridge that is closest to your school? What type of bridge is it?

Challenges:

❶ Design and build a drawbridge that will move out of the way of large ships.

❷ Use CAD software to design a bridge of the future.

❸ Research how bridges must be maintained so that they will be safe for many years.

❹ Contact the highway department, and ask how bridges are inspected.

Designing Communities

What effect does modern construction have on you? Do you move around your community easier because of a superhighway near you? How is shopping at the mall different from shopping in town? Your answers to these questions might be different from someone else's answers. Some people might move away from an area that all of a sudden is being developed. You might want to move into an area just because it is easy to get around on the freeway. You might like going to one place like a mall to see several different movies or shop.

Most people care about their community. They want their community to continue to meet their needs in the future. To make this happen, the community must have a development plan with laws to control what kinds of construction can take place. Planning boards, elected officials, and city planners are community people who plan for the future and guide community construction projects. They try to make the community a good place to live and work by planning what fits into the community. How would you feel about a fast-food restaurant being built next door to your house? Would you like the extra traffic?

Techno Teasers
Finishing

Techno Teasers
Answer–1

Techno Teasers
Answer–2

Techno Teasers
Answer–3

Ask a city planner or architect to talk to the class about community planning.

Proper community planning can make a city an enjoyable place to live and work. Poor planning or uncontrolled growth can make city living uncomfortable or annoying for some people. Do you think your town or city is planned properly? Can you suggest ways to make it better? (Courtesy of New York State Department of Economic Development.)

Community planning may require large construction projects like tunnels to prevent overcrowding of aboveground highways. (Courtesy of QA Photos Ltd.)

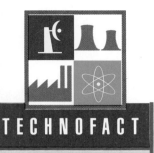

Most construction projects are built to satisfy a need. In a community, the needs might be houses, apartments, shopping centers, parks, waste management, and roads. Modern construction techniques are important in building airports, bridges, tunnels, overpasses, and interchanges that help traffic flow smoothly in and out of your community.

Designing a community is an enormous job. You have to make decisions that might not please all the people. This sometimes happens when a new highway is built through a city. Some people must be moved to new neighborhoods away from friends and familiar surroundings to make room for the road. A quiet neighborhood might change to a busy place with the new traffic. Sometimes old buildings must be torn down to make room for new construction. Again, people might have to move. Whenever possible, the structural engineers and architects should meet with community people and planners to talk over these problems before any construction starts. That way, the people in the community feel they are part of any construction decisions.

THINGS TO SEE AND DO: *Designing Communities*

Introduction:
Did you ever think about how towns or cities are designed? Sometimes there hasn't been much thought given to the way communities are laid out. Proper planning and efficient use of land can make our lives safer and more enjoyable. Communities must consider many factors in their design. In this activity, you will be a city planner. Your job will be to design an entire community that anyone would enjoy living in.

Design Brief:
Design and build a model of a community. Your design should consider how people work and play as well as how they shop and move from one place to another. Your community should provide space for at least 20,000 people.

Materials:

- Model materials, $\frac{1}{8}'' \times \frac{1}{8}''$ balsa wood
- Utility knife, masking tape
- Styrofoam
- $4' \times 4'$ plywood or particleboard
- 4′ roll of butcher paper
- 4-foot T square or straightedge and triangles
- Wood or hot melt glue

Equipment:

- Hot wire cutter
- Hot glue gun
- Scroll or band saw
- Computer with CAD or city planning software

Layers of recycled cardboard can be used to give the city model an elevation.

Procedure:

❶ In this activity, you will be part of a planning team to design an entire town or city. Your group size will be up to you. You can have as few as two or as many members as you would like. Your group will need a leader to coordinate its activities. Name your city, and decide where it is located in your state or county.

❷ Your group will have a $4' \times 4'$ square area to use as the base for your community. You should choose an appropriate scale so that the area covered in your model will be large enough to represent a city. Consider the size of an average home in your scale. For example, if you choose the scale of 1 foot = 1 mile, a house will be less than $\frac{1}{8}'' \times \frac{1}{8}''$. Discuss the size of your city, and choose an appropriate scale for your model.

❸ Make a rough sketch of how you would like your city to look. Some of the things you might consider in your design are:

Industrial Zones
- Light-industry area (small companies)
- Large-industry area
- Electric power generation plant

Residential Zones
- Single-family, low-density housing (individual homes)
- Multiple-family, high-density housing (apartments)

Public Areas
- Open space, parks, bike and jogging paths
- Schools, libraries, vocational training centers, colleges or universities
- Transportation access (major highways, railroad stations, airports, boat docks, and so on)

- Waste management

Commercial Zones

- Office space, fire departments, police stations
- Shopping centers, service areas, retail stores
- Other?

4 Cover a 4′ x 4′ square plywood or particleboard base with butcher paper. Lay out the streets and highways that your community will need. Include bridges or tunnels where they might be needed. Consider the location of industrial areas that need access to highways and of residential (home) areas that need quiet. Where should shopping centers be located? How much park space do you think is needed? Where should park space be located? Do you think one or two large parks are better than many smaller parks?

5 Cut balsa wood or Styrofoam models for the homes, apartments, shopping centers, schools, factories, and so on. Glue the model buildings to the paper layout. Name and label streets and highways. Put any finishing touches on your model.

Evaluation:

1 Where do people in your city live? How many people live in the following areas?

- High-density housing_____
- Low-density housing _____
- Total population _____

2 Would you like to live and work in your city? Explain.

3 What would happen if a disaster such as a hurricane or a flood required an emergency evacuation of your city? Would there be a problem? For example, is there only one bridge or tunnel for thousands of people to use if they have to leave quickly? How could your city be changed to make evacuation faster and easier?

4 In which areas of your community would you like to live? Why?

5 What special precautions do you think could be taken in case of a fire or explosion in the industrial area of your city?

Make a display of natural disasters that have caused loss of life or property.

Challenges:

1 Modify your city to hold two or three times the population. What would you have to do to schools, police and fire departments, or roads? Would high-rise apartment buildings solve the housing shortage, or would they create other problems? Why do you think there are zoning

laws that restrict the building of stores in residential areas?

2 Design a high-density living area that you think people would like to live in.

3 Take pictures of the storefronts in a business district of your town. Cut and glue the pictures together to show an entire block of storefronts. Place tracing paper over the pictures, and design a new front for each store that would make the business area look better. Show your design to a city planner or zoning committee.

4 With your teacher's help, contact a local city planner or a member of a local zoning committee. Ask him or her to visit your class and talk about your community models. Discuss the future of your community.

5 Modify your community model to show what it will look like in the future. How could the city be changed to help prevent air and water pollution? What could be added to your city to make it a place that people would come to on a vacation?

It is often easier to plan a new community than try to correct past mistakes. Sometimes entire towns must be moved to make room for highways, reservoirs, or airport construction projects. Here, a house is being moved to a new location. (Courtesy of Larmon House Movers Inc.)

Energy Conservation

Energy shortages are changing construction techniques in communities. New homes, schools, and other buildings have better insulation, so you don't have to use as much electricity or gas for heating or cooling. Many new homes are also being built with solar panels to collect **solar energy** (energy from the sun). Old homes are being **converted** (changed) to solar heat.

In cool northern climates, a great deal of energy can be saved simply by facing most of the windows of a structure toward the south. Of course, if you were in the Southern Hemisphere like Australia you would want your windows to face north! New technologies make it possible to build windows that are more energy-efficient. The windows have a transparent film that lets light pass through but reflects heat. The development of other technologies and materials will make construction even more energy-efficient in the future.

Ask a local building supply company to demonstrate energy efficient windows.

Building for the Future

The changes in technology to make life easier have affected the construction industry just as they have all of us. Products constructed with new materials and technology are generally better that those made the old way. The processes we use today are also more efficient. We have new lightweight, ultrastrong steel and reinforced plastic and glass materials to build things.

Computers have done much for the construction industry already. Structural designers and engineers can plan, revise, and even test new designs before any actual construction starts on a project. Using computers for these steps saves a company both time and money.

The need to use energy efficiently has definitely changed some parts of construction. Earth-sheltered construction is an example of a design meant to help save energy. Part of the finished building is belowground where the earth keeps the heat from escaping during cold weather. The ground also keeps the house cool during hot weather. Several new communities are being built with energy-efficient homes. These communities have no roads or cars. People walk to the surrounding shops or to neighbors' homes.

What can you expect building in the future to be like? The construction industry must meet the needs of an ever-growing population. You will notice many changes in the design of structures as new materials, building techniques, and construction methods are developed.

Better homes must be built at lower costs. Your new home might well be mass-produced in **modular** (separate) parts. Assembly of modular units is simpler and takes less time and money than building a home piece by piece at the site. You may live in a **smart house**. Smart houses will use electronics and computers to control energy uses in your house. You may have built-in computers that control lights, telephone calls, microwave ovens, televisions, and thermostats. You will probably speak to someone else in the house using special video and voice-activated devices. More people will work at home instead of going to the office thanks to computers. That will bring changes in design.

Plastics, reinforced concrete, and steel are just a few of the building materials used today. Attention must be given to the strength, appearance, cost, and fire resistance of all materials used in construction. (Courtesy of New York State Department of Economic Development.)

Ask a structural engineer to talk to the class about building designs.

Today, some of the mass-production techniques you learned about in Chapter 8 are being used to mass produce houses. This is an assembly line for house production. Have you ever seen a large truck pulling one half of a house down the street? Manufactured houses can often be made less expensively for the same amount of living space as a house built one piece at a time. (Courtesy of Clayton Homes.)

TECHNOFACT

Technofact 66

So you don't like the color of your car? Change it then! A special paint available in the future will allow you to change the color of your car's exterior (outside) by simply flipping a switch inside the car.

Visit a factory that manufactures homes.

T E C H N O F A C T

Technofact 67

If the present human birthrate continues, did you know that by the year 2600 there will be only 1 square yard of dry land surface available for each person? That's hardly more than the size of a telephone booth!

One possible solution to the over-crowded conditions in cities and on highways is to let people work at home. Communication technology makes it possible for many people to do most or all of their work in the comfort of their own home. Can you think of other benefits of working at home? (Courtesy of International Business Machines Corporation.)

Ask a local phone company to discuss how telecommunications will change in large buildings in the future.

Future cities may look very different from those of today. Maybe you will live in a completely enclosed, or **encapsulated**, unit. Your city would be completely air-conditioned and heated with purified air. Some researchers see the possibility of making a domed city underwater.

Office buildings will be connected to central computers. People in one office or at home can communicate with other people anywhere. Computers will continue to improve and thereby take over some office duties. You will have more free time!

How you get to the office is another matter. You might drive a car that rises **vertically** (straight up) to a parking spot on the roof! Construction in the future will need to use building space more efficiently. Buildings will need parking space for cars or, better yet, for other transportation systems that carry more people and are powered by non-polluting energy sources. Two possibilities are hovercrafts and maglev trains.

Space construction will create many different problems for structural engineers and designers. The construction industry will have to make the best use of resources in space. They will have to make people comfortable when they are living in space for months or years at a time. Space cities and moon bases are constructions planned for your future. How would you like to live in space or take a vacation on the moon?

Technology is making the idea of living and working in space or on the moon a real possibility. What special construction techniques or materials do you think would be needed on the moon? Would you like to live on the moon someday? (Courtesy of NASA.)

THINGS TO SEE AND DO:
Buildings of the Future

Introduction:
Now that you know a lot more about how buildings are made today it is time to think about the future. New materials and advancements in technology are making it possible to build buildings in new ways and new shapes. In this activity, you will design and build a model of a structure that uses the tensile properties and the compression resistance of materials.

Design Brief:
Design and build a tensile structure that could be built. Your design should include an interesting shape and a large unsupported floor area.

Materials:
- Wood
- Dowels
- Rubberbands or string
- Wood glue

Equipment:
- Scroll or band saw
- Power hand drill or drill press
- Hot melt glue gun
- Tape measure

Procedure:
1. In this activity, you will work with a partner to build a tensile structure. Study the following designs and brainstorm other possible designs with your partner.

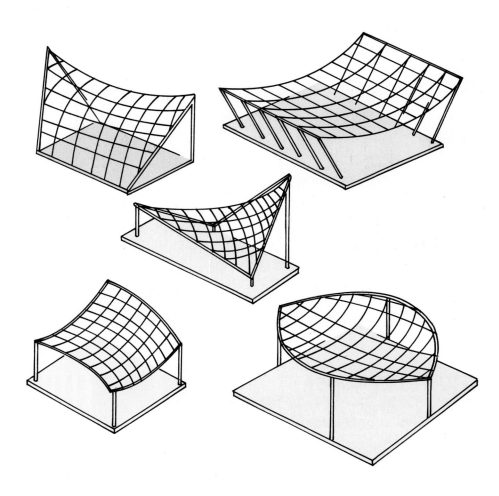

Tensile structures can be built faster and made stronger and lighter than structures built with traditional methods. How do you think the inside of a house would change if the roof looked like one of these designs?

SAFETY NOTE:
Follow the general safety rules for working in the technology lab area. Remember to wear eye protection. Take your time and work safely with machines and materials.

❷ Cut the base for your tensile structure. Notice that the roofs of the tensile structures are made of materials that are strong in tension. Steel cables are commonly used in these types of structures. The structural parts that hold up the tensile structure are made of concrete or structural steel to resist the compression forces. For your model, cut wood dowel rods to the length your design requires.

❸ Carefully measure and mark equally spaced holes for the compression parts of your structure. Drill holes in the base of your model to hold up the dowel rods. Cut a small notch in the end of each dowel to hold the string or rubberbands. Glue the dowels into the holes in the base.

④ Assemble the model using string or rubberbands as the tensile part of the roof. You might need to weave the string through the structure to make the design even stronger.

⑤ When the model is completed, gently press on the roof of your structure to test its strength. Fix any weak spots as needed. Real tensile structures complete the roof with a coating of concrete or a plastic fabric or membrane to keep out the weather. Can you think of a material that would simulate a finished roof on your model?

Evaluation:

① Would you like to live in a house with a tensile roof? Explain.

② What are the advantages of this type of construction? What are the disadvantages?

③ What materials are good for resisting compression forces?

④ What materials have good tensile strength?

Challenges:

① Design and build a pneumatic (air-filled) structure with the help of your teacher.

② Research pictures of tensile structures. Do you have any near your area?

③ Use a computer and CAD software to design other tensile structures.

④ Research one of the topics below and make a model similar to your findings:

- R. Buckminster Fuller's structures
- Geodesic dome
- Pneumatic structures
- Earthquake structural design
- Undersea structures
- Space stations or moon bases

Construction in the future will take place in new areas like space and undersea environments. Making people comfortable and meeting their needs in space and under the sea without ruining the existing environment will be challenging and certainly exciting! Maybe you will be the engineer who comes up with a plan for these future projects!

Pneumatic structures large enough for students to enter can be made inexpensively using polyethylene film. Be sure to provide an emergency exit.

TECHNOFACT

Technofact 68

Waste management is a community design concern. The Fresh Kills Landfill on Staten Island, New York, is a pile of garbage 140 feet high. It gets 22,000 tons of garbage daily. By the year 2000, it should reach an estimated height of 500 feet. That will make it the highest summit on the Atlantic Coast between Maine and the tip of Florida!

Summary

Construction is the part of production that deals with building structures such as houses, schools, roads, bridges, and skyscrapers. People have always needed shelter from weather. The materials used to build structures have changed from early people's use of animal skins, twigs, and mud. Today we use lumber, steel, and concrete for many building purposes.

Structures must be thoroughly planned before construction starts. Preliminary sketches are refined into plan, elevation, structural, and section views of the structure. A pictorial drawing is sometimes made to give a client an idea of what a structure will look like.

There are different forces that act on structures. These forces are called loads. Static loads either stay the same or change slowly. They include dead loads and live loads. The dead load of a building is the weight of the structure itself. Live loads are things that people move into the building, like furniture and even the people themselves. Dynamic loads are loads that change quickly like an earthquake or a tornado. The loads placed on a building create forces such as compression, tension, and shear. Compression forces squeeze objects together. Tension pulls objects apart. Shear tends to cut through objects.

Designing communities for people to live in comfortably involves many decisions. City planners and zoning committees make decisions based on the needs of a community. Some factors in designing communities include the location of industrial areas, residential areas, and commercial areas.

Changes in technology will affect communities and construction of the future. The use of computers to control traffic, new ultrastrong and lightweight building materials, and mass-produced homes are just a few areas where technology might help build the city of the future.

Challengers:

❶ Research the design and materials used in the first bridges. Make a model of a stone arch using Styrofoam blocks.

❷ Look at the plans for a large building such as your school. Find a plan view, an elevation view, a section view, and a detailed drawing. What is the scale of the drawing?

❸ With your teacher's help, ask an architect to talk to your class and share plans for a structure.

❹ Ask an architect to help you cut materials correctly to make scale models.

❺ Design and make a model of an undersea structure for people.

❻ Research how electronics and computers will be used in the smart houses of the future.

❼ Research and compare the construction techniques and materials used in different parts of the world to meet specific needs.

❽ Build 4-foot-long models of structural shapes out of wood. The shapes might include I-beams, angle iron, box beams, or wide flange beams. Place the beams so that they are supported on each end a few inches off the floor. Measure the distance from the floor to the bottom of the beam model without any load. Have someone stand in the middle of your beam and then remeasure it. Make a chart of the results of your experiment.

❾ Research a large construction project such as the building of the pyramids, the Great Wall of China, the Panama Canal, or the Statue of Liberty. What special problems did the construction of these large projects cause? How did engineers solve them?

❿ Find out why the Tacoma Narrows Bridge in Washington State collapsed. What did the engineers forget to plan for when they designed that structure?

See Teacher's Resource Guide.

Techno Teasers
Tacoma Narrows Bridge

Techno Teasers
Answer Segment

Chapter 11

Using Energy

Things to Explore

When you finish this chapter, you will know that:

▶ Potential energy is energy at rest. Kinetic energy is energy of motion.

▶ Electricity is produced by a generator turned by a turbine. The energy source might be moving water, steam made by burning a fossil fuel, or heat given off in a nuclear reactor.

▶ It makes sense to conserve energy rather than make more of it.

▶ Alternative energy sources might include sunlight, wind, biomass, or geothermal sources.

Chapter Opener
Using Energy

TechnoTerms

alternative energy sources
biomass
conserve
energy
fission
fossil fuel
fusion
generator
geothermal
hydroelectric power
infiltration

insulate
kinetic energy
nonrenewable
nuclear energy
photovoltaic
potential energy
power
precipitation
solar array
solar cell
turbine

Careers in Technology

Most people use energy, but they don't think about where it comes from or how it gets to them. Workers in many specialized fields are employed on this oil platform. Much of the energy you use comes from oil. Industrialized countries are dependent on energy from coal, oil, and natural gas. Millions of dollars are spent each year trying to find new sources of energy. Do you know where the electricity you use comes from?

Where Do We Get Energy?

Did you ever stop to think of the ways you use **energy** in a day? Every time you pick up a pencil or blink an eye, you are using energy from the food you eat. The bus or car you might ride to school in every day uses the chemical energy stored in gasoline. You flip on the light switch and you are using electrical energy to produce light. What is energy, and where does it come from?

What Is Energy? The definition of energy is the ability to do work. Some people confuse power and energy. They think they are the same, but they are not. **Power** is the amount of work done in a certain period of time. For example, if you run one mile, that is work. But if you run one mile in *4 minutes*, that's power.

Energy can never be lost or destroyed, but it can be changed from one form to another. For example, a battery changes chemical energy to electrical energy. A solar-powered car turns light energy into electrical energy. Technological processes often change one form of energy into another, more useful, form to do work for us.

TECHNOFACT

Technofact 69

Imagine a battery that is only 4/1000 of an inch thick. That's about the thickness of a piece of typing paper. Scientists at the Lawrence Berkeley Laboratories, University of California at Berkeley, have made such a battery. Not only is it half as heavy as other batteries, but it is also biodegradable. Besides that, once the battery is charged it stays charged almost forever. That's an efficient battery!

B

A

C

We use many different forms of energy every day. Can you describe the energy being used in each of these photos? (Figure A Courtesy of Toyota Motor Corporation. Figures B and C Courtesy of Bechtel Corporation.)

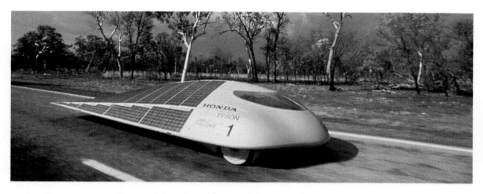

Energy is often changed, or converted, from one form to another. What types of energy are converted in an electric car? (Courtesy of Honda North America Inc.)

There are two kinds of energy. **Potential energy** is energy at rest waiting to do work. A compressed or coiled spring has potential energy. When you release a spring, like a spring on a pinball game, the spring has kinetic energy and moves the ball. **Kinetic energy** is energy of motion. A moving bicycle has kinetic energy. A book about to fall off the edge of a table has potential energy. Can you think of some other examples of kinetic and potential energy?

Kinetic energy involves motion. Potential energy is stored energy. It has potential to be used to do work. The compressed spring has stored energy. When the spring is released, it produces kinetic energy. Think of other examples of kinetic and potential energy.

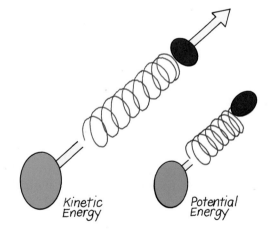

Kinetic Energy   Potential Energy

TECHNOFACT

Technofact 70
We're always looking for new alternative energy sources. A potential source of diesel fuel comes from the *Copaifera langsdorfii* tree resins found in the Amazon rain forest. The resins are collected much the way maple syrup is harvested, by hammering a spigot into the tree trunk. One tree can yield about 5 gallons of the resin, which can be used as an alternative fuel. The best part is that the resins can be gathered without destroying the trees or the surrounding forest.

Where Does Energy Come From? Did you know that nearly all of the earth's energy comes from the sun? Solar (sun) activities cause wind energy, water energy, and even fossil fuels.

Winds are a kind of kinetic energy caused by uneven heating of the earth's atmosphere by the sun. The hot air rises and the colder, heavier air moves in under it, creating winds. Sometimes these winds are strong enough to power electric generators.

Every living thing on earth needs solar energy. Plant cells store solar energy during photosynthesis. Animals like us eat plants for our own energy needs. Some of the decayed plant and animal materials become fossil fuels after millions of years. Burning is a way energy is released from plants.

Techno Teasers
Kinetic v. Potential

Techno Teasers
Answer Segment

Challenge students to think of energy sources that are not related to the sun.

A

Most of our energy can be traced to the sun. Can you explain how solar energy is used in each of these photographs? How long do you think we could last without the sun's energy? (Figures A and C Courtesy of New York State Department of Economic Development. Figure B Courtesy of United States Department of Energy.)

The sun also keeps the water cycle going. The sun heats water until it evaporates (changes to a gas). When the water vapor rises, it cools and forms clouds. The cooled vapor becomes liquid water again in the clouds and falls as rain or some other **precipitation**.

B

C

We depend on energy to be there when we want it. Without energy, our technology wouldn't be where it is today. Technology processes and systems depend on energy. Primitive people knew very little about energy. They used muscle power as their main source of energy. They could do only the things they were strong enough to do until they discovered that animals could be tamed and used as energy sources, too.

For thousands of years, animal energy was the greatest technological advancement around. Then people learned how to use other sources of energy and power such as the wind and falling water to do work. Eventually, adding power to tools and machines made it possible to do work faster and more easily. What would your life be like without the technological advances in energy and power?

Special Report
The History of Energy

Make a display of primitive attempts to use energy such as water or wind power.

Ask students if they know where their electricity is generated.

Without technology to harness energy, people must rely on their own energy to do work. Here, a girl is smashing grain into flour. How do we use energy to get flour for baking? What types of energy are used or changed to make wheat into the flour we buy in grocery stores? (Courtesy of UNHCR.)

Today's **conventional** (most common) energy supplies used in developed countries are:

▶ **Electricity:** Most electrical energy is made by an electrical **generator**, a machine that converts mechanical energy to electrical energy. You might have a portable generator on your bicycle that you power by pedaling!

Coal, oil, and natural gas are fossil fuels. We rely on fossil fuels for much of the energy we use. The problem is that they are being used at an alarming rate. Fossil fuels are not renewable. (Courtesy of CSX Corporation.)

We use electricity every day. Did you ever think about where electricity comes from? Generators change mechanical energy into electrical energy. Generators are turned by water (hydroelectric) or steam. The high-pressure steam is produced by burning coal, oil, or natural gas, for example. (Courtesy of Tennessee Valley Authority.)

Nuclear energy is released from atoms of radioactive materials. The radiation produces heat that directly or indirectly boils water into steam. The steam is used to turn a turbine that turns a generator to make electricity. (Courtesy of Tennessee Valley Authority/Ron Schmidt.)

▶ **Fossil Fuels:** Coal, oil, and natural gas are fossil fuels. These are the major source of energy today. They are made from decayed animals and plants. Burning fossil fuels produces a lot of heat energy that we can use. Fossil fuels are **nonrenewable** (cannot be replaced) in our lifetime.

▶ **Nuclear Energy:** Nuclear energy is the energy found in atoms. In **fission**, atoms of materials such as uranium are split, releasing huge amounts of heat energy that is used to heat water. The steam then spins a **turbine** (a wheel turned by the force of a gas or liquid striking the blades on the wheel). **Fusion** is another kind of nuclear reaction where hydrogen atoms are joined or fused. The reaction gives off lots of heat energy, just like fission does.

▶ **Hydroelectric Power:** This power is made when water stored behind a dam goes through a turbine. A generator working with the turbine produces electrical energy from mechanical energy.

Special Report
Nuclear Power Plant

Special Report
Energy Sources

Hydroelectric energy uses the potential energy of water stored behind a dam. As the water is released, it turns a turbine that turns a generator. (Courtesy of Tennessee Valley Authority.)

THINGS TO SEE AND DO: Where Does Electricity Come From?

Introduction:

What would your life be like without electricity? You wouldn't have light, heat, or air conditioning. You might not have hot water or even warm meals. Stereos, televisions, irons, toasters, and computers would be useless without electricity. Every day, we just flip a switch, push a button, or put a plug into an outlet without much thought about how electrical energy is produced.

Almost all of the electricity we use involves the use of a turbine blade turning a generator. A turbine is a type of fan that can operate under high pressures and sometimes high temperatures. In this activity, you will make your own turbine blade and put it to work making electrical energy.

Design Brief:

Build a working model of a turbine generator that will demonstrate how hydroelectricity and other conventional electricity are produced. Your model will include a turbine blade that harnesses the energy of moving water or air. The turbine will be connected to a generator to change mechanical energy into electrical energy.

Materials:

- Polycarbonate plastic ⅛" thick
- ⅛" steel welding rod
- Epoxy glue
- 1½–3 volt DC motor
- Hookup wire
- ⅛" ID (inside diameter) plastic shrink tubing
- Wood: 1" × 2" × 6" (optional)
- Rubber cement

Equipment:

- Drill press, ⅛" drill bit
- Scissors
- Scroll saw
- Measurement tools
- Plastic strip heater
- Pliers, wire cutters
- Digital multimeter
- Computer with CAD software (optional)
- Heat gun

Some materials for this activity may be obtained from the machine dissection done earlier.

Procedure:

1 In this activity, you will work with two or three other students to make a turbine blade and generator. Your first task is to draw the plan for the turbine blade on a sheet of paper. Use a compass and a 45° triangle or a computer and CAD software to make to make your plan.

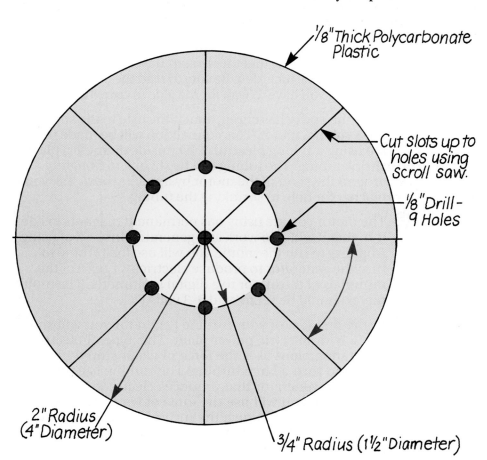

⅛" Thick Polycarbonate Plastic

Cut slots up to holes using scroll saw.

⅛" Drill – 9 Holes

2" Radius (4" Diameter)

¾" Radius (1½" Diameter)

Make a drawing that looks like this on a sheet of paper. Cut out your design and rubber cement it to a piece of ⅛" thick polycarbonate plastic.

Use pliers to bend the fins of your turbine blade to a 45° angle. Be careful not to burn yourself on the hot plastic. (Copyright © Pam Benham.)

2 Cut out the finished paper design with scissors. With rubber cement, glue the design to a piece of ⅛″ thick polycarbonate plastic (Lexan). Polycarbonate is a very strong thermoplastic. (Note: the proper way to use rubber cement is to coat both surfaces lightly and let them dry. When the glue is dry, press the two surfaces together.)

SAFETY NOTE:
Wear eye protection. Follow the general safety rules and specific rules for the machines you are using. Remember to ask your teacher for help.

3 Drill the eight ⅛″ diameter holes and another in the center. The holes will help prevent the plastic from cracking. Heat the plastic between the holes with a heat gun. Bend each fin of the turbine blade to a 45° angle.

SAFETY NOTE:
The plastic must reach a temperature of 350° to 400° before it will bend. Do not touch the hot plastic or the heating element of the strip heater.

4 Using wire cutters, cut an 8″ long piece of ⅛″ diameter steel welding rod. Be careful not to bend the welding rod because it will be the axle for your turbine. Glue the turbine blade to the middle of the welding rod using epoxy glue. Follow the directions on the tube of glue carefully. Mix the two parts of epoxy together to start the chemical reaction that will make it harden. Be sure to read the label to see how long you have to wait for the glue to **cure** (harden).

5 After the epoxy has cured, attach the axle to the shaft of a 1½–3 volt DC motor. This connection will be made flexible using a 1′ long piece of ⅛″ ID (inside diameter) plastic shrink tubing. The tubing can be shrunk to fit tightly around the shaft of the motor by applying heat. Ask your teacher for help in shrinking the tubing.

6 The motor you are using has permanent magnets inside it so that it will also work as a generator. Instead of using a battery to run the motor, you will use the turbine to turn the generator to produce electricity. Connect the terminals of the motor to a digital multimeter. The multimeter should be set to read 0–3 volts DC .

7 In the first test of your turbine generator you will simulate a hydroelectric power plant. This type of electrical generating plant uses the force of water stored behind a dam to turn a large turbine. The turbine is connected to a huge generator that produces electricity. In your simulation, you will use the force of water running out of a water faucet to turn the turbine blade. Hold the motor/generator so that it will be out of the water stream and low enough in a sink to prevent splashing.

Watch the multimeter reading and see what voltage your turbine will produce. If the meter shows a negative number, reverse the positive and negative wires.

8 The second part of this activity will simulate the use of a turbine being run by high pressure steam. In a real power plant, steam would be generated by burning coal, oil, or natural gas to boil water. Nuclear power plants use radioactive fuel rods to boil water and make steam. High-pressure steam is very dangerous. Instead of using steam to spin your turbine, you will use compressed air.

9 Design and build a stand to hold the turbine blade. Your design might be made of plastic or wood. Here are some possible designs:

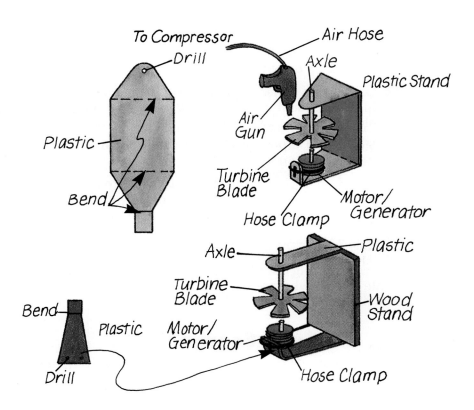

Design your support stand so that it will safely hold up your turbine. You might decide to use wood, plastic, or sheet metal. Ask your teacher to approve your designs before you start.

10 Connect the multimeter to the generator terminals as before. Use a compressed-air blow gun to spin the turbine blade.

SAFETY NOTE:
Do not point the blow gun at anyone or put your finger over the end of the gun. Compressed air can be dangerous. Keep your fingers and long hair away from the spinning turbine blade.

Students must be cautioned not to play with compressed air.

11 Watch the multimeter reading to see how many volts your turbine generator can produce.

Evaluation:

1 What was the voltage reading in the hydroelectric test? What was the reading using compressed air? Why was there a difference? Explain.

2 Make a sketch of a hydroelectric power plant. Label the turbine, generator, dam, and reservoir.

3 Where does the electricity you use come from? How is it produced?

4 How many generators like yours would it take to run a house (120 volts) outlet?

Challenges:

1 With your teacher's help, contact your local power company. Ask a representative to come to speak to your class. Ask questions about the future of the power plant. How will it keep up with the need for more electricity?

2 Design and make a modification (change) of your turbine blade so that you could use wind energy to make electricity.

3 How could superconducting power transmission lines help reduce energy uses?

4 Research the cost of electricity in your area. Find out how much the electricity used in your house costs every month. List ways you could reduce the electricity bill.

One of our biggest challenges today is to find ways to use these conventional energy sources more efficiently. Many of our resources are becoming scarce. Running out of energy sources is a tremendous problem for people like us who are used to using lots of energy without thinking about it.

TECHNOFACT

Technofact 71
Did you know that if we could figure out a way to better use energy from the sun, all our energy problems would be solved? The average yearly amount of solar radiation is equal to 178,000 Terawatts (10^{12} watts) a year.

How Can We Save Energy?

Did you ever leave a light on in a room when you weren't there? Have you seen people leave their cars running while they go into a store? Do you turn the heat up in your house but leave a window open for fresh air? Do you always wash your clothes in hot water? These are some habits that we need to change in order to **conserve**, or save, our energy resources.

Ask students to make a list of ways in which energy can be conserved.

Today, people are trying to find ways to conserve energy. That's because many of our resources are nonrenewable. Even ones like electricity, which we think will always be there, count on other energy sources for their production. In some places, power plants use coal or oil to make electricity. Both of these fossil fuels are running out. We need to conserve by using our energy supplies more wisely and efficiently.

You live in an **energy-intensive** world. That means most everything you use daily uses energy. Every day more products that use energy are being developed to meet your needs. Sometimes these products make life easier, but they are not needed for your survival. Do you always use an electric can opener? How about an electric pencil sharpener?

What can we do to save energy? One company has produced a new fluorescent light bulb that uses less energy than conventional bulbs and lasts longer. Automobile manufacturers are trying to make car engines more efficient. Half of the energy in fossil fuels is lost in the form of heat that doesn't do any useful work. This is true for car and other types of engines. Up to now, no machine can make total use of all the energy put into it. In other words, all products that use energy waste some of it.

So we need to change products so that they waste less energy. We can make cars that burn less fuel. Technology is also making products more energy-efficient by using plastics for many parts instead of metals. They don't use as much fuel because they are lighter. Furnaces are now being produced that are twice as energy-efficient as the ones made five years ago.

Technology is giving us ways to save electrical energy by making home appliances more efficient. For example, water heaters can be better insulated so that heat stays in. Your walls, ceilings, doors, and windows can be **insulated** to prevent heat loss during cold weather. Maybe your dishwasher has a special energy-saving setting that doesn't use energy to dry your dishes. Many large buildings and schools have special switches that automatically lower room temperatures and turn lights off at night or on weekends when most people are not working. That saves a lot of energy.

Have you been recycling cans, bottles, plastic, and newspaper? Recycling is another way to save energy because it takes less energy to recycle a material than to produce a totally new product. Throwing away one aluminum can is like throwing away one-half gallon of gasoline!

Ask students what products they would be willing to do without in order to save energy.

TECHNOFACT

Technofact 72

Besides being expensive, energy that powers your appliances and heats, lights, and cools our buildings produces 500 million tons of carbon dioxide a year. That's about 2 tons of carbon dioxide for every person in the United States. Think about what that does to your environment!

Ask a major appliance dealer to explain the energy labeling of appliances.

Automobile manufacturers are trying to make engines more efficient. Even an improvement of just a few miles per gallon in an engine's performance could save millions of gallons of gasoline. (Courtesy of Ford Motor Company.)

Recycling garbage can help save the energy needed to make new materials. It makes a lot of sense to recycle materials so that we can save energy, prevent overfilling of landfills, and help the environment all at the same time. Think about that the next time you throw an empty container away. (Courtesy of Waste Management Inc.)

THINGS TO SEE AND DO: *Energy Conservation*

Introduction:

If we could save or conserve energy, we wouldn't need to produce more of it. Conservation can take many forms. You have learned that recycling materials such as aluminum can help save a great deal of electrical energy. Some of the ways you can save energy are simple, but others are not so easy. For example, simply turning off a light when you leave a room is an easy way to save electricity. Other conservation methods have been more difficult to develop. Engineers and technologists have been trying to make gasoline engines more efficient for many years. In this activity, you will have a choice of ways to experiment with energy conservation.

Design Brief:

Design, build, and test a method for conserving energy. Develop a written plan of action that will help others understand how your research has helped to save energy. You have a choice of many different ways to demonstrate energy conservation. Here are a few:

▶ Investigate your house or apartment to point out wasted energy and suggest a solution.

▶ Test the effectiveness of different types of insulation used in house construction. Make recommendations based on your tests.

▶ Build or test a gasoline-powered vehicle for efficiency. Your vehicle must be designed for safety and energy efficiency with the help of your teacher.

▶ Design your own experiment that demonstrates conservation of energy. Ask your teacher for help in making a procedure for your idea.

Encourage students to think of their own experiment. Be sure they plan it with safety in mind.

THINGS TO SEE AND DO:
House Detective

Introduction:
In this activity, you will investigate your house or apartment for air leaks called **infiltration**. In cold climates, cool air infiltrating from outside can add a great deal to the heating bill. In warm climates, hot air coming in will add to the cooling bill. The places where air enters can be found and fixed easily.

Materials:
- Graph paper
- Weather stripping (optional)

Equipment:
- Thermometer
- Flashlight
- Screwdriver

A class project might entail weatherproofing the home of an elderly or physically challenged person.

Procedure:
1. Make a sketch of the floor plan of your house or apartment. Use graph paper with ¼″ squares. Use a scale of ¼″ = 1′. Use the architectural symbols from Chapter 10 to mark walls, windows, and doors.

2. You will be checking your home for infiltration. Infiltration is air leaking into your home through cracks and spaces around doors or windows. Turn off the lights at night. Ask someone in your house to shine a flashlight from outside around exterior (outside) doors and windows while you watch from the inside.

3. Mark any areas of light infiltration on your floor plan. Correct the problem by adjusting doors, using weather stripping, or installing storm windows if appropriate.

4. Measure the temperature inside and outside. Record the temperature difference.

5. Write (or use a word processor) a paragraph that explains your investigation and the steps you took to fix any problems.

Evaluation:
1. Did you find any areas of infiltration in your house? If you did, how did you correct the problem?

2. Find out what type of heat your house or apartment has. What does it cost to heat or cool your home each month? How could the heating bill be reduced?

3. What is the maximum temperature difference anyone in your house can remember between the inside and outside of your home?

Challenges:

1 With your teacher's help, organize a group to test and weatherize the homes or apartments of poor or elderly people in your neighborhood. Ask local building suppliers if they would either donate the materials needed or sell them at a discount to the disadvantaged.

2 Research how electronics will help conserve energy in the smart houses of the future.

THINGS TO SEE AND DO: *Investigating Insulation*

Introduction:

In this activity, you will test the effectiveness of different types of insulation used in building construction. Insulation is used to slow down the movement of heat through walls and roofs. The material used depends on many factors, such as cost, fire resistance, building codes, water resistance, and insulation value required.

Materials:

▶ Fiberglas, Styrofoam, polyurethane, or other types of insulating materials
▶ Gypsum wallboard
▶ Duct tape
▶ Graph paper

Equipment:

▶ Thermometers (2)
▶ 100-watt lamp
▶ Light fixture with clamp or stand
▶ Clock or timer
▶ Utility knife
▶ Framing square

Caution students to handle the insulating materials with care.

Procedure:

1 Work in groups of three or four. With the help of your teacher, gather the materials and equipment listed above. The experiment setup will look like this:

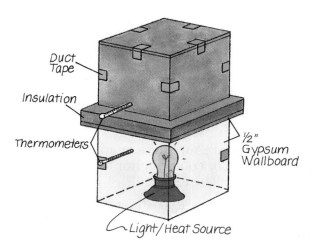

Duct Tape

Insulation

Thermometers

½" Gypsum Wallboard

Light/Heat Source

Set up two boxes of gypsum wallboard for the insulation efficiency test. Be sure that you don't leave the light bulb on when the test is over. The thermometers will tell you how much heat is coming through the insulation.

❷ With your teacher's help, cut eight pieces of ½" gypsum wallboard into 15" squares to make the walls of your experiment. Tape the sides of the "walls" together as shown in the drawing using duct tape. Cut two pieces of gypsum wallboard 16" square to make the ceilings in your test.

❸ Cut the pieces of insulating materials you will be testing 16" square, and make sure they are all the same thickness so that the test will be fair.

❹ Put the first test material into the tester. Poke a hole for the thermometer in the bottom and top of the test boxes as shown. Take the temperature of each side of the insulation. Turn on the light, and time how long it takes for the temperature to rise in each test chamber. Record the temperature every 10 minutes.

SAFETY NOTE:
Be sure to place a piece of gypsum wallboard between the lamp and the insulating material being tested. Some insulation is flammable. Do not leave the light on if you are not watching the test. Do not burn your skin on the hot light bulb.

Make a chart of the time and temperature for each insulation material.

❺ Write (or use a word processor) a paragraph that explains your experiment and the results you obtained.

Warn students not to leave the experiment unattended.

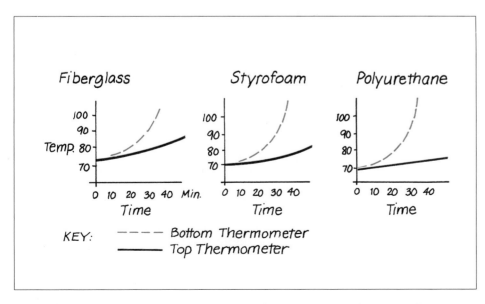

These graphs show how the temperature of the bottom thermometer went up rapidly as the bottom box heated up. The heat transferred into the top box had to go through the insulation sample. The flatter the line, the better the insulation.

Evaluation:

1 Which material was the best insulator?

2 Can you think of any advantages or disadvantages of the materials tested?

Challenges:

1 Research the measurement of insulation efficiency. What is meant by the R value?

2 Ask your teacher to show you plans for a super-insulated house. How are super-insulated houses built differently from normal houses? How long do you think it would take to pay for the additional cost of materials by saving energy?

THINGS TO SEE AND DO:
Energy-Efficient Car

Introduction:

There are millions of cars in use every day all over the world. Just think about the fact that we could save tremendous amounts of energy if car engines could be made just a little more efficient. For instance, if we could save just one gallon of gas per car every day, we would save billions of gallons of gas each year.

Test vehicles for this activity are available from technology suppliers. See Teacher's Resource Guide.

Materials:
- Different types of fuel

Equipment:
- Graduated cylinder
- Test vehicle
- Safety helmet

Procedure:

1 With the help of your teacher, discuss the possibility of purchasing or building a small energy-efficienct vehicle that could be used many times to test fuel consumption (use). You might be able to raise the cost of the vehicle by mass producing a product as you did in the activity in Chapter 9. Maybe a high school or vocational school technology class could help build your test vehicle.

2 Measure a course on the school grounds to use to test your vehicle. Your teacher will help decide what areas are safe. Remember, this is not a speed test. You will be trying to go as far as possible on 100 cc of fuel.

Instruct students on safe handling of flammable liquids.

3 Prepare your vehicle for the test. You might want to see if different factors influence the gas mileage of the test vehicle. Some factors to consider include:

- Weight of passenger and luggage
- Tire inflation pressure

- Wheel-bearing lubrication
- Engine-oil viscosity (thickness)
- Aerodynamic shape of the car (drag force)
- Driving technique (fast starts or slow-even style)
- Alternative fuels (alcohol, gasohol, propane)

4 Design your experiment to compare some of the factors listed. Carefully measure the fuel each time to make a fair test. Take turns testing your vehicle on the track you laid out.

SAFETY NOTE:
Even though you will not be traveling very fast, it is important to wear a safety helmet. Keep skin and clothes away from hot mufflers or exhaust pipes. Do not start the engine indoors. Exhaust fumes can be deadly. Be careful measuring and pouring the fuel. Keep all flammable liquids in the proper container and away from heat, sparks, electrical power tools, or open flames.

The energy efficiency of gasoline-powered vehicles can be improved through research and experimentation. Technology has helped to create the energy problems we face today. Careful use of technology can also help solve the problems. (Courtesy of Ford Motor Company.)

5 Write (or use a word processor) a paragraph that explains the test you did and the results.

Evaluation:

1 What can a driver do to improve the efficiency of a normal car?

2 List three ways that energy is wasted in normal driving.

3 Do you think you will drive an electric car some day?

TECHNOFACT

Technofact 73

How important is the wrapping on a candy bar to you? About 50 percent of the nation's paper, 8 percent of the steel, 75 percent of the glass, and 30 percent of the plastics produced today are used only to wrap products and make them look great. Then most of it is thrown away!

Techno Talk
Candy Bar Wrappers

Challenges:

❶ With the help of your teacher, convert an old riding lawn mower to run on electricity. Use a car starter motor to replace the gas engine. Provide a safe battery carrier. Test your design, and determine the advantages and disadvantages of electric-powered vehicles.

❷ Find out when your car was last tuned or when the tires were checked. Find out which service stations in your area sell gasoline mixed with alcohol. Ask a mechanic if there is a problem with burning alcohol fuels in cars.

Besides trying to conserve energy sources, we are looking at **alternative energy** sources. We call these *renewable resources* because we will not run out of them. They are constantly renewed through natural processes caused by the sun's energy. Can you name some alternative energy sources?

What Is Alternative Energy?

Alternative energy sources are important to you. They may replace or be added to nonrenewable energy supplies to do work for us. Renewable energy sources will provide energy with far less damage to the environment than nuclear or fossil fuel sources. They do not produce much waste or pollution. Solar energy, wind energy, biomass, tidal energy, and geothermal energy are examples of alternative energy sources that researchers are studying. We need to find ways to use them.

Solar Energy: The sun provides the earth with lots of energy, some of which can be used for heating purposes and to produce electricity. **Solar cells** or **photovoltaic cells** make electricity directly from sunlight. Solar cells were developed to use on satellites in the 1950s, and they were very expensive. Do you have a solar-powered calculator? Solar cells are now much cheaper to make, and calculators that contain them are inexpensive and powerful. Another direct use of solar energy is heating a home using solar collectors to heat water. Active solar heating is much more effective in sunny climates. Some solar collectors produce temperatures high enough to be used in industry and for generating electricity.

Solar energy can be used in many ways. These buildings have been constructed to use the sun's energy for heating. (Courtesy of United States Department of Energy.)

Wind Energy: Wind is one of the most promising alternative energy sources. Many countries, especially those that get a lot of wind, are developing wind power technology. The most important use of wind energy is to produce electricity. Wind energy can generate electricity at the same price as fossil fuels and nuclear power, but it is safer and doesn't cause pollution. The wind produces electricity by turning a turbine shaft that is hooked to a generator. There are more than 20,000 wind turbines producing electricity around the world. A wind-driven turbine depends on a steady supply of wind averaging 10 mph or more. Medium-sized wind-driven turbines have been the most efficient so far. There are "wind farms," or collections of wind generators, in California that have the power of two nuclear power plants but cost half as much as conventional power stations. In some places, batteries are used to store the energy produced for times when the wind isn't blowing.

Biomass: **Biomass** is living or dead plant or animal matter. The energy in biomass can be released and used in many different ways. Its main sources are wood, crops, animal wastes, and organic materials found in garbage. Garbage can be burned to produce lots of heat. Almost half of the world's population depends on biomass to supply energy for cooking, heating, and light. Most of the poorer nations get much of their energy from wood or from animal manure when wood is scarce or too expensive. Biomass can be used to produce biofuels such as methane, methanol, and ethanol. There are processes that can change garbage into petroleum.

TECHNOFACT

Technofact 74

Almost half of the electricity produced at a power plant is wasted as it is moved from the plant to someplace where you can use it. Experts say that if we did everything right and used the most efficient machines possible today, we could save as much as 44 percent more than we save today. The system still doesn't seem very efficient, does it?

"Wind farms" provide an inexpensive but plentiful energy source. (Courtesy of United States Department of Energy.)

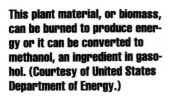

This plant material, or biomass, can be burned to produce energy or it can be converted to methanol, an ingredient in gasohol. (Courtesy of United States Department of Energy.)

Tidal Energy: Tides and waves have mechanical energy that can be changed to a form you can use. Turbines like those used in hydroelectric power stations can produce electricity from the rising and falling tides. When trapped tidal water is released, it makes electric power. The energy available depends on the difference between the heights of the high and low tides. Special generators can change wave motion into electric power. This source of energy has lots of potential, but scientists have some problems like high waves and strong winds to figure out first.

(Courtesy of Maine Office of Tourism/Joseph Devenney.)

Geothermal Energy: **Geothermal** energy is heat from beneath the earth's crust. When it is brought up to the surface as steam or hot water, it can be used directly to heat water or to drive generators or steam turbines. The geothermal resources in the upper three miles of the earth's crust are estimated to be more than all the world's natural gas and crude-oil reserves. Right now we have used only a small percentage of this energy resource.

You can see that many energy resources are available to us. We will have to start using some alternative energy supplies because the nonrenewable ones are running out. Could you live with less energy? Most of us can, because we waste a lot of energy. We need to find more efficient ways to use and conserve our energies and, at the same time, to keep looking for new energy supplies.

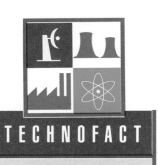

TECHNOFACT

Technofact 75

Even kindergartners can invent energy saving devices. Kacy White of Kentucky came up with the idea of an "afterglow" light bulb. It works by storing energy while it is turned on. Then when the light is turned off, the energy is released, and the bulb continues to burn or glow for another minute. That way you have time to get safely under the covers before the light goes out totally!

The natural geothermal energy of the Earth offers an enormous alternative energy source. (Courtesy of Coldwater Creek Operator Corporation.)

THINGS TO SEE AND DO:
Use Sunlight to Charge a Battery

Techno Teasers
Energy Alternatives

Techno Teasers
Answer Segment

Introduction:

We owe almost all of our energy to the sun. You know that even the gasoline we use in cars can be traced to the decay of vegetation millions of years ago. The decomposed vegetation would not have grown without sunlight. Unfortunately, we can't wait millions of years for the sun to make more oil and gas. Wouldn't it be great if we could change sunlight directly into usable energy? Well, we can, thanks to a thin slice of silicon called a photovoltaic cell. A more common name for this device is a solar cell. In this activity, you will build a solar battery charger. This simple electrical circuit will change sunlight directly into electricity that you can use while the sun shines, or store it in a battery for use at night or on cloudy days.

Design Brief:

Design, build, and test a photovoltaic battery charger. Your charger should produce approximately 1.5 volts DC. It must be able to charge a ni-cad battery.

Individual solar cells are available from technology and science suppliers. See Teacher's Resource Guide.

Materials:

▶ Silicon solar cells

▶ Silicon diode

▶ Hookup wire, rubber cement

▶ AA ni-cad battery, AA battery holder

▶ 1⁄16″ acrylic plastic (Plexiglas)

▶ Aluminum foil

▶ 1½ volt DC motor, 1⁄16″ steel welding rod, gears, belts, pulleys, from machine dissection activity in Chapter 6 (optional)

Equipment:

▶ Electronic soldering pencil

▶ Band or scroll saw

▶ Drill press

▶ Strip heater (optional)

▶ Digital multimeter

Procedure:

1 In this activity, you can work individually or in a group to make a solar battery charger. As a challenge, you might think about making your charger work as a solar car. You will need to use the problem-solving steps you learned in Chapter 1 to solve the problems of building a solar-powered car. Can you list the problem-solving steps in order?

2 A solar collector must be pointed at the sun to be able to produce electricity. You will need a way to adjust the collector so it will gather enough light to charge a battery or to run your solar car. You may also need to help gather sunlight by using reflectors to direct the sunlight to the solar cells. Brainstorm ideas for how you will meet this need. Some possibilities are illustrated for you.

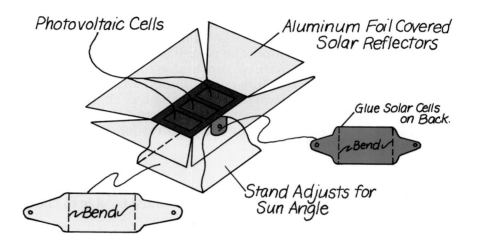

Photovoltaic Cells   *Aluminum Foil Covered Solar Reflectors*

Glue Solar Cells on Back.

Bend

Bend

Stand Adjusts for Sun Angle

The stand for your solar array needs to be adjustable so that it can be pointed toward the sun to gather the maximum amount of light. Reflectors help to gather the light and direct it toward the cells. Your design might be different if you are going to try to make a solar-powered car also.

❸ Refine your design and start construction. Think about the goals of your solar collector. It must:
- be adjustable,
- gather enough sunlight to operate at 1.5 volts or more, and
- be lightweight for a solar-powered car.

SAFETY NOTE:
Wear eye protection. Follow the general safety rules and specific rules for the machines you are using. Remember to ask your teacher for help.

Students will need assistance in soldering leads to the solar cell to prevent breakage.

❹ With the help of your teacher, solder the hookup wires to the solar cells according to the directions on the package. Handle the solar cells carefully. They are very thin and fragile. (They are also expensive.) Solder the hookup wires together to make a series circuit called a **solar array**. This arrangement will add the voltages of all the solar cells together. Solder the positive lead to a diode before it is connected to the battery holder. The diode will let electrons flow in only one direction.

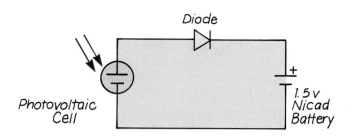

Diode

Photovoltaic
Cell

+

1.5 v
Nicad
Battery

Schematic diagrams use symbols to represent real electronic components. This schematic shows how to wire your photovoltaic array to charge a battery.

⑤ Carefully glue the solar cells to your collector using rubber cement. Do not press hard on the solar cells. They will crack easily if handled roughly.

⑥ Test your solar generator by pointing it toward the sun. Connect the leads of a digital multimeter to the positive (+) and negative (−) terminals of the battery holder. If you get a negative reading on the multimeter, you must reverse the battery holder. It is important that the polarity (+ or −) be correct for the charger to work.

A sun lamp can substitute for light on cloudy days. Warn students not to get too close to a light.

Evaluation:

❶ How many volts of electricity did your solar charger produce?

❷ How did you use reflectors to trap more light? Measure the voltage output with and without the reflectors. What is the difference?

❸ Why don't we glue photovoltaic cells to the roofs of houses to produce electricity?

❹ Can you think of a device you use that is powered by solar cells? Do you think other electronic products will have solar cells in the future? Explain.

❺ Why are solar cells used so often to make electricity for space satellites?

Challenge Activity:

The challenge for this activity is to use the solar collector as a part of a solar-powered car. You might decide to team up with another group to solve this problem. You will need to remember what you learned about mechanical systems in Chapter 6 to connect a motor to the wheel or axle of your car. If your first attempt does not make the car move, can you think of how you could connect two or more solar collectors together to get more power? If your class is successful in making more than one solar-powered car, set up a race to see which car goes the fastest or the farthest.

Challenges:

❶ Research the development of photovoltaic cells. What has happened to the efficiency and cost of producing solar cells? Where are they commonly used today?

· · · · · · · · · · · · · · · · · · · ·
Show a video on electric cars.
· · · · · · · · · · · · · · · · · · · ·

Special Report
The Sunraycer

TECHNOFACT

Technofact 76

Are you ready for an illuminating fact? Can you imagine how changing the size of a light used in a car's headlights by an inch would result in a totally new shape for cars of the future? The new light bulb that will be used is far more durable and energy-efficient than the bulbs used now. It doesn't use a heated filament. It uses an electronically controlled spark to heat gases. The smaller bulbs produce more light and less heat. Since the bulb itself is cooler, the rest of the assembly that holds the light can be made from plastics. The light assembly can be smaller and lighter. That means cars can be more streamlined, or aerodynamic, and therefore have improved gas mileage.

❷ How many volts could you produce if you connected all the solar collectors in your class together? What about connecting the solar collectors from your class to those of another class? Measure the total voltage, and with the help of your teacher see if you could power a portable radio or tape player.

❸ Research the solar-powered car made by General Motors called the Sunraycer. How is this car designed differently from an ordinary car? Do you think you will be able to buy a solar-powered car some day? Explain.

❹ With your teacher's help, contact a glass shop and ask them to save small pieces of mirror for a class project. Tape the edges of the mirrors to prevent cuts. Have each person in the class design and build an adjustable stand to hold the mirrors. Set all of the mirrors in the sun so that each one reflects sunlight toward a solar cell array. How does the voltage change when all the mirrors are adjusted properly? Calculate which would be most cost-effective, a large number of solar cells or a few solar cells and many mirrors.

SAFETY NOTE:
Caution: do not look directly into the reflections of all the mirrors.

Summary

Energy is very important to us. We depend on it for many purposes, from heating our homes to running our computers. Energy is the ability to do work. Sometimes we change energy from one form to another, more useful form to do work for us.

Energy is either resting or moving. Potential energy is energy at rest waiting to do work. Kinetic energy is energy of motion. It is doing work.

Most of the earth's energy comes from the sun. Some of the conventional energy supplies used in developed countries are electricity, fossil fuels, nuclear energy, and hydroelectric power.

One of our biggest problems today is to find ways to use energy more efficiently and to better conserve our energy resources. Some of our energy supplies, like fossil fuels, are nonrenewable. We are running out of them. We need to change products so that they waste less energy. Technology is helping find ways to make machines and other products more efficient so they don't use or waste so much energy.

Alternative energy sources are starting to be developed to replace the nonrenewable energy supplies. Solar energy, wind energy, biomass, tidal energy, and geothermal energy are some alternative energy sources that hold promise for the future.

Challengers:

1 Analyze your school for energy efficiency. Make a list of ways you could conserve energy.

2 Research a country in Africa or Asia. Find out what energy resources the people use most.

3 Make a list of all the electrical devices you use at home. Find out from your power company how much electricity each appliance uses per hour. Figure out how much electricity you use in your home in a 24-hour period.

4 Visit an electrical power plant. Ask what kinds of jobs people do and what skills they need for those jobs.

5 Choose a form of energy, and tell how it can be changed to another form of energy to do a specific job for you.

6 Research and list the advantages and disadvantages of alternative energy sources such as solar, wind, biomass, tidal, and geothermal energies.

7 Make a map of the United States showing where the alternative energy sources could best be used.

8 Research how nuclear power is made. Make a chart or diagram to illustrate your information. Write for information about the disadvantages and advantages of nuclear power as known today.

9 Human energy comes from the food you eat. List the food nutrients and how they are used in your body. Keep track for a week of everything you eat to see if you are using food energy efficiently.

10 Write a paragraph about how your life would be different if you lived in a world that did not have electricity.

See Teacher's Resource Guide.

Techno Teasers
Nuclear Power Plant Tour

Techno Teasers
Answer Segment

Chapter 12

Moving Things

Things to Explore

When you finish this chapter, you will know that:

▶ Transportation is the movement of goods or people from one place to another.

▶ Modes of transportation include airplanes, trucks, buses, conveyors, pipelines, trains, boats, and hovercraft.

▶ Composite materials are making transportation vehicles lighter and stronger.

▶ The transportation industry uses computer technology to make the movement of goods and people faster and safer.

▶ Future transportation modes might include faster and more efficient mass transport of people on maglev trains.

TechnoTerms

aerodynamic	hydrofoil
aileron	leading edge
airfoil	lift
angle of attack	maglev
barge	pitch
blimp	rudder
chord	symmetrical
dirigible	trailing edge
drag	transportation
elevator	vessel
hovercraft	yaw

(Courtesy of Pratt & Whitney.)

Careers in Technology

Have you ever flown in an airplane? It's hard to believe that the first airplane was invented less than a hundred years ago. The transportation industry employs many thousands of people who transport people and materials on land, air, water, and even in space. We often forget how products get from factories to stores or how thousands of people are moved through the air every hour of the day. What part of the transportation industry do you think you would find interesting as a future career?

Have students brainstorm various transportation modes.

TECHNOFACT

Technofact 77

One plan to reduce the number of landings and takeoffs at airports was to have passengers parachute from the plane just when the plane was over their destination. Don't you wonder what happened when they tried this one out?

Special Report
Introduction to Transportation

What Is Transportation?

Take a quick look around your classroom. Everything you see had to be transported to your room. **Transportation** is the movement of people or goods from one place to another. Transportation is a big industry that affects you in many ways.

Do you own a bicycle or a motorcycle? Does anyone in your house depend on a car or bus to get to work? How does luggage get from the airplane into the main airport terminal? How did the food you ate for breakfast get to the grocery store? Transportation is very important to your way of life. You depend on transporation such as trains, ships, automobiles, airplanes, or buses to get you where you want to go. Almost everything you use, from the food you eat to the chair you are sitting in, is there for you because of transportation systems.

A

Transportation technology helps you in many ways every day. Can you describe these modes of transportation? Think how the products you wear or eat get to you. (Figure A Courtesy of Grumman Corporation. Figures B and C Courtesy of United States Navy. Figure D Courtesy of French Government Tourist Office.)

B

C

D

Today's transportation technology is changing rapidly. The first airplane in 1903 could go only 9.8 miles per hour at top speed. Now we have the *Concorde* jet, which can fly faster than 1,200 miles per hour. Then, to top that, people travel faster than 18,000 miles per hour in space. Most transportation systems today use a lot of computer technology to control how an engine runs or the fuel-air mixture. Computers also control lights, door locks, and braking systems in many cars, trains, and airplanes. Even farms use mechanized transportation technologies today. Can you imagine how transportation systems will change in the next 100 years?

A

B

C

Air transportation technology has developed very fast. The Wright brothers' first flight flew less than seven miles per hour. Only 70 years later (one person's lifetime) the *Concorde* routinely flies at over 1,200 miles per hour. Spacecraft can travel over 18,000 miles per hour. What do you think will come next? (Figure A Courtesy of the Smithsonian Institution. Figure B Courtesy of British Airways. Figure C Courtesy of NASA.)

Industry and business depend on transportation systems to move goods. Raw materials are transported to a factory. Materials, parts, and finished products are then transported on conveyor belts through assembly lines. Finally the products are moved from the factory to places where we can purchase and use them. We want the fastest, most economical ways to transport products so that we don't waste products, time, or money.

There are different ways, or **modes,** of transporting, or moving, materials. The kind of transportation you use depends on whether you

want to move people or products on land, water, air, or in space. As you explore the different types of transportation, think about which ones you use a lot. Does where you live make a difference in the transportation systems you use most?

A

B

Transportation modes vary greatly. (Figure A Courtesy of USX Corporation. Figure B Courtesy of Raytheon Company. Figure C Courtesy of Pennzoil Company.)

C

Ask students if they know of underground utilities such as natural gas or electricity in their neighborhoods.

Hovercraft have an advantage in that they can travel over land or water. Hovercraft can carry heavy loads and people at high speeds. (Courtesy of United States Navy.)

Land Transportation

When you think of land transportation, do you usually think of some kind of vehicle such as a car, a train, or a truck? Actually any transportation that moves on or beneath the earth's surface is a form of land transportation. Land transportations such as railroads, pipelines, conveyors, moving sidewalks, escalators, and elevators move products or people from one point to another point.

Other land transportation vehicles, such as cars, buses, trucks, motorcycles, bicycles, or forklifts, might move along highways, streets, expressways, or even inside buildings. The hovercraft is a combination land and water vehicle which means it can travel on land or water.

Automobiles: Did you know that in the United States more than 120 million cars travel over thousands of miles of highways and streets? The automobile has become a necessity for many Americans. We use it to get to work or school every day. That has not always been the case. Before the automobile was invented, people walked, biked, or used animals to get places. Of course, most of the time they did not go very far from home. People usually worked close to their homes and sometimes even lived in the buildings they worked in. They really didn't need cars.

As automobiles were developed, people could travel greater distances in shorter times. We started to depend on the car to get us where we wanted to go. Do you know how fast the first car could go? The first car was built in 1771 in France. It used a steam engine invented by Nicolas Cugnot. It could go only 2.3 miles per hour (mph). You could actually walk faster than that car could move. The first gasoline automobile didn't show up until 1889. It was built by the Germans Gottlieb Daimler and Wilhelm Maybach. It could travel between 3 and 10 mph! Engines soon became more powerful. But building cars remained a slow process, and only the rich could afford them. Henry Ford's development of mass production of cars in 1913 made it possible to produce more cars faster. That made cars less expensive so more people could buy them.

The airplane and the automobile have developed steadily since the beginning of the 1900s. (Courtesy of Ford Motor Company.)

Today's cars are technological wonders. Almost all the cars made in the world today have on-board computers. The smart car has computers that use special sensors to help the different mechanical systems work right. Some cars have a computer that lets you know how far the car can travel before it runs out of gasoline or even your time of arrival at a certain place. Some computerized navigation systems in cars show local maps and even indicate the best **route** (path) to get to your destination.

Trucks: The trucking industry is big business in the United States. Have you ever noticed how many different trucking companies there are? There are over 40 million trucks carrying goods across the United States. Today, trucks might be owned by a company or by an independent owner-operator who usually has one truck. The big advantage to using truck transport is that goods can go directly from the producer of a product to where the product is going to be sold.

There are many kinds of trucks, such as dump trucks, delivery trucks, cement trucks, and garbage trucks. On our highways, most trucks are tractor-trailers. These trucks have a large tractor, which is the power plant, and one or more trailers, which carry the freight, or cargo. Tractor-trailers use powerful diesel engines that can usually go long distances before needing major repairs.

While Americans take automobiles for granted, residents of Tokyo must prove they have a parking place before they can even purchase a car.

Challenge students to research Henry Ford's production techniques.

T E C H N O F A C T

Technofact 78

How about a trailer that's also a houseboat? An Australian created the *Cummins Craft*, which is 22-feet long and 8-feet wide in the water. The craft sleeps six and has a stove, oven, refrigerator, shower, toilet, dining table, and even some storage space. Powered by an outboard motor, it cruises at about 8 knots. The neat part is that you can fold it up to 13 feet and take it on the road. It's easily towed behind a car. Even then you can use it to sleep as well as use the toilet and cooking areas! That's an efficient transportation system!

Ask a truck driver to bring a tractor-trailer to school so the students can appreciate its size and hauling capacity.

Many kinds of trucks are used to transport goods to stores or directly to you. Trucks also remove the garbage we make after using the products. (Courtesy of GMC Truck.)

Take a poll in the class. See how many students have taken a trip on a passenger train.

Challenge students to compare the energy usage of transporting goods by train versus other modes of transportation.

Have you heard the term **fifth wheel**? The fifth wheel is really a large, disk-shaped hitch that hooks the trailer to the tractor. Trailers today are made to carry special products, from refrigerated goods to melted chocolate! Sometimes trailers are **piggybacked** and carried on railroad flatcars to a location. Then a truck picks them up and takes them to their final destination.

Trains: Trains have been used for many years to move people and products. The first steam locomotive was the Trevithick, developed in England in the early 1800s. Its top speed was 13 mph when loaded. Over the next 80 years, many improvements in steam locomotives made trains that could carry more passengers and go faster.

In the United States, trains and railroads provided the fastest and safest way to travel before the automobile was invented. Trains were important in helping pioneers settle the interior of our country. They helped farmers by giving them a way to get their products from the farm to markets in larger cities.

In Japan and France, for example, passenger trains are still one of the most common transportation systems. Many of these passenger trains travel at speeds over 125 mph. Trains in the United States have lost riders to buses because buses are cheaper and to airplanes because they are faster. **AMTRAK** (*Am*erican *Tr*avel Tr*ack*) system provides all the long-distance rail service in the United States. It is owned by the federal government. The AMTRAK Superliner is a passenger train made especially for long-distance travel. It is designed to be a "rolling hotel" with passenger car bedrooms, a snack bar, and a diner. Each room has individual controls for heat, air conditioning, and music. This train brought more people back to train travel. The development of maglev trains, which don't use fossil fuels and can travel over 300 mph, will make train travel fast and efficient in the future.

In the United States, railroads are used mainly to move freight. Over two million freight cars carry goods such as oil, steel, automobiles, farm machinery, and home appliances across North America. The big advantage of trains is that they can move large loads over long distances economically and efficiently. Most locomotives today are powered by diesel-electric power or gasoline turbine engines like those used in airplanes.

Trains are efficient movers of people and products. Next time you have to wait for a train to cross a road, think about all of the products, materials, or people that are being transported. (Courtesy of AMTRAK.)

To make the freight move efficiently, computer systems are used to keep track of every freight car in the rail network. The computer system uses a database to show what each car is carrying and where it will be loaded and unloaded. A special system called **TRAIN** (*Telerail Automated Information Network*) keeps a step-by-step trail on a customer's cargo no matter where it goes by rail.

THINGS TO SEE AND DO:
Maglev Trains

Introduction:

Maglev is short for magnetic levitation, a fancy way of saying floating on magnets. Experiments are being conducted all over the world with ways to make trains that "float" on a magnetic field instead of rolling on wheels. Maglev vehicles can travel much faster than conventional trains. Speeds of over 300 mph are not uncommon. In this activity, you will experiment with designing and building a maglev train track.

Maglev is short for magnetic levitation. These experimental trains ride on a cushion of a magnetic field. Maglev trains can travel at speeds of 300 miles per hour of more. (Courtesy of Transrapid International.)

Design Brief:

Design, build, and test a maglev train simulation. Your train must levitate on a magnetic field. The train must move on the track, using a safe propulsion method.

Materials:

- Permanent magnets
- Styrofoam
- Acrylic plastic (Plexiglas)
- 2′ × 2′ × ½″ plywood sections (1 per group)
- Other, depending on design

Equipment:

- Aluminum angle
- Hot glue gun
- Scroll or band saw
- Hot wire cutter
- Plastic strip heater

Inexpensive permanent magnets are available from science and technology suppliers. Caution students to keep magnets away from computer disks or other magnetic media. The magnets are also somewhat brittle and easily broken.

Procedure:

❶ In this activity, the class will be divided into four groups. Each group will make a section of maglev train track and a train simulation. The sections will fit together to make a complete track.

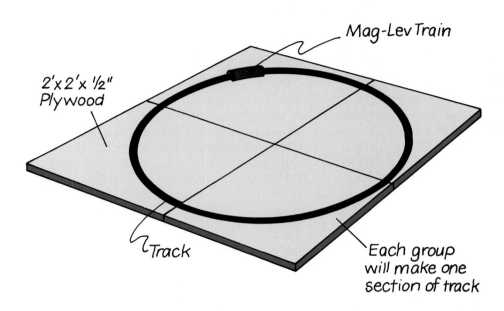

Mag-Lev Train

2'x 2'x ½"
Plywood

Track

Each group
will make one
section of track

This would be a good point to quiz students on the steps in problem solving.

❷ Each group should brainstorm ideas for a track design. Experimentation will be necessary to decide which design will work best for all four groups. Remember the steps in problem solving. The steps your group takes to solve this problem might look like this:

- Define the problem (see design brief).
- Gather ideas. Brainstorm with your group, ask your teacher, consult with experts, do library research, and so on.
- Try your best guess. Analyze the research, and pick out the idea you think is best.
- Test your idea. Experiment using magnets and other materials needed.
- Evaluate the results of your test. Did it work? How could it be improved?
- Retry after making modifications. Change the design to make it work the way you would like it to.

❸ Your design must consider the following specifications:

- The propulsion method must be safe for indoor use.
- The track should be made of nonmagnetic materials such as plastic, wood, or aluminum.

- The opposite poles of magnets attract each other. Like poles repel each other.
- Your teacher may add other design specifications to this list.

Some designs work better than others. Your group should try an idea you think will work. Here are some designs you might consider.

Pre-made maglev simulator tracks are available from science and technology suppliers.

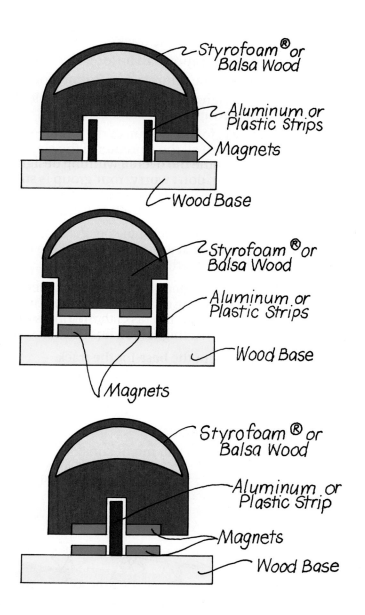

Styrofoam® or Balsa Wood

Aluminum or Plastic Strips

Magnets

Wood Base

Styrofoam® or Balsa Wood

Aluminum or Plastic Strips

Wood Base

Magnets

Styrofoam® or Balsa Wood

Aluminum or Plastic Strip

Magnets

Wood Base

❹ Build a test section of your track 2 feet long. Test a train design on your track to see if it will levitate. Experiment with a propulsion system.

❺ Each group will present its idea for the track design to the class. Evaluate each design as if you were a city council member deciding on spending millions of tax dollars on a mass transit maglev train for your city. Give a maximum of ten points for each of the following questions:

• Does the train work on curved sections of track?

• Is the track cost-effective? How many magnets per foot are required for the track?

• Does the design appear to be long-lasting?

• Could the idea be improved, or is it unworkable?

• Does the design team seem knowledgeable? Will they help other groups understand the design?

❻ Total the points for each team to determine your choice. Total the points of all the students in the class for each group to determine the overall winning design. If your group didn't win, don't worry. Your group is still in the running for the best train design. Each group will design and make a train to go on the class track.

❼ The class should make a list of specifications and make drawings so that each group will be making track sections that will fit together.

❽ Divide the tasks needed to complete your section of track and the building of a train simulator. Make a list of all of the tasks. Assign the tasks to members of your group. Your task list might look like this:

• Measure and cut the base for the track.

• Measure and cut the alignment rails according to the track specifications.

• Glue the magnets to the track with the proper polarity.

• Make the train simulator.

Here are some ideas you might consider for your train.

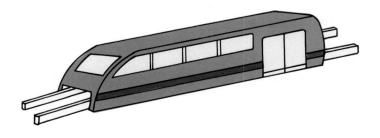

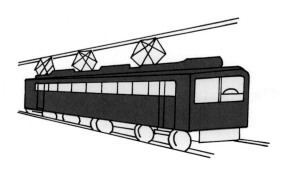

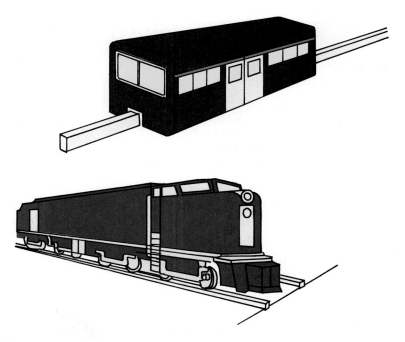

9 Start the construction of each track section.

SAFETY NOTE:
Remember to follow the general and specific safety rules for working in the technology lab. Keep magnets away from computer disks, video tapes, or any other magnetic information storage. Magnets will erase the information. Wear eye protection at all times while working around machines.

10 Assemble the sections of track together to make a complete track. Each group will have three chances to have its train make it around the track. Time each of the tests and take an average. The group with the best average time wins!

Differences in track size may require adjustments. Discuss real-world situations where parts made by different companies must fit together.

Evaluation:

1 Why do you think the winning train design went the fastest?

2 How could your train design be improved?

3 What do you think it would be like to ride on a maglev train? Explain.

4 Can you list three advantages and three disadvantages of maglev trains compared with conventional trains?

Challenges:

1 Design a maglev train that uses superconducting magnets. How could the superconductors be kept cold?

A

B

C

Trains aren't the only modes of rail transportation. Can you identify a streetcar, subway, and monorail? (Figure A Courtesy of San Francisco Convention & Visitors' Bureau. Figure B Courtesy of Busch Gardens, Tampa 1991. Figure C Courtesy of San Francisco Bay Area Rapid Transit (BART) District.)

❷ Research the current designs being considered for actual maglev trains. Make a bulletin board display of your findings.

❸ Think of a name and logo for your maglev train. Use computer graphics software to make your design. Glue your finished design to your train.

❹ Design a way to make your track using electromagnets. Try to design a method of using electromagnets to propel the train.

Mass Transit Rail Vehicles: Many types of mass transit rail systems are used to move people today. They can be located underground, above-ground, or at ground level.

▶ Streetcars in San Francisco are ground-level transportation.

▶ Subways, found in large cities, are underground rail systems. Subways often use tunnels, which are expensive to build.

▶ Monorails are transit systems that are sometimes elevated. They use a single rail.

Other mass transit rail systems are totally automated and do not have a driver. These **automated transit systems** (ATS) are used at airports, remote parking areas, and shopping centers, for example.

Buses: There are many kinds of buses used to move large numbers of people around. City buses, school buses, and motor coaches are the most common ones. Motor coaches are usually large buses that provide long-distance transportation.

Most transportation systems use vehicles to carry people and products. Other transportation systems, such as pipelines and conveyor belts, do not use vehicles to move things.

Pipelines: Certain kinds of materials can be moved by pipeline very economically. Examples are natural gas, coal, grain, oil products, wood chips, and gravel. Most pipelines are one-way and are buried underground.

There are over 1 million miles of pipelines crossing Canada and the United States. Most of these carry natural gas or oil products. The Trans-Alaska pipeline moves crude oil from Prudoe Bay to Valdez. Can you find out how far that is?

The cost of building pipelines is very high. The amount of material they can move is also very high. Pipelines are used to move great quantities of oil and natural gas all over the country. (Courtesy of The Coastal Corporation.)

Conveyors: Many other land transportation systems use conveyor belts. One of the most popular is the "people mover," or moving sidewalk, found in major airports. Have you ever walked on one of these? They really help to move people along quickly over long distances, for example, from one airport gate to another.

Special Report
Transportation on Water

Water Transportation

Did you know that early people sewed animal hides together, filled them with air, and then used them to float on water? Using water for transportation is nothing new. Early people used floating logs. Later, after people developed tools, they hollowed out the log to make a dugout.

The first boats used muscle power only. People moved boats with poles, paddles, and oars on short trips close to shore. Many of the first large cities developed along waterways because it was easier to trade goods between towns that had access to a waterway than over land. In the United States, 19 of the 20 largest cities today are located on inland waterways or an ocean.

As people learned how to use wind power to move boats, or **vessels**, they could travel longer distances. Explorers from many countries especially Portugal, Spain, England, and France set out for new lands using two-masted or three-masted sailing ships.

Before the steam engine was invented, ships had to depend on wind, currents, or muscle power from people or other animals. Steam-powered ships could sail any time, whether the wind was blowing or not. Larger ships could be built and that meant larger cargoes could be carried. The *Clermont*, the first steam-powered ship to be used regularly, was built by Robert Fulton in 1807. After that, many other steamship designs were used to carry people and products.

Sailing ships were used by explorers to discover new lands. How are spacecraft of today similar to the sailing ships used by early explorers? (Courtesy of Greater Boston Convention & Visitors' Bureau.)

Challenge students to research the use of satellites in navigation.

Today's modern ships have lighter-weight steel hulls and more efficient, powerful engines. Some commercial ships have a double bottom, so the ship can safely carry liquid cargo or fuel for the ship engines. Modern ships are built in prefabricated sections and then are welded together at an assembly area.

Computers also play a big role in water transportation. Computers determine a vessel's position and plot a course for the ship. The computer can store a complete set of charts that take up less space than paper. The charts are easier to update using the computer. The captain can see easily where his ship is and what other ships nearby are doing. It is also easier to plan long-term voyages using the data from the computer.

Water transportation vessels have design features to fit their uses. One thing all water vessels have in common is that they are designed to be watertight!

Ocean Liners and Passenger Ships: Small ocean liners or cruise ships are still popular for vacations. Many have on-board swimming pools, stores, movie theaters, and exercise rooms to entertain people on long cruises. Since most people today want the fastest transportation possible, the number of passengers on ships has **declined**, or gone down.

Technology has helped to make modern sailboats more efficient and safe. The white "sails" of the *Alcyone* are visible here. The *Alcyone* is an ocean research vessel of the Cousteau Society. (Courtesy of The Cousteau Society.)

Ocean liners and passenger ships provide transport for thousands of people all over the world. (Courtesy of Royal Caribbean Cruise Lines.)

Hydrofoils skim over the water thanks to the lift created by the hydrofoil "legs." The hydrofoil is similar in shape to an airfoil on an airplane wing. (Courtesy of Boeing.)

Hydrofoils: Have you ever seen a boat that looks like it's flying? That's a hydrofoil. A **hydrofoil** is a passenger ship that moves above the surface of the water. A hydrofoil has wings called foils. When it is not moving, a hydrofoil rests in the water like any other ship. Once it reaches a certain speed the foils lift the ship out of the water. Hydrofoils can go very fast because there isn't much friction or resistance from the water.

Hovercrafts: Also known as **air cushion vehicles** (ACVs), **hover-crafts** ride on a cushion of air. High-speed fans driven by gas turbines push air under the boat. The air is trapped around the outside edges of the ship so that the vehicle is actually lifted above the surface of the water. You get a very fast, smooth ride. Hovercrafts can travel over ice, snow, land, or marshes, too!

Show a video of air-cushioned vehicles, hydrofoils, and other water transportation vehicles. Have students bring in toy hover-crafts to demonstrate.

Submarines: Submarines are ships that work either on or beneath the surface of the water. By changing the amount of air in their tanks, submarines can float at any level below or on the surface of the water. Submarines have to be very strong in order to take the tremendous pressure, which increases as they go deeper in the water. The newest submarines are powered by nuclear energy. They can stay underwater, or **submerged**, for long periods of time. Smaller robotic submersibles are being used to explore deep-water areas. They are controlled by computers on other ships.

Challenge students to research the development and use of submarines in history.

Cargo Ships: Most ocean-going ships in use today are cargo ships. There are several specialized kinds of cargo ships that carry products and people from one port to another. Some supertankers are over 900 feet long. That's about the size of three football fields! Supertankers can carry large amounts of oil but if they are damaged at sea, the resulting oil spills really affect the environment for a long time. Just ask the Alaskans about the 1989 *Exxon Valdez* accident that dumped 11 million gallons of crude oil into Prince William Sound. It was only one of thousands of oil spills a year all around the world. Can you imagine the damage done to birds, fish, sea otters, and the tiny marine animals that are part of the food chain by these oil spills?

Cargo ships transport millions of tons of products and raw materials over the oceans.
(Courtesy of New York State Department of Economic Development.)

Submarines can stay underwater for many days. They are a part of our national security. (Courtesy of United States Navy.)

***Barges*: Barges** have flat bottoms and blunt ends to haul cargo on inland waterways such as canals, rivers, and lakes. A barge might carry other transportation vehicles such as trucks or railroad cars. Because the barge has a flat bottom, it can operate in shallow water.

***Tugboats*:** Tugboats pull barges or ocean liners into and out of harbors. They need powerful engines in order to move such large ships.

Tugboats and barges are the workers of water transportation. Raw materials such as coal or oil can be moved easily on rivers, lakes, or oceans. (Courtesy of Garry Nelson.)

THINGS TO SEE AND DO:
Testing the Design of Boat Hulls

Introduction:

The amount of fuel that a large ship uses to transport materials or products is a major cost of water transportation. If the shape of the ship's hull can be made so that it would reduce the drag through the water, the ship could go faster and save fuel. The same is true of pleasure boats and speed boats. In this activity, you will build a water tunnel to test the shape of boat hulls.

Design Brief:

Design, build, and test a boat hull testing device.

Materials:

- Plastic rain gutter, end caps
- Clear acrylic (Plexiglas) plastic or polycarbonate
- $\frac{1}{16}''$ steel welding rod
- Garden hose, faucet-hose connection
- Drinking straws, rubber cement, rubberband
- $\frac{3}{4}''$ plywood, $2'' \times 4''$ blocks
- Silicone sealant, abrasive paper (60 grit sandpaper)

Equipment:

- Band or scroll saw
- Plastic strip heater
- Power hand drill
- Water faucet and sink
- Belt or disk sander

Procedure: Building a Water Tunnel

1 In this activity, you will work in a large group to build a water tunnel and individually to test a boat hull design. Work with your teacher to decide on a design for your water tunnel. Keep the following design specifications in mind while you design your water tunnel.

• The water flow of the tunnel should be adjustable and made to resist splashing or spilling.

• A method of measuring the drag force on the test hull should be made so different designs can be compared.

• Straws should be used to straighten the flow of water and to prevent **turbulence** (twisting currents of water).

SAFETY NOTE:
Be sure to follow the general safety rules and wear eye protection while you build the water tunnel. Keep extension cords out of the way. Keep all electrical power tools and extension cords away from water. Clean up any spills immediately to prevent slipping.

Caution students about the danger of using electric power tools near water.

Your design might look like this:

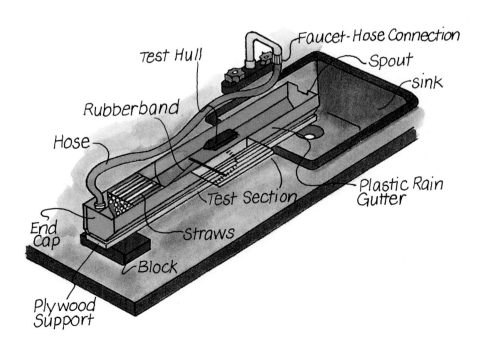

❷ List the tasks needed to complete your water tunnel. Assign individuals or small groups to complete each task. Your task list might look like this:

- Cut two lengths of plastic rain gutter for each end of the water tunnel.
- Design, cut, and bend acrylic or polycarbonate plastic to fit the shape of the rain gutter. Be sure to leave a flat area for the drag scale as illustrated.
- Glue drinking straws near the end of one half of the rain gutter to straighten the flow of water.
- Measure and cut a piece of plywood to support the water tunnel.
- Cut out an outlet spout on the output end of the rain gutter.
- Bend a piece of welding rod to work as a pointer for the drag test. Your test section might look like this:

If a sink is not available, use a 12-volt DC water pump and a five-gallon bucket to circuit the water through the tunnel.

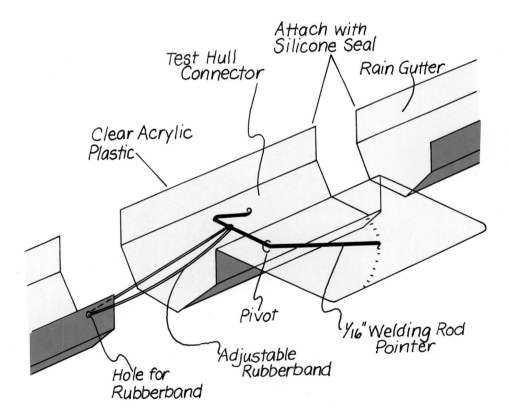

❸ Assemble the parts of your water tunnel. Connect the plastic test section to the rain gutters with silicone sealant to prevent leaking. Mount the assembly to a piece of plywood for support. Glue the end caps to the gutter sections. Attach the hose to a faucet and run the end into the input side of the tunnel. Place the output end over a sink. Support the input end with 2″ × 4″ blocks so the tunnel flows downhill.

4 Test the flow of water through your tunnel. Adjust the flow of water through the tunnel so that it is constant and will not overflow. Adjust the size of the output spout if necessary.

Procedure: Testing a Boat Hull

1 Design a boat hull to test in the water tunnel. Make your hull 1½″ wide and 5″ long. All the hulls should be the same width and length to make a fair comparison of shapes. Your hull design might look like one of these:

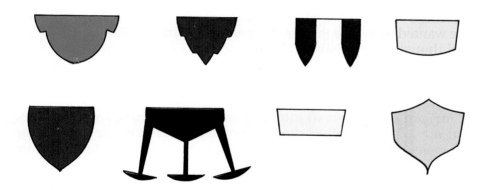

2 Sand the surface of your hull to make it smooth. Attach the hull to the tester section of your water tunnel. Mark equal divisions on the test sections and number them, starting at the point of no resistance.

Sanding Styrofoam can be messy. Have students use waterproof abrasive paper. Sand the designs in the sink.

Evaluation:

1 Make a sketch of your hull design and mark the number showing its drag. Make a bulletin board display of each hull design and its resistance.

2 Which design had the most resistance? Which design had the least resistance? Why?

3 What other factors could influence the drag resistance of ocean-going ships? Explain.

4 What is the shape of the hull of high-speed racing boats? Why do some of the racing boats have a double hull?

Challenges:

❶ Design and test a method of testing submarine shapes in the water tunnel.

❷ Connect two or three hulls together with thread or fishing line. What happens to the drag force?

❸ Research the America's Cup race. Why is the design of a hull so important to racing sailboats?

❹ Research hydrofoils, and explain why they can go faster than boats.

There are many other boats, such as motor boats, canoes, kayaks, and paddle boats, that we use for recreation as well as transportation. In general, water transportation is slow, but it is also less expensive than air or land transportation.

Special Report
Air & Space Transportation

Air Transportation

Probably the most inventive and imaginative transportation ideas have come in the field of flight. That's because since early times, people have wanted to fly like the birds. Even great inventors like Leonardo da Vinci thought people should be able to fly by flapping some kind of a wing device. What really started air flight were dreams of going up into the sky, not a special need to transport people or things somewhere else. People didn't think about how aircraft could help them in the beginning.

You can divide aircraft into **lighter-than-air** vehicles and **heavier-than-air** vehicles. Lighter-than-air vehicles float in air depending on their weight and volume. Heavier-than-air vehicles must supply power to fly.

The first really successful aircraft of any kind was the hot air balloon designed by Frenchmen Joseph and Etienne Montgolfier in 1783. They did not know what really made their balloon go up. They thought maybe some unknown, mysterious gas was released from burning wood! Do you know why hot-air balloons go up?

This principle of lighter-than-air flight was used in many more designs as people experimented with hydrogen-filled and hot-air-filled balloons to carry people up. The first human flight was also in 1783, when two people stayed up 25 minutes at an altitude of 3,000 feet.

Other lighter-than-air vehicles, called **dirigibles**, carried passengers and freight around the world in the early 1990s. The *Hindenburg* was the biggest one. It was more than 800 feet long and could carry 100 passengers. The problem with the dirigibles is that they used hydrogen as the lightweight gas. Hydrogen burns very quickly when it is ignited. The *Hindenburg*, like many other dirigibles, came to a tragic end when the hydrogen exploded and the dirigible burned.

Today, **blimps** use helium gas, which is safer than hydrogen. Blimps don't have a rigid structure like the dirigibles. Most of them are used for advertising or taking camera shots. Have you ever seen the Goodyear Blimp? Sometimes, blimps do some freight lifting, too.

Ask students to make a list of places they have flown to in an airplane. Calculate the total mileage traveled by air in their class.

Challenge students to make a small working hot-air balloon using tissue paper or a plastic bag.

After people got one foot off the ground, there was no stopping new flight ideas and technologies. Do you know who made the first takeoff in an engine-powered plane? It was a Frenchman named Clement Ader, in 1890. The wheels only came off the ground a few inches and the flight was very short (only 160 feet), but it was a start!

The Wright brothers' flight in 1903 was the first long, controlled engine-powered plane flight. They had experimented with gliders and had even built a wind tunnel to test different aerodynamic shapes. They had to experiment with how to control a plane in flight. This meant knowing the forces that work on an airplane, and how to control them.

Challenge students to make a display of facts relating to the Wright brothers' first flight.

A

TECHNOFACT

Technofact 79

Are you a nervous flyer? A company in Long Beach, California, is making special safety compartments that fit inside commercial airplanes. If there is an emergency, the passengers inside the individual safety enclosures will be safe from shock, changes in air pressure, and temperatures up to 2,000 degrees Fahrenheit. The system also has a communications system so you can communicate with search-and-rescue people.

B

Aircraft are divided up into lighter-than-air craft and heavier-than-air craft. (Figure A Courtesy of Garry Nelson. Figure B Courtesy of Alaska Air Group Inc.)

Bernoulli's Principle: What Makes Airplanes Fly?

Over 200 years ago, a Swiss mathematician named Jakob Bernoulli discovered a scientific principle that every airplane depends on. Bernoulli's principle states that as the speed of a fluid increases, its pressure decreases. You're probably asking what that has to do with an airplane's flying. First, a fluid can be a liquid or a gas. Air is a fluid. Bernoulli didn't know it at the time, but his scientific principle would help the Wright brothers and every aircraft developed since.

As air flows over an airfoil (wing), its shape and its angle of attack makes a low-pressure area above the wing. The shape of the wing and its angle of attack makes it necessary for air to speed up above the wing surface. Bernoulli's principle says that when a fluid (air) speeds up, its pressure goes down. This low-pressure area creates lift for the aircraft to overcome gravity and to fly. Bernoulli would be amazed to see how technology has put his scientific discovery to use.

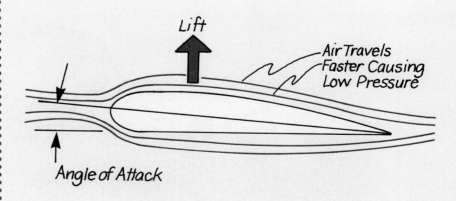

Lift

Air Travels Faster Causing Low Pressure

Angle of Attack

THINGS TO SEE AND DO:
Testing Aerodynamics in a Wind Tunnel

Introduction:

When the Wright brothers were designing their airplane, they tested wings in a wind tunnel. Today, there are huge wind tunnels used by NASA to test aircraft at supersonic (faster than the speed of sound) speeds. Wind tunnel testing is a method used by aircraft and automobile companies to test the aerodynamics of the shapes of airplanes and cars. **Aerodynamics** is the study of the motion of air. Models are often used to simulate the real product. Many design problems can be solved by wind tunnel testing before production starts.

Design Brief:

Build and use a wind tunnel to test a wing shape for lift.

Materials:

- Cardboard or hardboard
- Clear acrylic (Plexiglas)
- Permanent magnets
- Duct tape, drinking straws
- ¹⁄₁₆″ welding rod
- Stiff paper

Equipment:

- Window fan
- Plastic strip heater
- Lab balance
- Power hand drill
- Protractor
- Hot wire cutter
- Scissors

Wind tunnels are commercially available from science and technology suppliers.

Procedure: Building a Wind Tunnel

1 In this activity, you will work individually to make an **airfoil** (wing) and with the entire class to make a wind tunnel. You and your class can think of other uses for your wind tunnel. Your class should decide whether this activity will be taken apart after everyone has tested their airfoils, or left together to experiment with other shapes. Your decision will determine what materials are used to build the wind tunnel.

2 There are many ways to build your wind tunnel. Your design will depend on the fan that you are going to use and the materials available to you. Work as a large group to brainstorm ideas. Here are a few guidelines to follow as a general design:

- All fan blades and any moving parts must be covered with a guard to prevent injuries. All electrical connections must be safe, and extension cords must be the proper size wire for the fan being used.

The dust collecting system in a technology lab might be used as part of a wind tunnel.

- The wind should be drawn through the tunnel rather than blown in. This design reduces the turbulence (twisting air currents) from the fan blade.
- A venturi design will greatly increase the speed of the air flow. A venturi is a narrowed area where the air must speed up to flow through.
- A method of measuring the lift of the airfoil shape must be considered in the wind tunnel design.

Your design can be adjusted to fit the materials you are using or the fan shape. Here is one design that you might consider.

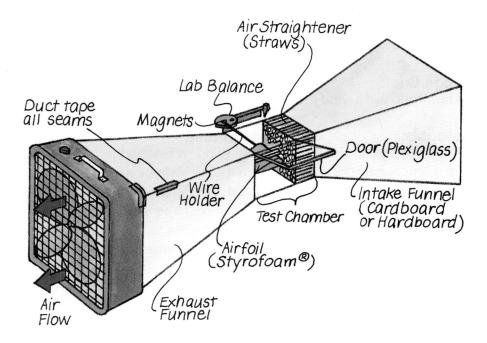

3 With help from your teacher, list the tasks that need to be completed to make the wind tunnel. Divide the tasks among smaller groups to make the building process go faster. Here are some tasks that must be completed:

- Mark and cut the cardboard or hardboard used for the intake and exhaust funnels.
- Design and build the test chamber so that it will open and close easily. The test chamber must be clear to let you see your airfoil test. Acrylic (Plexiglas) or polycarbonate plastic would be a good choice.
- Design and build an air straightener that will prevent any turbulence from entering the test chamber. A simple method would be to pack the intake side of the test chamber with drinking straws. The air straightener

Be sure all students are cooperating and helping with this activity.

produces **laminar flow**. Laminar flow is air that flows smoothly in layers without turbulence.

- Design and build a method for testing the lift of your airfoil using a lab balance.

  Your design might look like this:

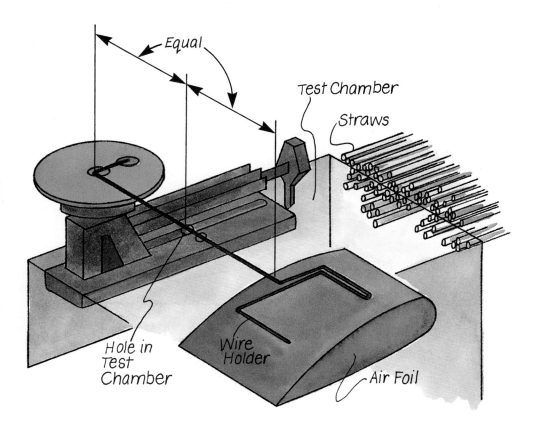

④ Assemble your wind tunnel using duct tape to seal all the connections. Ask your teacher to inspect your completed wind tunnel.

Procedure: Using Your Wind Tunnel

① Now that you have completed your wind tunnel, it is time to design and test an airfoil. Airfoil shapes vary with different aircraft. Some airplanes must be able to fly at supersonic speeds or even upside down in acrobatic stunts or air combat. In this part of the activity, you will design an airfoil shape and test it in your wind tunnel.

② Airfoil design has been studied since the first airplane flight in 1903. Your test will show how airfoils create lift to make aircraft fly. **Lift** is the force created by the airfoil that must be greater than the force of gravity so that an aircraft can fly. To design your airfoil, you need to know a little about the terms used by aeronautical engineers:

- **Leading Edge:** The point on the wing that is farthest forward. The spot where wind hits the airfoil first.
- **Trailing Edge:** The back edge of the airfoil.
- **Chord:** A straight line from the leading edge to the tail.
- **Angle of Attack:** The angle between the chord and the direction of the wind.
- **Drag:** The resistance of an object to the flow of air, often determined by its shape.

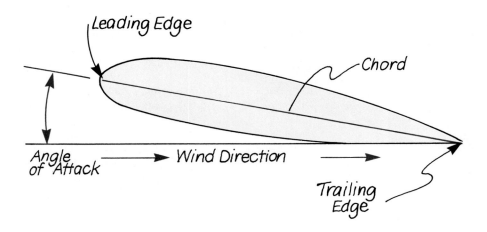

❸ Design an airfoil shape on a piece of stiff paper. Cut out your design with scissors. Pin the pattern to a piece of Styrofoam. Use a hot wire cutter (see Chapter 6) to cut your airfoil shape. Follow the paper pattern carefully to make a smooth cut.

❹ Locate the leading edge and trailing edge of your airfoil. Draw a straight line between the leading and trailing edges to show the chord of your airfoil. Mount your airfoil on a wire holder made from a ¹⁄₁₆″ welding rod. Place the wire holder through a hole in the test chamber of your wind tunnel. Attach the wire holder to a lab balance with permanent magnets.

❺ Adjust the wire holder so that the angle of attack is zero. Make a data table to record all the measurements you are going to make. Your data table might look like this:

Angle of Attack		0°			5°		10°	
Wind Tunnel Test Data		A-Balance setting (fan off)	B-Balance setting (fan on)	C-Difference (B–A)	D-Balance setting (fan on)	E-Difference (D–A)	F-Balance setting (fan on)	G-Difference (F–A)
Fan Speed								
Low								
Medium								
High								

6 Adjust the lab balance so that it counterbalances the weight of the wing and holder. You are now ready to test your wing for lift. Turn the fan on to a low speed. See if the wing goes up in the test chamber. If it does, adjust the balance so that the weight of the wing, wire holder, and lift force are balanced again. Write the new balance setting down.

36A sensitive spring scale could be used instead of a lab balance.

7 Continue to test your airfoil at medium and high speeds. Readjust the lab balance, and mark the new weight measurement on your data table.

Evaluation:

1 Did your airfoil fly?

2 Do you see any relationship between the lift and the wind speed? Explain.

3 Is there a relationship between the angle of attack and the amount of lift? Explain.

4 Can you explain why early airplanes had two and sometimes three wings?

5 What is the meaning of leading edge, trailing edge, chord, and angle of attack?

Challenges:

1 Make a graph that illustrates the effect of wind speed on the amount of lift. Make another graph that illustrates the effect of angle of attack on the amount of lift.

2 Design and test a method of safely making "smoke" to see any turbulence around the test shape. You might try a vaporizer or humidifier to make a safe "smoke" source.

❸ A symmetrical wing is the same shape on both sides of the chord line. **Symmetrical** means the shape is identical on both sides of a line. Why do you think some stunt planes and jet fighters have symmetrical wing shapes?

❹ Design and test other wing shapes. Make a data table for each shape. Can you predict the performance of an airfoil by looking at its shape? Explain.

❺ Design a method for measuring the drag force (air resistance) of an object such as a model airplane, car, or rocket. Try your idea in the wind tunnel. Can you make a general statement that would apply to all shapes and the amount of drag they produce?

Challenge students to bring in a model airplane and identify its parts.

How Planes Fly: In order to fly, an airplane has to overcome two major forces: gravity and drag. Wings that provide lift can overcome gravity. Once in the air, the plane is held back by drag. Drag is overcome by the thrust of the engines and the streamlined shape of the airplane. The less drag an airplane has, the less power is needed to propel it.

In the air, the plane also uses **ailerons** to control the **roll** of an airplane. If the wing tip begins to dip or rise, the pilot operates the flaplike devices on the wing edges. The **rudder** controls the **yaw**, or the movement of the plane to the left or right. The rudder is a flap on the vertical (upright) section of the tail. When the nose of a plane begins to **pitch** (dip or rise), the pilot operates the **elevators**, which are horizontal flaps on the tail to control the plane.

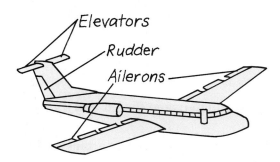

Elevators

Rudder

Ailerons

The pilot operates the control surfaces of an airplane from the cockpit. The control surfaces make the pilot able to control the flight of the airplane.

Many technological advances in airplane design and manufacture came in World War II. One of the most important was the development of the jet engine.

Today, new materials for building planes also make new designs possible. One of the first jet-powered passenger planes was built by the British in 1952. Unfortunately, these planes (called the British Comets) had many bad accidents, but we learned from those mistakes. Planes had to be made safer. They had to be built with lightweight yet strong materials that could take the stress put on them as they flew higher and faster.

Show a video of developments in the aeronautics industry.

Today there are many different kinds of planes, and each is built for a special job. Airliners can take off and land on short runways. Jumbo jets carry hundreds of people over long distances but need a long runway.

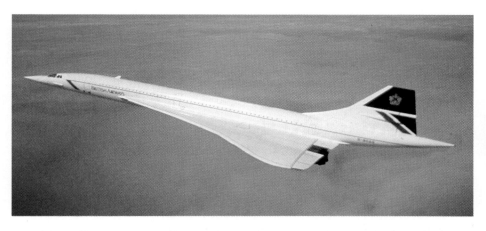

The *Concorde* SST is the fastest commercial airplane in service today. SST stands for supersonic transport. (Courtesy of British Airways.)

The *Concorde* Supersonic Transport (SST) can fly from New York to Europe in 3½ hours. Many people own small planes that they use for short trips.

As in other transportation systems, computers play a huge part in the air industry. Computers are used to ticket passengers, order and stock food on planes, list flight routes, and most important, control landings and takeoffs at airports. Computerized radar helps controllers keep aircraft a safe distance apart. All airliners are equipped with computerized weather radar that lets a pilot see even through clouds. That way they can go around thunderstorms. Computers also handle many jobs inside the airplane cockpit, from signaling fuel levels to controlling lights, heat, braking systems, and air pressure.

Helicopters are another form of heavier-than-air transportation. Helicopters are able to land people and supplies in places where other types of transportation can't go. They can fly straight up while taking off or straight down while landing. Have you ever seen a helicopter land on the roof of a building? Helicopters can also hover (fly in one place) in the air. They can also change directions very quickly. Helicopters are used a lot today for traffic control in large cities. They can easily send word about traffic jams or accidents. They are also used in the construction industry to move lumber or even large prefabricated sections.

TECHNOFACT

Technofact 80

Did you know that when you fly you travel along superhighways in the sky? In order not to run into other planes, pilots travel along numbered air routes, or corridors, in a three-dimensional traffic network. A corridor is about 9 miles wide, and at lower altitudes each corridor is 1000 feet above or below the next corridor. Keeping track of as many as 15 planes at one time is a challenge for air traffic controllers even with the use of computers and guidance systems!

Visit an airport. Ask a helicopter pilot to give the class a tour of a helicopter.

Arrange a tour of the local airport.

Helicopters have the ability to hover and to move forward, backward, and from side to side. (Courtesy of United States Navy.)

Air transportation has changed so much in the last hundred years. Creative thinking and the development of new technologies have made airplane transportation safer and more convenient. Even the idea of a stewardess or steward on an airplane, which was an American idea, has made airplane travel more comfortable for you. What can you expect in the future?

Future Transportation

Transportation is continually changing to meet the needs of people. In the automobile, trucking, train, and airline transportation industries, aerodynamic designs are making vehicles more fuel-efficient. Design engineers are also looking for ways to reduce the weight of vehicles, using aluminum, plastics, or lighter-weight steel.

Plastics and **composites** (which are fiber-reinforced plastic) are used to make smaller aircraft that are light and strong. The trend to use more plastics in cars and trucks will soon make many exterior parts recyclable.

Computers and other electronic devices are also changing transportation. Computers will continue to control many functions in all forms of transportation, from scheduling trains to reducing air pollutants in vehicle exhaust. Satellite communications systems will help air traffic controllers monitor planes on long-distance flights so that planes don't run into each other.

Transportation systems that use alternative energies will be explored further to see if they are practical and economical. Maglev trains and hovercrafts hold much promise as possible transportation systems. Wouldn't you like to ride a hovercraft to school or work? Another alternative energy, wind energy, is being used to supplement electricity on large ships. Photovoltaic cells, which change sunlight into electricity, are being used to power motors in cars and airplanes.

TECHNOFACT

Technofact 81

How about an all-terrain dirtboard that lets you ride on dirt hills, grass, and mud just as if you were riding a regular skateboard on smooth concrete? What makes this dirtboard special are the wheels. They are air-filled tires that are 8 inches in diameter! Big wheels are safer. On the dirtboard, the wheels rise above the platform, so the board is still close to the ground.

Computer and radar technology helps air traffic controllers keep track of aircraft to avoid accidents. (Courtesy of G. M. Hughes Electronics Corporation.)

THINGS TO SEE AND DO:
Make a Hovercraft You Can Ride

Techno Teasers
New Energy Sources

Introduction:

Hovercraft are vehicles supported on a cushion of air. They can move over land or water at high speeds. Hovercraft today are limited to small recreational vehicles or very large commercial or military craft. Hovercraft are used to transport people quickly across open water in some parts of the world. In this activity, you will make a hovercraft that you can actually ride on.

Design Brief:

Design, build, and test a hovercraft that will support the weight of a person.

Students will actually build and ride a small hovercraft in this activity. Be sure their design and use of the craft is safe.

Materials:

- ⅜″ plywood
- Polyethylene (Visqueen) plastic (6 mil)
- Duct tape, wood screws (flathead)
- Abrasive paper (60 grit sandpaper)
- Plastic coffee can lid

Equipment:

- Saber saw
- Stapler
- Utility knife
- Vacuum cleaner

Procedure:

1 In this activity, your class will be divided into two groups. Your teacher will help each group with its hovercraft design. Keep the following ideas in mind as you design your hovercraft:

- Your design must be safe. Electrical extension cords must be the proper size for the vacuum motor. Your hovercraft will not be tested over water or on a wet surface to prevent possible electrical shock.

- The area of the base of the hovercraft must be large enough to support the weight of a person.

- A switch must be provided so that the rider can stop the vacuum motor and stop the hovercraft.

- The bottom of the hovercraft must be smooth and free of nails or screw points that could cut the plastic or scratch the floor.

- A chair or seat must be mounted to the hovercraft for people to sit on while riding. No one should attempt to ride the hovercraft while standing. Only one person should ride at a time.

Some possible ideas look like this.

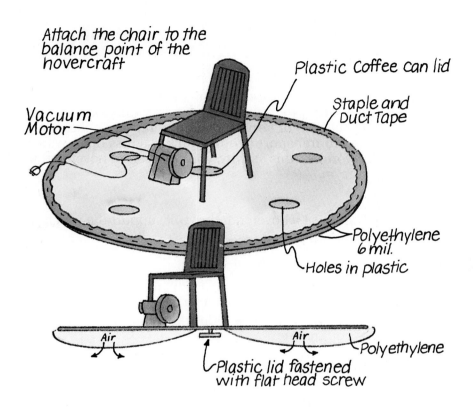

Attach the chair to the balance point of the hovercraft

Vacuum Motor

Plastic Coffee can lid

Staple and Duct Tape

Polyethylene 6 mil.

Holes in plastic

Air

Air

Polyethylene

Plastic lid fastened with flat head screw

Inspect the bottom of the hovercraft for protruding staples or screws that might scratch the floor.

Be sure all students in the class are helping to complete the assignment.

SAFETY NOTE:
Be sure to follow the general safety rules and wear eye protection while you build the hovercraft. Be careful with sharp tools. Keep extension cords out of the way.

2. Make a list of the tasks needed to complete your hovercraft. Assign the tasks to individuals or small groups on your team. Some tasks might include:

- Mark the plywood for cutting. Cut the plywood shape and sand the edges to remove sharp splinters.

- Drill a hole large enough for the vacuum motor exhaust port or hose.

- Measure and cut the polyethylene sheet to the desired size using a utility knife. Mark and cut the holes to release the air. Be careful not to scratch the workbench or floor while cutting.

- Staple the polyethylene sheet to the plywood. Screw or staple a plastic coffee can lid to the center of the plastic. This reduces the friction of the plastic against the floor.

- Design and build a mount for the vacuum motor.

❸ Assemble the hovercraft using the parts built by each group. Place duct tape around the edges to help prevent rips in the plastic. Attach the emergency stop switch with your teacher's help. Attach the vacuum motor to the base with wood screws. Be sure the screws do not go through the plywood and cut the plastic. Seal around the motor and plywood with duct tape.

❹ With your teacher's help, test your hovercraft without anyone riding it. If everything is working, have the first test driver climb aboard. Clear a path for the hovercraft test run. Be sure to keep the extension cord out of the way. Start the vacuum motor and give the rider a gentle push. The test driver should practice balancing on the hovercraft so the weight is distributed evenly. Test the emergency stop switch so the driver will have brakes to stop the hovercraft. Take turns test driving your hovercraft, and suggest ways to make it better and safer.

Evaluation:

❶ What are the advantages and disadvantages of hovercraft compared with cars?

❷ Do you think you will have a choice of buying a car or a hovercraft in the future?

❸ What is the relationship between the area of the hovercraft and its carrying capacity? Explain.

❹ How could the hovercraft be made so it would not need to be plugged into an extension cord?

Challenges:

❶ Research other designs of hovercraft. See if you can find out which hovercraft is the fastest and which is capable of lifting and transporting the most weight.

❷ Design and test a safe method of propelling your hovercraft forward and backward.

❸ Design and test a way to steer your hovercraft.

Transportation in the future will be safer, more economical, and more convenient. New frontiers in space and underwater exploration will also bring new challenges to transportation systems as we try to move people and products. Just imagine the possibilities for future transportation systems as we explore space and new frontiers in technology!

TECHNOFACT

Technofact 82

Have you heard the term "Mach 1" before? Mach 1 is the speed of sound, or about 700 miles per hour. When an airplane approaches the speed of sound, the air it pushes ahead of it makes a shock wave called the sound barrier. Just as a plane breaks through the sound barrier, it becomes very hard to handle. Chuck Yeager was the first person to fly at more than the speed of sound. The *Concorde* cruises at Mach 2, or twice the speed of sound. There is a hypersonic plane planned that will reach speeds of Mach 6 and up! Now that's flying!

Techno Talk
Speed of Light

Techno Talk
Answer Segment

Special Report
Future Transportation

Summary

Transportation is the movement of people or goods from one place to another. You depend on transportation to get you where you want to go. You also depend on transportation systems to bring products you use to markets.

Today's transportation technology is changing rapidly. Most transportation systems have computers controlling many parts.

There are different modes of transporting, or moving, materials. Land transportation includes any forms of transportation that move on or beneath the earth's surface. Some examples of land transportation include cars, buses, trucks, pipelines, conveyors, railroads, escalators, hovercrafts, or even forklifts. Mass transit systems such as streetcars, subways, and monorails carry many people at one time.

Water transportation includes passenger ships, cargo ships, hydrofoils, hovercrafts, submarines, barges, and tugboats.

Air transportation is divided into two groups: lighter-than-air vehicles and heavier-than-air vehicles. Hot-air balloons, dirigibles, and blimps are examples of lighter-than-air transportation systems. Airplanes and helicopters are heavier-than-air vehicles that need power to fly.

Transportation vehicles in the future will need to be lighter, safer, and more economical in order to meet our needs. Computers and other electronic devices will play a big part in future transportation systems.

Challengers:

1 Try to make a lighter-than-air, helium-filled balloon neutrally buoyant by attaching weights to it until it will stay in one place without moving up or down.

2 With the help of your teacher, arrange a visit to an airport. Investigate the maintenance of aircraft and how air traffic control works.

3 Research the design of the Wright brothers' plane, and make a scale model of it.

4 Ask one of the following resource people to visit your class:

- Airline pilot
- Train engineer
- Air traffic controller
- Automobile race driver or test driver

5 Contact a local natural gas supplier. Find out how natural gas gets to your home through pipelines.

6 Find someone with a radio-controlled scale model hovercraft. Ask her or him to visit your class and to demonstrate the hovercraft.

7 Write to a major oil company and ask them what they are doing to prevent oil spills.

8 Visit a flight museum, and chart the progress of manned flight.

9 Make a video of how planes fly that could be shown to elementary school students.

10 Make a bulletin board display of jet fighter aircraft and their specifications.

See Teacher's Resource Guide.

Chapter 13

Finding and Using Information

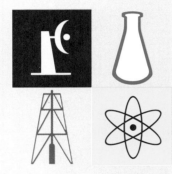

Things to Explore

When you finish this chapter, you will know that:

▶ Communication is the process of exchanging information either by sending information or by receiving it.

▶ A communication system includes a sender, a receiver, feedback, and interference.

▶ Computers are changing communications fields such as technical writing, illustrating, photography, and electronic communications.

▶ Electronic communication systems include radio, television, fiber optics, modems, satellite communications, networking, telephone, and facsimile (FAX) machines.

▶ The future of communications is unlimited.

Chapter Opener
Finding & Using Information

TechnoTerms

broadcasting
communication
design elements
electronic bulletin
  board
electronic commu-
  nications
electronic mail
electronic noise
facsimile (FAX)
feedback
fiber optics
geostationary

geosynchronous
  orbit
hologram
interference
modem
network
orthographic
  drawing
pictorial drawing
technical writers
  and illustrators
video
videoconferencing

Careers in Technology

When you watch the evening news on television, you're used to seeing live, on-the-spot coverage of major news stories for the day. Broadcasters for the major networks rely on satellite news-gathering services like GTE's Spacenet to provide that information. Working for one of the news-gathering services, you would control communication "traffic" on satellites and monitor transmissions to be sure they are top-quality.

Techno Teasers
Communication Systems

Answer
Segment

How Do We Communicate?

Can you imagine your world without television, telephones, books, signs, newspapers, tape recorders, or computers? It would be pretty hard to go through one day without using any form of **communication**. What is communication? Communication is the process of exchanging information either by sending information or by receiving it.

In order to make sure your communication works, the message has to be sent, received, and understood. Sometimes your message is sent to another person, an animal such as your dog, or even a machine such as a robot or a computer. Sometimes, machines communicate with machines such as computer to computer.

A communication system is like any other system. It has input, which is the **sender** of a message, a process, which is how the message is sent, and output, which is the **receiver** of the message. **Feedback** from the receiver lets you know if your message was understood. There is always a chance for **interference** anywhere in the communication system. Interference is anything that gets in the way of the message being

C

B

A

Which of these communication technologies can you identify? (Figures A and C used by permission of Gannett Company, Inc. Figure B Courtesy of Scotsman Photo/Rockwell International.)

understood. Sometimes it is hard for people to communicate if they do not understand the same language. Other forms of interference might be **electronic noise** such as static. A scratch on a record or a CD can cause noise interference. Have you ever not heard what your teacher told you to do because you were thinking of something else? That's also interference!

People have been communicating with each other since early days when they first used signs and symbols. Just like most everything today, communication is changing very quickly. That's because we are developing new technologies that help us communicate in many exciting ways. Did you ever think you could send your voice on a laser beam?

Techno Teasers
Early Communication

Techno Teasers
Answer Segment

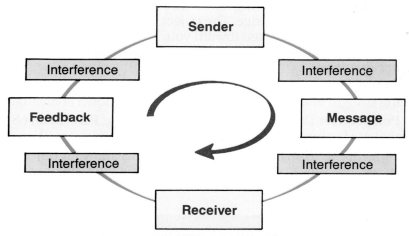

Few people really understand the process of communication. The entire process makes a complete circle to tell the sender that the message has been understood. How does a teacher get feedback from students?

Writing and Drawing Instructions

Have you ever tried to put together a model of a rocket or use a computer software program only to get frustrated trying to follow the written directions? It isn't easy for **technical writers** to write instructions or manuals that everyone can follow. Adding graphics or pictures is one way that technical writers have tried to make directions clearer for you.

Sometimes when you are writing directions, they seem crystal-clear to you, but when other people read them they get confused. Even writing complete, clear directions for making a peanut butter and jelly sandwich is not as easy as you may think. Try it. If someone had never seen a sandwich before, your directions would have to include how to put the two pieces of bread together to make a sandwich. Did they? In order to make sure your message is clear, you should have someone else **edit**, or check, your work. Technical writers are trained to write technical manuals and instructions. They have people test their instructions to make sure they haven't left out a step that they assumed everyone else already knew.

TECHNOFACT

Technofact 84

Have you ever used a laser? Sure you have! Anytime you buy an object that the checker scans at the checkout line, you are using a laser inside an optical scanner. The scanner is reading a bar code. The bar code is a series of black lines and white spaces. The scanner laser light senses the lines and spaces and changes the information into on or off pulses of electricity. Each product has its own code. The code is sent to the computer, which then gives the price for that item. Bar codes are pretty handy things for store owners. The computer can use the bar codes to keep track of price changes, how many of that item are in stock, and even to let the store know when it's time to order more!

Work with the language arts instructors to help students develop technical writing skills.

Show examples of technical illustrations.

Challenge students to make a display of various drawing types.

See Teacher's Resource Guide for orthographic projection techniques.

Computers have made the editing process much easier than it used to be. If there are things that need changing in your instruction manual, a word-processing program will let you change it easily as well as check spelling and grammar.

Technical drawings or **illustrations** help people understand the sizes and shapes of objects. **Technical illustrators** are people skilled in making illustrations. Technical illlustrators show objects in either pictorial drawings or orthographic drawings.

A **pictorial drawing** shows a three-dimensional view of an object. You learned about pictorial drawings in Chapter 10. There are three kinds of pictorial drawings:

- **Isometric:** These drawings show an object as though you are looking at it from an edge. The object looks as though it is angled and slightly tilted toward you. Lines that show the width and depth of an object are drawn at 30° angles from the length or horizontal.
- **Oblique:** These drawings show one surface as if you were looking straight at it. The other two surfaces are shown at an angle.
- **Perspective:** This is the most realistic drawing. Parts of an object that are farthest away look smaller.

Orthographic drawings are a second way to show an object. Here three views (the top, front, and side) of an object are shown as if you were looking straight at each one.

Technical writers try to write directions that people will understand. The written copy is edited and proofread carefully. Word processing has made it very easy to make changes. (Courtesy of International Business Machines Corporation.)

Technical illustrators make it easier to understand written instructions by making drawings. You may have even seen instructions that are all illustrations and no words. (Courtesy of Aldus Corporation.)

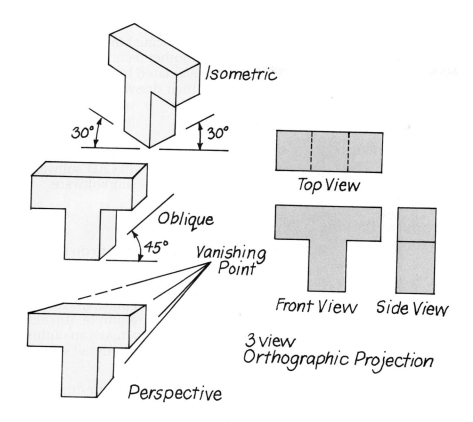

Isometric

30°   30°

Oblique

45°

Vanishing
Point

Top View

Front View   Side View

3 view
Orthographic Projection

Perspective

Illustrations can be made using
different types of drawings.
The drawing types that make
the object look like a picture
are called pictorial drawings.

Computer-aided design has improved the field of technical drawing. CAD allows technical designers to create very precise, or accurate, drawings that are easier to edit than hand drawings before final printing. Making drawings that are realistic is becoming more and more important in communicating an idea. Computers help that process. Desktop publishing also lets you put the text and graphics together easily. Anytime you can use pictures or drawings to back up the words in a manual, it will make more sense to more people.

THINGS TO SEE AND DO:
Writing and Drawing
Technical Information

Introduction:

You know how frustrating it is when you don't understand directions or the directions don't make sense? It isn't easy to write directions that everyone will be able to follow. Sometimes it is easier if the written directions refer to a drawing. Drawings used in this way are called **illustrations**. Technical illustrators are skilled artists who make drawings easier to understand. Technical writers are skilled at writing explanations or procedures clearly and completely. In this activity, you will be working both as a technical illustrator and a technical writer.

Design Brief:

Write a procedure for drawing an object located in your classroom. Your written instructions will be given to another technology student in another class. Your instructions will be evaluated by having someone else follow your procedure and compare their drawing with your original drawing.

Materials:
▶ Paper, pencil

Equipment:
▶ Drafting equipment
▶ Computer with CAD software, word-processing software (optional)

Procedure: Making a Drawing

1 In this activity, you will work individually to do two tasks. Choose an object that is visible in your classroom. The object should be recognizable from a front view.

2 Use drafting tools or CAD software to make the front view of the object. Keep your drawing simple. You may need to leave out some detail and just make an outline of the object. You should draw just enough detail so that the object can be recognized easily.

3 Ask someone else in the class to look at your drawing. Can it be identified easily? If not, you need to refine your drawing or put in more detail so that there won't be any confusion.

Procedure: Writing Directions

1 In this part of the activity, you will write directions for another student (in another class) to make a drawing exactly like yours and to identify the object.

2 Start your list of directions carefully so that both drawings will be identical. You can write the directions on paper or use word-processing software. Sometimes a direction can have more than one meaning. Then it is called **ambiguous**. As you follow these sample directions, check for ambiguous directions. Compare your drawing with another student's work.

Sample Directions

- Tape the paper to a drawing board so that the long edge is parallel to the T-square.
- Starting from the upper left corner, measure 5½″ to the right and 1″ down. Mark this position as point A.
- Draw two 75°-angle lines, ½″ long, down to the left and right of point A.
- Join the ends of the lines in step 2 with a horizontal line, forming a triangle.

- From the lower corners of the triangle, draw two parallel vertical lines 4¾″ long.
- Join the lines with a short horizontal line to complete the outline of the object.

What directions needed to be more specific? How would you rewrite them so they are not ambiguous?

3 If you used CAD software to make your drawing, your directions will be slightly different. In this case, you should give directions such as scale and screen position. The **hard copy** (printout) of both drawings should be identical if your directions were clear and accurate.

Evaluation:

1 Ask your teacher to give your directions to a student in another class. Compare the results of the drawing made from your directions with your original. Were they the same?

2 Your teacher will give you the directions from a student in another class. Make a drawing by carefully following the directions. Make a note on the paper when you think you know what the object is. Complete the drawing, and return it to your teacher so the author of the directions can compare the results with the original.

3 Were there any ambiguous directions in your list? Explain.

4 How do you think directions could be made simpler for very complex tasks? How could other forms of communications technology help people learn new things? Explain.

Challenges:

1 Make a pictorial drawing of your object rather than the front view. Was it easier or harder to write the directions for a pictorial drawing? Explain.

2 Use walkie-talkies to give directions to another person instead of writing them or typing them. Was this method easier or harder? Explain.

Ask students to explain the steps in the communications model in this activity.

3 Use communications software to **upload** (send) your instructions to a student in another school using a modem. **Download** (receive) the directions from another student in another school. Exchange the resulting drawings using a fax machine.

4 Send your written directions to students in a foreign country. Get a set of directions from a foreign student. Use a translation dictionary to understand the instructions. Exchange the drawings through the mail. Were there more errors in understanding the instructions when you had to change languages? Why?

The activities in this book are an example of technical writing. They are written in a step-by-step procedure to make it easier for you to follow the directions.

Communicating with Pictures and Symbols

Communicating with pictures and symbols is a common way of getting a message across to another person. We've been doing it since prehistoric times. With the changes in technology, you still communicate with simple pictures and symbols but you can also use photography, animation, and video to name a few.

Photography: What catches your eye first when you look at a billboard, a magazine ad, or a picture in a book? You look at so many different things in a day that it is really a challenge for graphic artists to catch your attention. Did you know that there is a way of placing objects or text in pictures that makes you notice them more? If you're going to communicate well with pictures and symbols, you need to know these **design elements**.

A skilled photographer can capture images that give you a feeling of what is happening. These photos show speed, drama, happiness, and worry. Can you tell which photo goes with each of these descriptions? (Figure A Courtesy of General Dynamics. Figure B © M. Mannix 1992. Figure C Photo by Doug Persons. Figure D Courtesy of Amoco Corporation.)

▶ **Line of golden proportion**: Also called the **rule of thirds**, you can find the line of golden proportion by dividing the height of a paper into three equal sections. The third line from the top is the line of golden proportion. Half of any text or graphic should be above this line, and half should be below this line. If you arrange your pictures or writing this way, it makes them much more interesting.

Ask the art teacher to suggest ways of illustrating design elements.

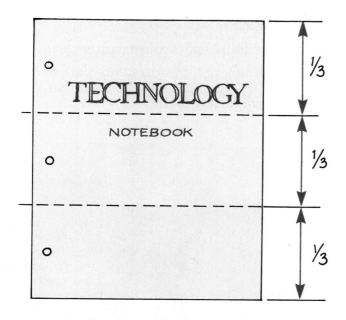

When you put the title on a page, use the rule of thirds. Most people just put titles or graphics in the middle. This is not a strict rule, it is a design suggestion that most people find more pleasing to look at.

▶ **Rhythm**: Another important design element, rhythm, guides the viewer's eyes. Sometimes you might number items to guide the eye. Other times you might use arrows to guide the eye to a specific spot in the picture. In a photograph, shadows can guide the eye to different areas.

▶ **Unity**: This is a very important element. For example, if you use too many different type styles or sizes in one ad it looks confusing. If you are adding graphics, you need to think about how they best fit together to get your message across.

These design elements are all part of the composition of a picture. The **composition** (where objects are placed) of a picture or photograph is very important. You want to have an interesting picture. Here are some general things to think about to make your composition good.

❶ Select the subject to be photographed. (See Figures a and b.)

❷ Determine the main figure, form, or area. Use the rule of thirds or line of golden proportion. (See Figures c and d.)

Special Report
Front Page Photos

Special Report–1
Designing Messages—Audience

Special Report–2
Designing Messages—Content

Special Report–3
Designing Messages—Form

❸ The other parts of the picture should not take attention away from the main area of interest. (See Figures e and f.)

❹ Get close enough to your subject so you can get rid of unwanted background.

❺ Choose a good background that is not cluttered with objects such as telephone poles and fences. Plain areas or solid areas such as clear sky or bushes make good backgrounds.

❻ Use contrasts like light against dark or small among large.

❼ In color photography, warm colors like reds and oranges jump out at you, while cool colors like dark blue stays in the background more.

A

B

E

F

C

D

Photographic composition is mostly common sense. These photos give examples of proper composition. Remember, these rules apply to video production, too. (Figures A,B,E,F Photos by Michael Bombard. Figures C, D Photos by Doug Persons.

Producing pictures with good composition takes time. You will need to develop your own style for what you like!

Animation: Animation is just graphics or pictures in motion. Have you ever seen a flip book or a piece of film from a movie camera? You can see that a film is really made up of a series of still pictures, each slightly different from the one before it. Animation relies on **persistence of vision** to make you think those pictures are in motion. Persistence of vision is the way your brain naturally holds onto an image or picture longer than your eye sees it. So when several images are flashed in rapid order, as a movie projector does, and when the images are only slightly changed from one image to the next, you see them as continuous motion. Have you ever watched an old cartoon where the motion is jerky or bumpy? Animation technology has improved greatly since then thanks to computers!

Challenge students to make a small flip book showing an animated figure.

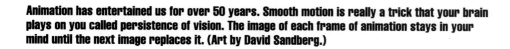
Animation has entertained us for over 50 years. Smooth motion is really a trick that your brain plays on you called persistence of vision. The image of each frame of animation stays in your mind until the next image replaces it. (Art by David Sandberg.)

TECHNOFACT

Technofact 85

How would you like to "play" your photo albums on a compact disk? Kodak is planning a new system that lets you use your 35mm camera to take pictures that you can show on a TV screen. Here's how it works. You will still take your film to your photofinisher. But instead of getting slides or prints you will have the images converted to digital data and put on a compact disk. Then you will use a CD player that lets you see pictures as well as have sound. Even better, you can control your images on the CD and even add special effects. Finally, one CD will hold thousands of pictures!

Video: Video photography technology keeps changing fast. The video cameras you can use today are smaller and able to add special effects. With a still video camera, you can take a picture that is recorded on small magnetic disks, like computer disks, instead of film. The camera not only records your image but plays it back for you on your television.

To make a good video, you still need to follow the design elements for regular photography. By the way, there is a difference between taping and filming. Filming is a photographic process. Taping is the magnetic process of putting images and sound on tape.

Not all video cameras record moving pictures. This is a still video camera that records images onto a small computer disk. The camera plays the pictures back through a television or a computer. (Courtesy of Canon USA Inc.)

THINGS TO SEE AND DO:
Photographic Communication

Introduction:
You have probably heard the saying "a picture is worth a thousand words." We live in a world of visual images. Advertisements use eye-catching photographs or bold graphics to get our attention. You probably found that it took a lot of directions to draw a simple object in the last activity. Graphic illustrations and photographs are used in text-books, newspapers, and instruction manuals to give the reader a visual image. Many people learn faster by studying visual images than by reading about the same subject. In this activity, you will work as a photographer and graphic artist to make a poster that teaches about safety.

Design Brief:
Design and make a poster that will effectively teach or remind students to work safely in the technology lab. Use photographic and graphic arts materials to produce a poster that can be displayed in your technology room.

Special Report
Video Animator

A Polaroid camera could be used in this activity.

Materials:
Photographic materials:

▶ Black and white film, chemicals

'ic arts materials:

▶ Poster board, rubber cement
▶ Dry transfer letters, markers, etc.

Equipment:
Photographic equipment:

▶ Camera, enlarger, etc.
▶ Darkroom
▶ Paper cutter

Ask the art teacher or yearbook advisor if a darkroom is not available in the technology lab.

Procedure: Staging Your Photo

1 In this activity, you may work individually or with a partner to make a safety poster.

2 Brainstorm ideas about what your poster will look like. Consider the following ideas:

- Keep the topic simple and to the point. Don't try to cover more than one idea.
- Photograph the subject of your poster so the topic is obvious.
- Keep the text (words) that explain your safety topic very short. You might not really need words at all.

3 The objects you use in your photograph are called **props**. Make a list of the props you will need. Ask your teacher to check your prop list. The way you arrange the items or people in your photo is called **staging**.

4 Your teacher will give you instructions on using a camera. The actual **photo shoot** (picture taking) will take place after all of your props are in place and you have removed any distracting items from the picture area.

5 Adjust and focus the camera according to directions from your teacher. Make at least six **exposures** (pictures), experimenting with different camera angles and lighting conditions.

Procedure: Developing Film

1 Let other groups use the camera to finish the roll of film.

2 Your teacher will demonstrate the development procedure to the class. Developing film is easy and fun. It is important to follow the steps carefully and use the proper time for each chemical. The steps needed are:

- Developer
- Stop bath
- Fixer
- Wash

Some students may be sensitive to the photographic chemicals.

SAFETY NOTE:
Many photographic chemicals can be poisonous or cause skin or eye irritation. Follow the directions on the package. Do not eat or drink around photo chemicals. If chemicals splash in your eye, tell your teacher immediately. Wear chemical goggles, and ask your teacher for help when mixing chemicals.

3 There are different types of films that can be used to make photographs. Each type of film requires a special time for the developer. Your teacher will help you find the correct time for the film you are using.

4 Handle the film only by the edges. Hang the film in a clean area to keep dust from sticking to the wet surface. Put a weight on the end of the film to keep it from curling up as it drys.

Procedure: Making a Photograph

A television and VCR can be used in the darkroom to show how to develop a print. Place red ortholith film over the television screen to make its light safe with black-and-white photographic papers. Get the film from graphic art suppliers.

1 Cut the film into sections for each group. Inspect the developed film to see which frame you will **print** (make into a photograph). The film is called a **negative** because the light images are dark and the dark areas are light.

2 Put the negative in an enlarger according to your teacher's instructions. **Compose** (position) the enlarger and a photographic **easel** (paper holder) to make the best possible photo. Make a test print to determine the best

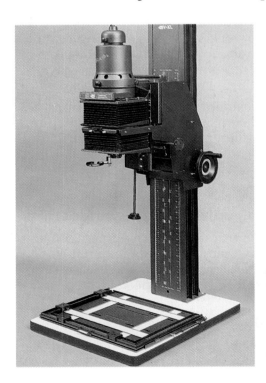

A photographic enlarger is used to project the image of a negative onto photographic paper. The easel is used to hold the paper in place during the exposure. Be careful not to expose black-and-white paper to white light. The red safe light lets you see what you are doing without ruining the paper. (Courtesy of Beseler Corporation.)

exposure time. The photographic chemicals used for developing a print are different from those used for film. Your teacher will help you mix the chemicals and give you the proper time for each. Photographic prints use chemicals in this order: developer, stop bath, fixer, and wash.

3 Photographic paper is sensitive to white light. Be careful not to spoil the photo paper by turning on a white light in the darkroom. Red light does not expose black-and-white photo paper. Darkrooms have red **safe lights** so you can see what you are doing without spoiling the photo paper. Cut a piece of photo paper to the proper size for your poster.

4 Put the photo paper in the easel, and make the final exposure according to your test print. Your teacher will give you the proper times for each chemical. Place the print on a drying rack. Be careful not to put your print on top of another one.

Procedure: Graphic Arts Layout

1 When your print is dry, it is time to make the finished poster using graphic arts materials and methods. Review the design elements listed in the reading. Get the materials you need such as posterboard, rubber cement, and so on.

2 Cut the posterboard to the size you need. Remember to leave room for the text. Use a paper cutter to keep the edges of the poster square and straight.

SAFETY NOTE:
Only one person at a time should be near the paper cutter. Use the guard, and keep your fingers away from the cutting blade.

Ask a graphic artist to talk to the class about their projects.

3 Your teacher will show you how to use dry transfer letters or make your text on a computer. Your finished poster should look professional. Markers can be used to draw borders or make special illustrations. Text should not be produced freehand.

4 Use rubber cement to attach your print to the posterboard. Make careful measurements, and make light pencil marks to help position the print.

Evaluation:

1 Trade your safety poster with another student or group. Write a paragraph explaining the message of the poster. Was there any misunderstanding or interference in sending the message? Explain.

❷ List three types of interference that could prevent the receiver of a photographic message from understanding the message.

❸ Why do road signs need to be designed carefully? Explain.

❹ Have you ever misunderstood or been confused by a poster or advertisement? Explain.

Challenges:

❶ Look at a black-and-white photograph printed in a book or newspaper with a magnifying glass. Describe what you see. How are shades of gray printed when only black ink is used?

❷ Look at a color photograph printed in a book, magazine, or newspaper with a magnifying glass. Make a list of the colors of dots you see. How many colors of ink does it take to make all of the colors in the photograph? Explain.

❸ Ask your teacher if your darkroom could be used to develop and print color photographs. Research the steps in developing color film and making color prints.

❹ Research ways to make your poster using a computer and avoiding the photographic process. Does your school have the technology to do this yet? How could you use this process to make photographs for your school newspaper?

Where does the computer fit into graphic communications? Graphic artists used to have to work long hours to put photographs, artwork, and text together. Today's graphic artist uses a computer workstation that allows the artist to add a tree or move a fence. A regular photo can be changed in a special effect and then added to another picture by a special machine called a **digital image processor**. You could combine a picture of you with a picture of the beach even if you've never been there! Graphic artists can electronically create exciting posters or magazine layouts all on the computer screen and make any changes easily before the final printing. In animation, graphic artists only need to create a character or scene once, save it on the computer, and call it up for changes in each scene.

Electronic Communications

Did you talk long distance today to anyone? Or did you watch a news program live from another country by satellite? Communications technology has changed rapidly because of new improvements in electronics. Today you use many **electronic communication** systems such as the radio, television, telephone, fiber optics, modem, satellite communications, and FAX machines. These systems all use electronic signals or electromagnetic signals to carry messages through cables or through the air.

A

B

Electronic communication has made it possible for people to send or receive messages almost instantly anywhere in the world. Can you identify a FAX machine, communication satellite, and fiber-optic communication systems? (Figure A Courtesy of US Sprint. Figure B Courtesy of Canon Corporation. Figure C Courtesy of TRW Inc. © 1991.)

C

Radio: How would you like to have been Guglielmo Marconi, the first person to transmit and receive a wireless message? That first wireless signal crossed the Atlantic in 1901. Right away, people could see how radio communication was going to help them.

How often do you turn on the radio to listen to music or to hear the latest news or the weather report? Have you ever picked up a station hundreds of miles away? Today's radio communication lets you have two-way communication between your home and an airplane, a car, or a boat. Even people on the moon or in the space shuttle can communicate with Earth by radio signals.

In radio communication, information is sent through the air. Transmitters send the messages into the air at a certain **frequency**, which is measured in cycles per second. One cycle per second is called one **hertz**. The signals have different wavelengths and frequencies. Each transmitter uses a different frequency to send information to a receiving antenna someplace else. That is why you can choose from so many radio stations. It also happens very fast. Just like light, radio waves travel at about 186,000 miles per second no matter what the wave frequency!

Visit a local radio station for a field trip.

Radio broadcasting provides news and entertainment for millions of people around the world. It is easy to see the electronic communication technology in use here. Have you ever visited a radio station? (Photo by Michael Bombard.)

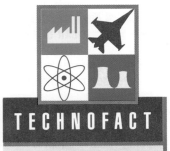

Visit a local television station for a field trip.

Radio can be used in different ways. **Broadcasting** is where a single transmitter sends out a signal to many listeners. Most of you probably have a favorite AM or FM station already. Sometimes though, remote places such as forest ranger stations or doctors' paging systems need a different kind of radio transmission. This service is called **point-to-point**. It is used for communicating with one receiver instead of many. There are lots of "**ham**," or amateur, radio operators who talk to other people point-to-point around the world. Many ham operators have even saved lives and helped others out during disasters like hurricanes.

Television: Americans watch an average of 3 hours of television every day. Some experts say that average is rising. How does television affect you? While you're watching television, you are **potential** (possible) consumers for the products advertised during the shows. How many commercials are there in a regular 30-minute program? Television is a big business involving many people, from camera operators to sponsors.

Television communication changes **visual** (things you can see) information into electrical, or **video**, signals. The signals are transported to another place and then changed back to pictures you can see on a screen. In television, the camera is part of the transmitter that changes, or **converts**, the visual information to a video signal. Then the signal is stored in a video recorder or VCR, sent directly by cable, or broadcast using a television transmitter. The monitor changes the video signal to a picture on the screen.

Broadcast television is the most commonly used form of transmission. The signals are sent on special carrier frequencies that have assigned TV channels. TV transmitters also send sound along with the picture.

In places where television reception is not good, cable TV systems deliver all the channels received by their up-dated equipment to subscribers on a single cable. Do you have cable TV in your school or home? An advantage to cable TV is that extra programming can be added to the cable. That way you can pick up movies, special concerts, and other events that regular broadcasts might not have. If you could set the program scheduling, what kinds of programs would you like to see on TV?

How does a television signal get into your home? Television signals can be transmitted from antennas and received by an antenna on your house or television set. Signals can also be sent on local cable systems or received directly from a satellite in your back yard. (Photo by Michael Bombard.)

THINGS TO SEE AND DO:
Radio and Television Production

Introduction:

Millions of people watch television or listen to the radio more than 4 hours every day. While watching or listening, we are the target of dozens of commercial messages. We can even shop at home by calling a toll-free phone number and using a credit card to order hundreds of products. Thousands of people are involved with the production of radio and television programs. In this activity, you will be the **talent** (actors or actresses on camera or on the radio) and the production personnel such as camera operator, sound engineer, script writer, and so on.

Design Brief:

Produce a video or audio (or both) program that is both informative and entertaining. Your program should teach the audience some aspect of technology. Your program should be from one to ten minutes long.

Materials:
- Props as required
- Videotapes
- Audiotapes

Equipment:
- Video camera and recorder, or camcorder
- Stereo, tape deck, CD player, mixer, microphone (optional)

Challenge students to help set up a television network that goes to each room in your school.

Procedure: Getting Organized

1 In this activity, you will work in a group to help make an audio (sound only) and video (picture and sound) program. Your finished program will be shown to other technology classes to teach them what you have learned about the subject of technology. You will need between three and five people in your group. Your group will team up with another group for the actual production. While your group is performing in front of the camera, the other group will be operating the equipment. When your group is finished, you will switch places.

Be sure all students have a chance to hear their voice on audiotape.

2 Brainstorm ideas that you could use for your program. Remember, it must be instructional and entertaining. Here are some possible topics and themes:

Technology Topic
- The fast growth of technology
- How has technology affected our world?
- Making a pneumatic robot arm
- Technology quiz

Presentation Theme
- News program
- Soap opera
- How-to program
- Game show

❸ Write a script for your idea. Everyone in your group should have a part.

Procedure: Producing Audio

❶ Your television program should include an audio (sound only) commercial for a product. While the commercial is playing, a computer-generated graphic or hand-drawn image should be shown on the video.

❷ Commercials usually run for 1 minute. Some commercials are 30 seconds or up to 2 minutes long. You should time your commercial to fit one of these times.

❸ The commercial message can also be educational. Use stereo equipment to record your audio commercial. Save the audiotape for the final television production.

Stereo equipment can be set up easily to produce good-quality audio (sound). A mixer can add two or more sounds together. If you wanted to talk with music in the background, a mixer would let you control the volume of each sound.

Procedure: Television Production

❶ You are now ready to rehearse your program. Your teacher will demonstrate the proper use of the video equipment. Gather or make any props that you will need. Some things you might include are:

- Graphic title for the introduction
- Graphic for audio commercial
- Credits (who were the actors, actresses, production staff, and so on.)
- Costumes or backgrounds to make your production look real

Be sure all students have a chance to be videotaped.

❷ Your goal should be to make the video look and sound like real television programming. Rehearse your program while another group practices using the video and audio equipment. Play the tape back while both groups watch. Discuss how the production could be improved, and make changes for the final production.

❸ Make the final tape of your program. Keep track of where it is on the video tape. Write down the starting and finishing time or index number so that your production can be found easily in the future.

Evaluation:

❶ Do you think that you look and sound the same on television as you do in real life? Explain.

❷ Many professional actors and actresses get very nervous just before performing. Why do you think some people get "stage fright"?

❸ What is the difference between filming and taping?

❹ Ask your parents to come to school to see your production, or take the tape home.

Challenges:

❶ Ask your teacher about the possibility of making a video studio in your technology room to improve the quality of your programs. See if it would be possible to send the video to other rooms in the school.

❷ Research how long it takes to produce one television show. Make a list of the jobs related to television production. You can see the titles in the credits at the end of each show.

❸ How do the television programs you watch reach your home? If a program is taped in Hollywood, California, how does it get to your television?

❹ Research high-definition television. How is it different from what you usually watch? What do you think television will be like 50 years from now?

Telephone: What would you do without a telephone? It's hard to imagine not having a phone to call a friend or order a pizza. Alexander Graham Bell never knew that his invention would end up being one of the most used electronic communication tools today.

The telephone's mouthpiece changes sound to electrical signals. The earpiece changes the electrical signals back to sound. Since many telephone conversations happen at the same time, switches with lots of wires have to be connected from town to town. Today, telephones are using fiber-optic cable because one fiber-optic cable can carry over 10,000 conversations at one time!

The new advances in telephone communications make it possible to have **cordless** telephones. These telephones have a small radio transmitter that sends a signal to the base of the telephone connected by wire to the telephone lines. **Cellular** phones, like cordless phones, also send radio signals. But cellular phones can transmit over an entire city, where cordless phones work only over small distances. Telephones are also

TECHNOFACT

Technofact 89

Everyone knows how to use a telephone. The problem is that lots of people don't know how to use computers and are afraid of them. AT&T wants to help solve that problem by designing a computer that looks like a telephone. The Smart Phone is a computer that links households to several services through a special information network. You will be able to program each button on the telephone computer so that it obeys a certain command. Maybe one button will order groceries for you, while another button will transfer money from one bank to another. If you travel a lot you would enter instructions into the computer to make it dial the travel agency and always ask for an aisle seat and the lowest possible rate. Then the computer makes the phone call and asks the questions for you!

A

B

Telephone technology has changed a lot with advancements in technology. Miniature electronic circuits, rechargeable batteries, and fiber optics have made a big difference in the way we use telephones. (Figure A Courtesy of AT&T Archives. Figure B Courtesy of Ball Corporation.)

Ask the local telephone company for a sample of fiber-optic cable.

connected to computers. In some places, you don't have to look up a number in a telephone book anymore! The computer does it for you.

Fiber Optics: Fiber-optic systems use light to carry information. What's so special about fiber optics? For one thing, fiber-optic cables are smaller, weigh less, and cost less than the many copper wires used on long telephone routes.

A fiber-optic cable is made out of many thin strands of glass fibers, sometimes thinner than a strand of your hair! This cable can carry light for long distances without losing power. A strong light source such as a laser or light-emitting diode (LED) is the transmitter. The fiber-optic cable carries the information that is received by a device that converts light energy into electricity. The cable can carry over 1000 circuits at a time.

In computer communication, fiber optics will really speed up how fast computers communicate with each other using modems. Right now, using fiber optics would increase the communication speed by a multiple of 41,667! By the time the airline agent punched in your ticket request, you'd already have your reservation!

Satellite Communication: You now have instant access to any part of the world with satellites. Communication satellites are relay stations for television and radio. Have you ever heard of the Clarke Belt? It is an orbit 22,300 miles over the Earth's equator where many communication satellites are located. These satellites are in **geosynchronous orbit**, which means they seem to stay in the same place above the Earth at all times. They are moving just fast enough to turn with the Earth so it seems they are fixed, or **geostationary**.

Techno Teasers
Fiberoptics

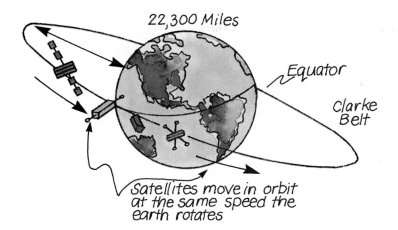

22,300 Miles

Equator

Clarke Belt

Satellites move in orbit at the same speed the earth rotates

Geostationary and geosynchronous mean the same thing. Satellites in this type of orbit rotate with the Earth. In this way, the satellite seems to stay in the same place in the sky all the time.

Satellite signals are microwave signals. Like light, microwaves travel in a straight path from the satellites to the Earth's surface. Antennas on board each satellite focus the microwaves onto a special part of the Earth's surface. An Earth antenna pointed at a satellite doesn't have to move either. A transmitter station sends a transmission on one frequency called the **uplink** to the satellite. Inside the satellite, a receiver takes in the signal and changes it to another frequency called the **downlink**. The microwave signals can be picked up over a very large area called the **footprint**. Satellite footprints show where each satellite's power levels are the best. If you have a satellite receiver, you can pick up any transmissions. Some uplink sites purposely scramble transmissions so that you cannot receive them for free.

If your school has a dish, show how a satellite dish is maneuvered to pick up satellite transmissions.

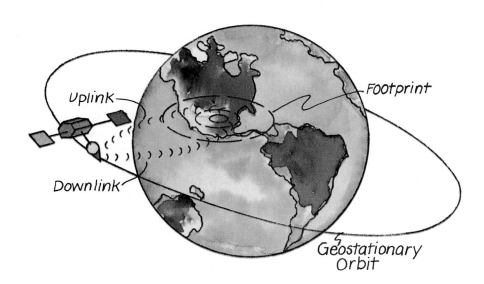

Uplink

Footprint

Downlink

Geostationary Orbit

Electronic signals are uplinked (sent) from Earth to communication satellites in space. The signals are downlinked (sent back to Earth) from the satellite, and cover an area of the Earth called the footprint.

THINGS TO SEE AND DO: Satellite Communications and Fiber Optics

Introduction:

Much of the television and long-distance telephone communication that we take for granted are sent from the ground to satellites and back again. Satellite communications can be simulated in a technology classroom using a laser. Actual voice and even television signals can be sent and received in the lab using special equipment. In this activity, you will work as a communications engineer to set up and test a satellite communications system and a fiber-optic communication system.

Design Brief:

Design, build, and test a satellite and a fiber-optic communication system.

Materials:

- Mirror, front surface
- Fiber-optic cable
- Magnifying glass

Equipment:

- Modulated helium-neon laser
- Receiving module, microphone
- Humidifier, vacuum cleaner hose (optional)

Procedure: Fiber-Optic Communication

❶ In this activity, you will work with a partner to set up two types of communications systems. You will be using a laser to send information. Special safety precautions must be taken when using any laser.

SAFETY NOTE:
Laser light can be very dangerous to your eyes. Do not look directly into a laser beam or a reflection of the beam. Warn others of the hazards of laser light. Do not point the laser at anyone, through windows, or outside. Be aware of where the laser light is going at all times. Turn off the laser when you are not using it. Be careful of electrical shock hazards such as wet floors or wet hands. Do not use lasers with an output greater than 5 milliwatts (mW).

❷ Point a HeNe (helium-neon) laser across a table toward a receiving module. Adjust the laser to point to the receiving port. The laser light from a HeNe laser is visible only if it reflects off of something. If you watch the beam carefully, you will see bits of dust sparkle as they move through the beam. Remember, do not look directly into the laser.

Use lasers with caution. Do not look directly into a laser beam or its reflection. In this activity, you will point a laser toward a receiving module to send and receive voice signals over the laser beam. (Photograph by Pam Benham.)

❸ Other materials that enter the laser beam will help make it visible. Dust is often too messy to use around lasers and other equipment. A simple way to view the laser is to attach a vacuum cleaner hose to the output of an ultrasonic humidifier. Ultrasonic vaporizers produce very tiny droplets of water vapor. The laser light reflects off the water droplets, so you can see it easily. Adjust the output of the humidifier so that it will not make the work surface wet. Remember to keep the table, floor, and your hands dry to prevent the possibility of electrical shock.

An all-plastic replacement vacuum hose is available at most hardware stores.

❹ When the beam is adjusted to point into the input port of the receiver module, try talking into the microphone connected to the laser. Make any adjustments needed so that you can hear your voice through the receiver module. You are now talking on a laser beam. What is happening is that the sound waves are changed to pulses of laser light.

❺ Sending your voice through the air on a beam of light might seem to be a practical way to communicate. Can you think of some possible interference with this technique?

❻ Without changing your setup, place a coil of fiber-optic cable in line with the laser and the receiver. Try talking on the microphone again. This time, your voice is going through a thin glass fiber around in circles to the receiver module. Many long-distance and sometimes even local

Bend a glass stirring rod using a propane torch or Bunsen burner. Demonstrate the property of internal reflection by directing a laser through the bent rod.

phone calls use fiber optics instead of wires to send and receive messages.

Procedure: Satellite Communications

1 Satellite communications systems commonly use geo-stationary satellites 22,300 miles in space. Signals are uplinked (sent) to the satellite and downlinked (sent back) to Earth. In this activity, you will make a simulated satellite that you will hang from the ceiling of your classroom.

2 Study pictures of different communication satellites. Make a plan of the design you would like to model. Instead of a receiving antenna, your satellite will use a front-surface mirror. Most mirrors, like the one in your bathroom, are rear-surface. You look through glass to a reflective coating on the back of the glass. The laser beam will reflect a more powerful signal if you use a front-surface mirror. Work with your partner to complete your model.

3 With help from your teacher, hang your model from the ceiling to simulate a geostationary orbit around the earth. Carefully point the laser toward the satellite.

SAFETY NOTE:
Remember to warn others of the hazards of looking into a laser beam or into a mirror reflection.

4 Your partner should adjust the receiving module so that the reflected light will enter the input port. If the beam has diverged (spread out), use a magnifying glass to refocus it into the input port. Test your setup by talking through the microphone. Make any adjustments needed.

5 This simulation uses laser light to send your voice to a satellite. In actual use, **microwave** energy would be sent and received. The satellite dishes you have seen gather the microwave energy and **amplify** (enlarge) it back into a television picture. Satellite communications has made it possible for very remote parts of the world to share the same information we use.

Evaluation:

1 Name three different kinds of interference that could interrupt a message sent on a laser beam through the air.

2 What is internal reflection? Make a sketch that illustrates a laser beam going through a glass fiber and around curves.

3 What does HeNe stand for?

4 List three safety rules related to the use of lasers.

Ask students to apply the communications model to this activity.

Lasers

Mention lasers to most people and they think of high-power weapons used to shoot down satellites or missiles. Some lasers are used that way, but most of them are used in factories and hospitals to help make products or to help doctors perform operations. The word **laser** is also often misunderstood. It is an acronym that stands for: **l**ight **a**mplification by **s**timulated **e**mission of **r**adiation. The radiation is in the form of light energy.

Lasers were first suggested by Albert Einstein in 1917. Einstein predicted that atoms could be excited to a higher energy state. He predicted correctly that when the atoms returned to their normal state, they would release energy in the form of a **photon** of light. The first laser was made by Theodore Maiman in 1960 using a ruby rod with mirrors at both ends. Lasers have been around for over 30 years. Some people say lasers are a tool looking for a way to be used. In recent years, lasers have become common in many applications: supermarket checkout stands, laser disk players, CD players, optical data storage for computers, computer printers, and laser surgery.

Lasers are made in many different sizes and strengths. Different colors or wavelengths of light are produced by different materials used in the laser. Lasers used in schools commonly use helium and neon gas. They are called HeNe lasers and produce a red-orange light. Other lasers use materials such as argon, carbon dioxide, or organic dyes to produce different wavelengths or colors of light.

The future of lasers is very bright. There are hundreds of uses that people haven't thought of yet. Maybe you will think of a new way to use lasers someday. Maybe you can think of a new way to use them today!

Techno Teasers
The Workings of Light

Techno Teasers
Answer Segment

Challenges:

1 Research the history of lasers. Who first came up with the idea of lasers? How long has it been since the first laser was made?

2 Shine a laser at a rear-surface mirror. Use an ultrasonic humidifier to make the beam visible. How many beams are reflected from the mirror? Can you explain why?

3 Research the names of the communication satellites in geostationary orbit. Make a bulletin board display of a drawing or photo of each satellite, and list the names of programs available.

4 Use a satellite dish and receiver to downlink a satellite television program. Consider how communication technology has changed through history. What do you think the phrase "global village" means when used to describe the ease of communicating around the world with satellites?

Modems: You learned about **modems** in Chapter 3. It is important to realize that modems are communications systems that change computer signals to audio tones so that they can travel over telephone lines. At the other end of the line, another modem turns the audio tones back to computer signals.

Connecting computers to each other through modems and phone lines lets you get information easily and quickly from **electronic bulletin board** services. You just call up a computer using your computer, telephone, and modem. Then you can leave messages for others to read or access other information left by someone else. **Electronic mail** is different in that you can leave messages in private electronic mailboxes. That way, only the person you want to get your message can. You also have your own electronic mailbox.

Networking: If you have purchased airline tickets, you know that the ticket agent makes your reservation on a computerized reservation system that connects thousands of agents with computers to a central data **network**. That's why they can tell you immediately if you can get tickets to almost any place in the world. Computer networks are systems that have many smaller computers connected to a larger computer by telephone lines. In many offices, computers in each office can share the same information or data. Even automatic banking machines use computer networks after banks have closed.

FAX (Facsimile): **FAX (facsimile)** machines send graphics or pictures electronically. FAX communication is becoming more important today when we want to send graphic information quickly. Did you know that since the 1920s, weather maps have been sent by FAX? A FAX works by an optical scanner moving across a page. It changes information into electrical impulses that are sent over the telephone wires to another location and then changed back to a picture. Corporations and businesses use FAX communications to order materials and to send letters instead of using the post office.

Computers used with a modem should be equipped with virus detection software.

Ask students if their parents use FAX machines.

FAX machines are a common piece of electronic communication equipment used in offices and businesses. Does your school or someone you know have a FAX machine? (Courtesy of Canon Corporation.)

Other communication systems use electronic devices to get pictures and graphics. Sometimes two or more different communication systems are used together.

THINGS TO SEE AND DO:
Capturing a Video Image with a Computer

Introduction:
In this book, you have learned that computers are being used in hundreds of ways. Today it is possible to combine more than one technology to make a new way of doing things. For example, you have used a computer and you have used a video camera for separate tasks. In this activity, you will combine a video camera with a computer to capture an image. Graphic artists, video production engineers, and publishers use computers to capture video images for newspapers, magazines, books, and television production.

Design Brief:
Use a computer and a video source such as a camera to capture and print a video image.

Materials:

Equipment:
▶ Computer with capture board (interface), capture software, printer
▶ Video source (camera, VCR, camcorder, laser disk player, still video camera, etc.)

Still video cameras are available through mail order or through computer suppliers. See Teacher's Resource Guide.

Procedure:
① The process of capturing a video image on a computer is quick and easy. In this activity, you will work with a partner to capture each other's picture.

❷ Ask your teacher to demonstrate the proper use of the equipment you will be using and the software you will need.

❸ Have your partner adjust the video camera (or other video input). Use the computer software and the capture interface to grab a frame of video into the computer's memory.

❹ Some capture software lets you adjust the picture before printing it. You might be able to use special effects to enhance the picture or to make it look different. Ask your teacher what options you have after you capture the picture.

❺ If your computer has the ability to change the captured picture, experiment with things such as erasing the background or adding your name or a bubble caption similar to a comic strip character.

❻ Print the picture using the proper commands for your computer.

Challenge students to design and make a picture I.D. card using this technology.

Evaluation:

❶ What are three possible video input sources you could use to capture an image on a computer?

❷ What happens if the subject moves when the capture process is started?

❸ How does the printed output of the captured image compare with a real photograph?

❹ How could news photographers use this technique to get photos into the newspapers quickly?

Challenges:

❶ If it is almost impossible to detect a computer-altered photograph, what do you think could happen if photographic evidence were changed in a trial? How could this problem be prevented?

❷ If you could take a picture and enter it into a computer, how could you send that picture overseas quickly?

❸ Try to exchange computer pictures with another school using a modem.

❹ What do you think the home camera of the future will be like? Will you need to send film out to be developed? What will happen to the size of cameras in the future?

Communications in the future

Just think what will happen in the future as we develop machines that understand our voice. Right now there are voice-activated software

TECHNOFACT

Technofact 90
Did you know that you will soon be using pocket-sized disk players that use minidisks? These disks are only 2.5 inches in diameter and can play 74 minutes of audio. That's as much as a standard compact disk.

programs. As the programs become more advanced, computers will be able to recognize voices and complete sentences without being trained to respond to a certain voice. You will just speak to a computer, and it will respond in some way. Maybe it will print out your words or make something else, such as a robot hooked to it, respond.

Other appliances such as lamps, dishwashers, clothes dryers, and televisions will turn on and off at voice command. Your telephone might be holographic so you can see who you call in three-dimensional full color. Holographic phones use laser beams to create the three-dimensional effect called **holograms**. You might have to start worrying about how you look before you answer the phone! Already television sets are being produced to give you sharper, more detailed pictures.

Satellites in the future will bring us in more contact with people in other parts of the world. **Videoconferencing** (using television signals beamed by satellite from one location to another) is already opening up ways to communicate with others face-to-face. You can transmit voice and picture through videoconferencing. NASA also has plans for sending a communication spacecraft beyond our solar system. The project will send a probe that will travel for 50 years after it is launched, sending back information and pictures. Hopefully, you will learn more about the sun and other planets. Maybe you'll be part of that project.

As always, computers will bring big changes in the way things work and how we do things. As computers become better, their uses in communications will grow. Artificial intelligence uses computer networking. Computers cannot think like people. Through artificial intelligence, computers will make decisions and solve problems that are commonly solved by people like you.

There is a trend toward home shopping done by connecting phone lines to computers. You can also pay bills or order things through the computer. In the future, you might do your homework from a computer networking database. Even your entertainment might be provided by a computer network. Because networking computers helps people share information, more businesses and schools will start using networking to

Videoconferencing can save time and travel expenses when people need to discuss business decisions. It makes sense to use technology in this way. Can you think of some possible ways communications technology might affect your future? (Courtesy of Sony Corporation of America.)

improve communications, give you more information resources, keep track of records, and make things run more smoothly.

The future of communications depends a lot on what you as the consumer need and expect from communications. Who knows? Maybe you'll be able to turn your own room at home into a live TV studio using the telephone lines and a video camera. Won't it be fun to send a "live" greeting to a friend somewhere far away!

Summary

Communication is the process of exchanging information. A communication system is like any other system. It has input, process, and output. To make sure your communication works, a message has to be sent, received, and understood. Feedback from the receiver lets you know if your message was understood.

There are many ways to communicate. Today, technical writers and technical illustrators use computers and other technologies to write manuals and instructions explaining how to do things. They add pictures and graphics to make the message clearer for you.

Photography, animation, and video are other ways to communicate a message in addition to writing. To communicate well in any of these areas, you need to use the following design elements: line of golden proportion, rhythm, and unity. All of these elements are part of the total composition of your photograph or picture. Animation is just graphics or pictures in motion. It relies on persistence of vision to make you think pictures are moving. Video technology is quickly changing. Technology is helping us make smaller cameras that can do more things. It is important to know that taping and filming are different processes. Taping is a magnetic process, while filming is a photographic process.

Improvements in electronic technologies have made it possible for you to use many other kinds of communications systems. Television, radio, telephones, computers, FAX machines, fiber optics, modems, and satellites are just a few that we use today. In each case, these systems use electronic signals or electromagnetic signals to carry messages through cables, through glass fibers, or through the air.

Computers are used in hundreds of ways in communications technology. Networking computers is a good way to share the same information among many people. You can also connect with an electronic bulletin board or have your own electronic mailbox to receive and send messages.

The future of communications is unlimited. The computer will continue to get better. As it does, there will be more uses for it in communications. You will see more machines that respond to voice command. Satellite communications will bring people around the world even closer together as well as let us learn more about outer space.

Challengers:

1 Use a small pad of paper to make your own flip book animation series.

2 Choose an object. Make a three-view (orthographic) drawing and a pictorial drawing of the same object.

3 Find out if there is a ham radio operator in your area. Ask for a demonstration.

4 Visit a local radio or television station. Find out how programs are produced and aired.

5 Contact your local phone company. Ask them for samples and information about fiber optics.

6 Use computer-aided drafting software to draw the satellites in the Clarke Belt.

7 Research how to make a hologram and try to make one.

8 Research how to network the computers in your school or classroom.

9 Use a modem and computer to access an electronic bulletin board service.

10 Ask your teacher to set up a group activity with other schools that lets you share information using communications technology such as modem, FAX, telephone, and so on.

See Teacher's Resource Guide.

Chapter 14

Space Technology

Things to Explore

When you finish this chapter, you will know that:

▶ There are many milestones in space history.

▶ Modern rocketry began with three people: Konstantin Tsiolkovsky, Robert H. Goddard, and Hermann Oberth.

▶ Things do not work the same in space as they do on Earth because of microgravity and the near vacuum conditions.

▶ Putting objects in space takes lots of energy to overcome Earth's gravity.

▶ Living in space for long periods of time presents new challenges to people and equipment.

▶ Plans for building space colonies on the moon and Mars are in the near future.

▶ Space spinoffs are new technologies and products that were developed during space projects but can be used for different purposes.

▶ Future plans for the space program depend on how taxpayers want their money spent.

TechnoTerms

alloy	*Skylab*
gravity	space spinoffs
launchers	*Sputnik*
mass	thrust
microgravity	vacuum
NASA	vector
Newton's laws of motion	zero-g
self-sufficient	zero gravity

(Courtesy of NASA.)

Careers in Technology

Can you imagine how exciting and challenging a career as an astronaut is? Training to be an astronaut is vigorous and demands you be in excellent physical shape before you go into outer space. Astronauts have many different jobs. Some are trained specialists in biotechnology, physical science, or astronomy. Others are computer and electronic specialists. All astronauts must be team players in order to live and work together in such close quarters. Many new discoveries are made by astronauts during experimentation in the space environment. Just viewing the Earth from space must be an amazing adventure!

Astronomers estimate there are 100 billion stars in the Milky Way Galaxy we live in. They also estimate there are over 100 billion other galaxies! Have you ever wondered what's out there? The "final frontier" is there for you to explore. Here are the Earth and moon, comet Kohoutek, and the Lagoon Nebula. (Courtesy of NASA.)

Space History

Do you ever just look up at the sky at night and watch the stars and planets? Do you wonder what it's really like out there? For a long time, space travel was just a dream. Now it is a real thing. Space is sometimes called the "final frontier." We've already explored the earth's continents and oceans, and so it seems natural to go beyond earth. The challenges are great. How do we send scientific instruments, machines, and people into space and bring them back safely?

The first big problem was just getting a spacecraft off the earth. Today's rockets are a combination of many different ideas tried through the ages. Stories of rocketlike devices come from as far back as 400 B.C. In Rome a wooden bird moved using rocketry principles that weren't even described until the late 1600s by Sir Isaac Newton. Even the Chinese in the thirteenth century used a form of gunpowder to launch a simple form of a solid-propellant rocket they called fire-arrows.

Modern rocketry really began with three people: Konstantin Tsiolkovsky, Robert H. Goddard, and Hermann Oberth. They are called the first true pioneers of space travel. They took the ideas and dreams of others such as Jules Verne, the science fiction writer, and tried to make them work.

Challenge students to research early rocket designs and scientists.

▶ In 1898, Konstantin Tsiolkovsky, a Russian schoolmaster, suggested using liquid propellants for rockets. Even though he never launched a rocket, he explained the principles by which rockets could fly in space.

Challenge students to make a picture display of early rockets.

▶ An American, Dr. Robert H. Goddard, launched the first liquid-propellant rocket in 1926. The flight, rising only 41 feet, lasted 2.5 seconds, but it was a start. His experiments in liquid-propellant rockets continued for many years.

▶ Hermann J. Oberth also worked with long-range liquid rockets, but he was more an idea man than an inventor. His writings were important because they got scientists and engineers excited about building rockets.

All three of these people believed space travel was possible, and they were right.

Show a NASA video on space history.

New developments in technology quickly moved people from dreaming about space travel to making it happen. The launch of

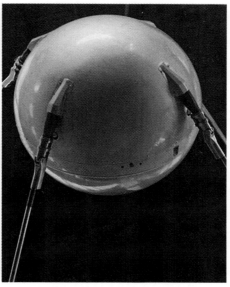

Robert Goddard was a pioneer of rocket science. His experiments proved that powered rocket flight was possible in space. (Courtesy of the Goddard Collection/Clark University.)

The space race really started with the Russian satellite called *Sputnik I* in 1957. The satellite weighed 184 pounds and was only 23 inches in diameter. *Sputnik I* was the first artificial satellite to enter low Earth orbit. It circled the Earth every 90 minutes. (Courtesy of NASA.)

Sputnik by the Soviet Union (now the Commonwealth of Independent States) in 1957 was an exciting moment for the entire world. We now had entered the Space Age. *Sputnik* stayed in a low Earth orbit, where it circled the Earth once every 90 minutes. For the first time, people realized that a human-made object really could stay up in space and orbit the Earth. This was a giant step! Then *Sputnik 2* carried the first dog, Laika, into space. This experiment proved that living things could survive in outer space.

The next step was to put people into space. In 1961, a Russian cosmonaut, Yuri Gagarin, made the first manned space flight. Two Americans were right behind him that year in Mercury missions: Alan Shepard, Jr., in *Freedom 7* and Virgil I. "Gus" Grissom in *Liberty Bell 7*. Throughout the 1960s, more and more space missions were made. Finally, in July 1969, Neil Armstrong and Edwin "Buzz" Aldrin, Jr., set foot on the moon.

TECHNOFACT

Technofact 92

Without an atmosphere or weather to disturb them, the footprints left on the moon by *Apollo 11's* Aldrin and Armstrong in 1969 will stay there for thousands of years.

One of the most exciting moments in history happened when the first people, Neil Armstrong and Edwin "Buzz" Aldrin, set foot on the moon. All of the knowledge gained by earlier astronauts and scientists was put to use to get the Apollo mission to the moon. The research that went into this project greatly increased our technology knowledge base. (Courtesy of NASA.)

Technology has grown rapidly thanks in part to the exploration of space. Advancements in the exploration of space and hundreds of benefits called spinoffs have greatly expanded our knowledge. Here are a few of the milestones in the space program. Can you identify a Redstone rocket and *Skylab*? (Courtesy of NASA.)

Challenge students to research and compare the Soviet space station *Mir* with *Skylab*.

Show NASA video on the story of the Space Shuttle.

Since then space has been an important area for research and development in the United States. In 1973, ***Skylab*** went into space. Three different crews worked and lived in *Skylab* as it orbited 270 miles above the Earth. The unmanned *Voyager* satellites launched in 1977 have visited all the planets except Pluto and sent back fantastic images of the moons, the atmospheres, and the planet surfaces. The *Voyagers* will leave our solar system by the end of the twentieth century and will probably wander through space for thousands of years.

Until 1981, **NASA**'s (National Aeronautics and Space Administration) manned programs such as Mercury, Gemini, Apollo, and Skylab used launch vehicles and spacecraft that went up only once. The development of the Space Shuttle was a technological wonder. The first one, *Columbia*, became the first reusable space vehicle. Even the solid rocket boosters are rebuilt after every mission.

The shuttle is a remarkable flying machine that takes off like a rocket but lands like a glider on a runway. The former Soviet Union's *Buran* works just like the United States' space shuttle. Every space shuttle mission set new records and accomplished "firsts" in manned space flight. The first U.S. woman in space was Sally Ride. All kinds of people, from doctors, U.S. senators and congressmen to many different scientists have been part of the shuttle crews. Then, in 1986, the *Challenger*, carrying a crew of 7, exploded just seconds after it was launched. That disaster helped us remember that there are still many problems to solve in making space travel safe.

There are many **milestones** (important events) in space history. In your life so far, what events do you think have affected space travel most? Now you have robot spacecraft that travel to the planets. Satellites launched by rockets and the shuttles enable scientists to investigate your world, forecast weather, and make it easy for you to communicate instantly with other people around the world.

The Space Shuttle is sometimes called a "space truck" for its ability to deliver payloads to Earth orbit. This sequence of photos shows the major steps in preparation, launch, orbit, landing, and reuse of the orbiter and solid rocket boosters. (Courtesy of NASA.)

THINGS TO SEE AND DO:
Rocket Technology

Introduction:

You have read about Robert Goddard's early experiments with rockets. You probably have seen old movies of early attempts to fly. People learned about rocket and airplane flight mostly by trial and error. They tried their idea and if it didn't work, they learned a valuable lesson. They knew what not to try again. In this activity, you will work as an aerospace engineer to design a launch system and a rocket.

Design Brief:

Design a safe launch system, rocket, and altitude gauge that demonstrate Newton's third law of motion. The rocket will be powered by compressed air and water.

The launch of bottle rockets using compressed air must be supervised by the teacher. Use only bottles designed to hold carbonated soda as they can withstand the pressure.

Materials:

▶ Wood, plastic, fasteners, and so on as needed

▶ 2-liter Plastic soda bottle

▶ Rubber stopper, hose connection

▶ Cardboard, duct tape, string

Equipment:

▶ Band saw or scroll saw

▶ Drill press

▶ Air compressor, hose, regulator

▶ Calculator (optional)

Procedure: Building the Launch Pad

1 In this activity, you will work in a group to design and build your launch system, rocket, and altitude gauge. Your group should be large enough to complete the following tasks quickly:

• Design and build the launch pad.
• Design and build the rocket.
• Design and build the altitude gauge.

SAFETY NOTE:
In this activity, you will use everything you have learned about the proper selection of materials and the safe use of power tools. The design of your rocket launch system must be made with safety in mind. The rocket will leave the launch pad quickly, and it will reach altitudes of over 100 feet. The use of your launcher must be supervised by your teacher. All sharp edges and other hazardous parts of your rocket must be removed. The materials used for the rocket must be safe for the launch and return of your rocket.

2 Brainstorm the design of your launch system and rocket. The first design consideration must be safety. You will be using compressed air to build pressure in the rocket. If the rocket releases too soon, it could hit you. Your teacher will advise you about the safety of your design. Consider the following design specifications:

- The bottle (rocket) must be held firmly to the rubber stopper while air pressure is increased.
- A guidance system must be used to hold the rocket upright and direct the rocket launch vertically.
- Launch controls should be able to be at least 20 feet away from the launch system.
- The bottle-holding mechanism must release the rocket smoothly without catching.

The illustrations on pages 367 and 368 show some possible ideas you might use.

3 Make a detailed sketch of your ideas. Ask your teacher to inspect your plans for safety.

If time does not permit students to build their own launch system, the teacher might want to build one launcher for the class.

Inspect student launcher plans to ensure an even release of the bottle rocket to prevent it from tipping over.

Hinges can be used to hold the neck of the bottle rocket down on the rubber stopper. A locking mechanism is used to release both hinges at the same time.

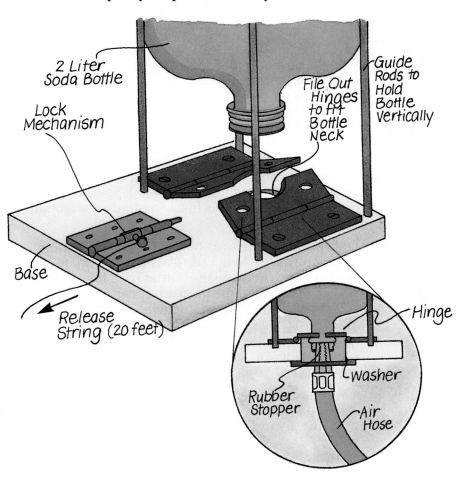

2 Liter Soda Bottle

Lock Mechanism

Base

Release String (20 feet)

File Out Hinges to fit Bottle Neck

Guide Rods to Hold Bottle Vertically

Hinge

Washer

Rubber Stopper

Air Hose

A simple sliding mechanism can be used to hold the neck of the bottle rocket while air pressure is added. File the edges smooth so it won't stick to the plastic bottle.

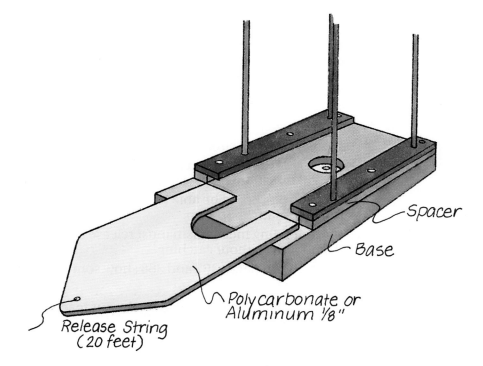

Spacer

Base

Polycarbonate or Aluminum ⅛"

Release String (20 feet)

4 Divide the tasks of building the launch mechanism among your group members. Assemble and test each part without air pressure to test their operation.

Procedure: Building the Rocket

1 While part of your group is building the launch pad, others can be building the rocket and altitude gauge. The best launch system will not help a poorly designed rocket. Keep the following ideas in mind as you build your rocket:

- Only plastic 2-liter soda bottles in good condition should be used. Do not use glass bottles. Do not use water bottles. Because the bottles used for soda are made to withstand the pressure of carbonation in the soda they are ideal for this activity.

- The materials used for the fins, nose cone, or any other part must be made of cardboard or other material that will collapse easily on impact. Wood, metal, or plastic parts should not be used.

- The method of attaching the nose cone, fins, or other rocket parts to the plastic soda bottle should not weaken the bottle. Tape or silicone sealant are good choices. Hot melt glue or other adhesives might weaken the bottle.

Here is a possible design:

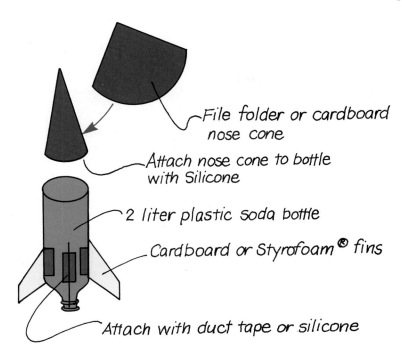

File folder or cardboard nose cone

Attach nose cone to bottle with Silicone

2 liter plastic soda bottle

Cardboard or Styrofoam® fins

Attach with duct tape or silicone

The bottle rocket must be made with safe materials. Paper, cardboard, or Styrofoam are acceptable. Wood, metal, or hard plastic materials should not be used.

❷ Make a sketch of your rocket plans. Have your teacher inspect your plans for safety. Build the rocket and have it ready for testing when the launch system is complete.

Procedure: Building the Altitude Gauge

❶ While others in your group are building the rocket and launch system, someone should build an altitude gauge to determine the maximum height your rocket reaches.

❷ The design of the gauge is very simple. It will be used to measure an angle, so you can use trigonometry to calculate the altitude.

❸ Roll a piece of paper into a tube, or use a cardboard tube from aluminum foil. Tape two pieces of thread to one end to make cross hairs for accurate sighting. Use tape or glue to attach a protractor to the side of the tube. Attach a string and weight as shown in the figure on page 370.

Work with the math teacher to cover basic trigonometric functions that students will use in calculating the rocket heights.

Procedure: Rocket Launch

❶ Ask your teacher to inspect your work. Fill the bottle rocket halfway with water. Mount the rocket on your launcher.

❷ Take the launcher outside. Do not put air pressure in the bottle until your teacher tells you. Attach the air hose. Run a string at least 20 feet away from the launch pad to operate the release mechanism. Point the rocket so that it is vertical.

The altitude gauge is used to find the maximum altitude the rocket reaches. Trigonometry will be used to calculate the height when you know the angle from the ground and the distance from the rocket launch pad.

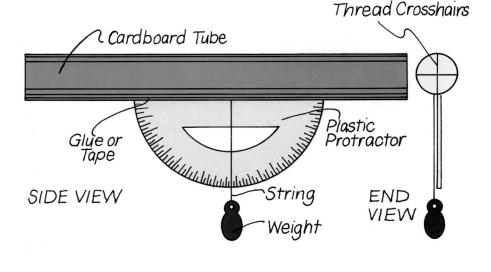

Thread Crosshairs

Cardboard Tube

Plastic Protractor

Glue or Tape

SIDE VIEW

String

Weight

END VIEW

Air pressure should not be applied to the bottle rocket while students are in launch area. Be sure to warn bystanders to stay at least 20 feet away.

SAFETY NOTE:
Warn others in the launch area to move away. No one should be within 20 feet of the launch area when air pressure is turned on. If the rocket does not launch, release the pressure at the hose <u>before</u> going near the rocket. Do not use pressures greater than 80 psi (pounds per square inch). Your teacher must be present during setup and launch.

❸ Measure 100 feet away from the launch pad. Have someone in your group use the altitude gauge at this point and be ready to sight the rocket's highest point. The altitude gauge will give you an angle that will be used in measuring the maximum height of the rocket.

❹ This is the final test! Be sure everything is ready. Go over the following checklist to be certain:

- Launch area clear of people.
- Teacher present.
- Air pressure ready.
- Launch string at least 20 feet away.

If everything is ready, start your countdown; 10 - 9 - 8 - 7 - 6 - 5 - 4 - 3 - 2 - 1- pull the launch string. Sight the maximum altitude of your rocket. Measure the angle using the protractor on the altitude gauge.

Evaluation:

❶ The height of your rocket will be measured using trigonometry. This sounds hard, but it is easy. First, subtract the angle you measured on your altitude gauge from 90°. This will give you the angle from the ground to the highest point your rocket reached.

Angle from ground = 90° − Angle on altitude gauge

To calculate the height, you will use a trig (trigonometry) function called tangent (TAN).

The formula is easy:

$$\text{TAN} \underline{\quad}° = \frac{\text{Opposite side (altitude)}}{\text{Adjacent side (100 feet)}}$$

The tangent of the angle you measured can be found in the trig table in the appendix of this book. Just find the angle you measured and read the tangent value to plug into the formula. To make it even easier, this formula can be changed to look like this:

TAN ___° × 100 feet = _____ altitude (feet)

What was the altitude of your rocket?

❷ Why do you need to subtract the angle you measured on the altitude gauge from 90°?

❸ How could the design of your rocket have been changed to make it go higher?

❹ What do you think would happen if you put more or less water in the rocket?

Students may need help in finding the tangent of an angle using the trig chart in the Appendix.

Challenges:

❶ Experiment with different amounts of water in the rocket. Keep the air pressure constant for each test (less than 80 psi). Make a graph of the results of your test.

❷ Contact a local bottling plant. Ask them for the maximum pressure the soda bottles are designed to withstand. Ask what pressure is used in the carbonation process. Explain why there is such a difference.

❸ Redesign your launch pad to operate electrically using 12 volts DC or less. Think of ways you could accomplish this. Ask your teacher to approve your design. Keep electrical equipment and connections dry. Test your idea using two 6-volt lantern batteries and a 20-foot length of wire so you can stay clear of the launch pad.

❹ Use your altitude gauge and the trig formula above to measure the height of objects around your school. Birds flying, flagpoles, trees, or even the maximum height of your school could be measured.

Be sure students do not use more than 80 psi air pressure to launch their bottle rockets.

Electrically operated launch systems must use safe low voltage current.

We want to know whether people can live for long periods of time in outer space. Many men and women astronauts go beyond earth's boundaries. They want to find out about life in outer space and how things work. Eighteen nations, working together and independently, have satellites in orbit above the earth today. By the early 2000s, over fourteen nations expect to have sent people into orbit. How would you like to be the first student in space?

Weightless conditions can be simulated on Earth in water. (Courtesy of NASA.)

Teacher in space, Barb Morgan, tries out weightless conditions in the KC-135, or "vomit comet." Can you guess why the KC-135 is called this? (Courtesy of NASA.)

Ask the students to calculate how much they would weigh on the moon.

TECHNOFACT

Technofact 93

How about a ride on the MMU (manned maneuvering unit)? Some people call it a "space scooter." Most people think you can just tear around in space on an MMU. But that's not really true. You have to accelerate (speed up) and decelerate (slow down) very slowly or you'll totally miss where you're going, or worse yet, you might smash into something! Astronauts learn to first estimate how far they need to go. Then they slowly accelerate to about a normal walking pace. When they think they've reached that speed, they just float to their target.

Space Physics

Now if things worked the same in space as they do on the Earth, space physics would be simple. But that's not the case. In outer space, there is almost no air. It is a near **vacuum**. There is also **microgravity** (very little gravity). Some books call microgravity **zero gravity** or **zero-g**. It really doesn't mean there isn't any gravity. It just means gravity's force is very, very small. **Newton's laws of motion**, which explain how things move on earth, still work for the space environment. Sometimes the results are different though! You can do tumbling stunts in outer space that would be impossible to do on earth!

What does it take to get objects into space anyway? Putting objects like satellites and space shuttles into orbit takes lots of energy to overcome the Earth's **gravity**. What is gravity? No one knows exactly. It's one of four basic forces that we have discovered in nature. Gravity is not a very strong force compared with the other basic forces. But it's strong enough to keep you stuck firmly to the ground, to keep the moon orbiting the Earth, and to keep the planets orbiting the sun.

Gravity is the force of attraction between objects. The Earth's gravity pulls you and everything toward the center of the Earth. Things have weight because of gravity. Just how much you weigh depends on how strongly the earth's gravity is pulling on you. Have you seen the films of astronauts leaping around the moon like they were on a trampoline? On the moon, the gravitational force is only one-sixth the pull of gravity on Earth. So you can have the same **mass** (amount of matter) but weigh only one-sixth as much on the moon! How much would you weigh on the moon?

Have you seen the pictures of astronauts or things they are using such as pencils floating weightless in the shuttle? Some people used to think that this happened because the astronauts and the shuttle were outside of Earth's gravity. That's not true. Earth's gravity extends for a very long distance out into space because the Earth is so big. The astronauts appear to be weightless because the space shuttle along with the astronauts is really "falling around" the Earth. The astronauts really still have weight, but they are in "free-fall." The astronauts can't press against the "floor" of their shuttle because the floor is falling at the same speed toward the Earth as they are. Does that sound strange to you?

Newton's Laws of Motion

Sir Isaac Newton (1641–1727) described physical motion with three scientific laws. The laws explain the motion of all objects, whether on earth or in space. In simplest form, Newton's laws of motion state:

❶ *Objects at rest will stay at rest, and objects in motion will keep moving in a straight line unless acted on by an unbalanced force.* A rocket on the launch pad is balanced and at rest. The surface of the pad pushes the rocket up, while gravity tries to pull it down. As the engine starts, the rocket thrust causes things to become unbalanced. The rocket travels upward. In space, a spacecraft will travel in a straight line if all the forces stay balanced. If it gets close to a larger body such as a planet, the planet's gravity will pull on the spacecraft, and it will curve the path of the spacecraft.

❷ *Force is equal to mass times acceleration (F = ma).* To make something as large as a rocket move quickly, you need a lot of force. The amount of force or thrust produced by a rocket engine is determined by how much (mass) rocket fuel is burned and how quickly the gas escapes from the rocket.

❸ *For every action, there is an equal and opposite reaction.* A rocket can lift off from the launch pad only when it pushes gas out of its engine. The rocket pushes on the gas, and the gas pushes on the rocket. The **action** is the gas coming out of the engine. The **reaction** is the movement of the rocket in the opposite direction. In space, even tiny thrusts will cause the rocket to change direction because of microgravity.

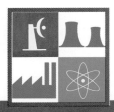

T E C H N O F A C T

Technofact 94

If you want a close look at the stars, visit the Keck telescope on top of Hawaii's Mauna Kea volcano. It is the largest in the world. Its main mirror is around 400 inches in diameter. Each separate mirror segment has an accuracy of four billionths of an inch. Astronomers say the Keck telescope will let you look as far as two-thirds of the way toward the edge of the universe. That's far out!

Show NASA video of astronauts in microgravity.

Work with the science department in the discussion of gravitational pull.

Special Report
Life in Space

Begin
Segment

Swing a ball on a string in a circle to demonstrate the principles.

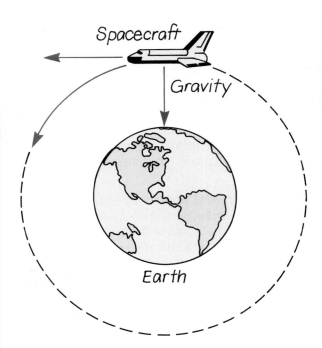

Spacecraft

Gravity

Earth

Weightless conditions are not caused by being away from the pull of gravity. Weightlessness is caused by the fact that spacecraft are "falling" around the Earth. It is like being in a free-fall through space.

All orbiting spacecraft are actually "falling" toward the Earth. Why don't they all fall straight down to Earth? The answer is because of the spacecraft's speed and the shape of the Earth. If a satellite or spacecraft stopped moving, it would fall down to Earth. When a spacecraft is launched into space, it keeps moving in a certain direction. Remember, there isn't much air to slow it down. Gravity pulls on the spacecraft, so it doesn't keep going straight out either. The spacecraft is traveling forward fast enough to fall toward the Earth just as fast as the Earth curves away from it. Have you ever tried swinging a piece of string or a chain with something such as a key or a ball on the end of it in a circle? Your hand acts like the Earth. The string or chain represents the force of gravity keeping the object from flying off when you swing it. How fast (acceleration) you keep it moving makes the difference between whether it falls back to your hand or continues in a circle.

THINGS TO SEE AND DO: *Using Vectors*

Introduction:

A **vector** is a graphic way of showing a quantity such as force. Vectors are fun and easy to use. They make it possible to find the answer to complicated problems. Forces are measured in pounds. Your weight, for example, is the force you put on the Earth due to gravity. Forces can be measured with scales. You could hook a spring scale to a chair with wheels and pull. The amount of force needed to move the chair would show on the scale. In this activity, you will work as a NASA technician to help determine the direction of a stranded astronaut on a space walk.

Design Brief:

Use vectors to determine the result of five forces pulling on a stranded astronaut.

Materials:

▶ Paper

Equipment:

▶ Computer with CAD software (optional)

▶ Drafting tools

Procedure:

1 In this activity, you will work individually to try to help a stranded astronaut. The microgravity conditions of space make it easy to move when you push off of something like a space station. If you are "floating" free in space, there isn't anything to push off of.

2 This simulation shows what might happen to an astronaut in an MMU (manned maneuvering unit) if propellant ran out for the operation of the maneuvering jets.

In this activity, we will simulate an astronaut in trouble, unable to return to the space shuttle. The good news is there are five other astronauts waiting to help pull him in.

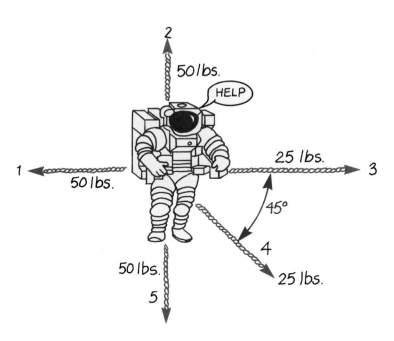

3 You can see from the illustration that there are five other astronauts pulling on ropes from different directions with different forces. Your problem is to determine exactly which way the astronaut will move and with what force.

4 The solution to this problem is easy if you use vectors. We can give a length measurement to represent the forces. In this example, you will use a scale of one inch equal to 25 pounds of force. This way, a 50-pound force would be drawn two inches long (50/25 = 2). The other important thing we need to know when using vectors is their direction. In the astronaut example, you can use a 45° drafting triangle and a T square to draw all the vectors you need. If you are familiar with CAD software, you could solve this problem on a computer.

⑤ There are many ways to get the answer or result of this problem. Basically, you draw each vector the proper length and direction, but you "add" them graphically. You just start with one vector and add the next one by drawing it where the first one ended. This "tail-to-tip" method is used until all the vectors are drawn. The answer or result is the distance and direction from where you started to where you ended. Here are two ways of solving this problem. Try to draw them to scale on paper or on the computer.

Help students understand the scale used to represent the force of a vector.

Vectors are lines drawn to represent the strength and direction of forces. Vectors can be added to find the effect of all the forces working on one object. The answer is called the result. The vectors are added tail-to-tip in order clockwise or counterclockwise. The results are the same in length (strength) and direction (angle).

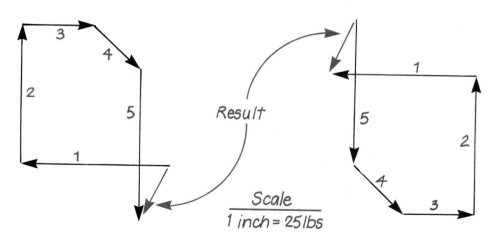

Scale
1 inch = 25 lbs

Evaluation:

❶ What two things must you know to draw a vector?

❷ What other things besides forces can be drawn as vectors?

❸ Which of these quantities—temperature, time, velocity, acceleration, weight—could be drawn as a vector?

❹ What scale could be used to draw a 1200-pound force on a piece of paper?

Challenge students to research the difference between a vector and scalar quantities. Have them make a list of these examples.

Challenges:

❶ Research the term *vector* in a physics book. Make a list of the quantities that can be represented as vectors. What are quantities other than vectors called?

❷ Explain how navigators on airplanes might use vectors when they know the speed and direction of winds.

Graphically (using vectors) show how a head wind would slow the speed of an airplane.

❸ Place a student on a chair with wheels. Have three other students pull on the chair in different directions. First predict which direction the chair will move. Measure the forces and their directions, and find the answer using vectors.

❹ Design a problem that could be solved using vectors. Some ideas might be:

- A lion on six ropes with six people pulling in different directions.
- A truck stuck in mud with five other trucks pulling it out.

Space Travel

In order to get into space, you must travel more than 17,000 miles per hour. This is called the **escape velocity**. Space vehicles reach orbit at that velocity. **Launchers**, such as the solid rocket boosters and the motors in the space shuttle, are used to give enough power and speed to overcome gravity. Even so, some people think the space shuttle can go to the moon. That's not true. Even the powerful space shuttle engines can only put the shuttle in low Earth orbit, or about 300 miles out. The moon is 244,000 miles away. The farther out an object is going, the more energy you need to get it there. If a space vehicle wants to move farther away from the Earth, it will need an extra push, or **thrust**, from its engines. Rockets actually work better in space, where there's a near vacuum, than they do in air. The escaping gases don't have to push away much air to get out. Pushing against air takes up rocket energy.

THINGS TO SEE AND DO: *Space Conditions*

Introduction:

Gravity attracts the air in our atmosphere and holds it near the surface of the Earth. In fact, if we could weigh the amount of air in the atmosphere at sea level, it would weigh 14.7 pounds for every square inch of the Earth's surface. The higher you go in altitude above sea level the less air there is in the atmosphere. Scientists have measured the number of air molecules at different distances from the Earth. At sea level, in one cubic centimeter there are a million, million, million air molecules. When you go farther out away from Earth the number gets smaller. At 20 miles above the Earth, the air becomes too thin for jet engines. At 50 to 600 miles away from Earth, there are one million air molecules per cubic centimeter. When you get to 1,200 miles in altitude, there is only one air molecule per cubic centimeter. People who say that there is nothing in space are not quite right.

A hand vacuum pump and plastic bell jar are available from science and technology suppliers. See Teacher's Resource Guide. A motor-operated vacuum pump and glass bell jar may be available in your science department.

In this activity, you will work as a research technician to investigate the effects of reduced air pressure. We call this condition a vacuum, but that doesn't mean there is no air at all. The pressure will be much less than normal atmospheric pressure, however.

Design Brief:
Demonstrate the effects of a vacuum.

Materials:
- Marshmallows, beaker
- Carbonated soda
- Battery-operated buzzer
- Balloon, tennis ball

Equipment:
- Vacuum pump
- Bell jar

Procedure:

1 In this activity, you will place different objects or materials in a vacuum. You will work in groups of three or four.

2 Your teacher will demonstrate the proper use of the vacuum pump and bell jar. Gather the materials you would like to test in reduced pressure. Ask your teacher to approve your test materials.

3 Make a chart of the materials you are testing, your prediction, and the effects of reduced pressure you observe. Your chart might look like the one shown below.

4 Place each object in the bell jar and draw a vacuum. Watch and listen for the effects of reduced pressure.

A bell jar and a vacuum pump can be used to simulate the reduced air pressure of space. (Photograph by Pam Benham.)

Evaluation:

1 What happened to the balloon when you placed it in a vacuum? Explain.

2 What do you think would happen to your body if you didn't have the protection of a spacesuit in space? Explain.

Make a chart like this to keep track of the data you gather in this experiment. Ask your teacher to approve other materials for testing in the vacuum.

Use a water/moisture trap when using a motor-operated vacuum pump.

Effects of Vacuum		
Materials	Prediction	Actual
Marshmallow		
Balloon		
Tennis Ball		
Buzzer		
Soda		

3 What caused the marshmallow to swell?

4 Why are tennis balls sometimes packaged in pressurized cans?

Challenges:

1 Are there sonic booms in space from the rapidly moving space vehicles like the space shuttle? Explain.

2 Make a list of similarities and differences between space suits and deep sea diving suits.

3 Why do mountain climbers at high altitudes have to boil food longer than they would at sea level? Why does water "boil" at cold temperatures in reduced pressure?

4 NASA researchers have found that metal parts sometimes weld themselves together when they touch in a vacuum. Can you think of how this could happen? Can you design an experiment that will illustrate "cold welding?"

Living and Working in Space

Once you're up in space, what would it be like to live and work there? What's so different about living in space for a long time? In order to find out, both the United States and the former Soviet Union have built space stations where people could stay for longer times. Living in space for long periods brings new challenges to both people and equipment.

Space Stations: *Salyut 1*, launched by the Soviet Union in 1971, was the first space workstation. *Salyut* was about 44 feet long. Though it was used only once, it was the beginning for many other successful *Salyut* missions.

Skylab, launched on May 14, 1973, was America's first space station. It became the home for three crews of three people Each. *Skylab* was about the size of an average three-bedroom house. On earth, it weighed almost 100 tons. The entire *Skylab* station was almost 120 feet long. It had a workshop area that included living and working quarters.

TECHNOFACT

Technofact 95

Did you know that astronauts occasionally smuggle food into space? John Young and Gus Grissom during the *Gemini 3* mission were the first. They wanted something better than packaged "spacefood," so Young smuggled two corned beef sandwiches inside his leg pocket! One of the better meals in those early days!

Show NASA video on *Skylab* operation.

Special Report
Preparing Food for Space

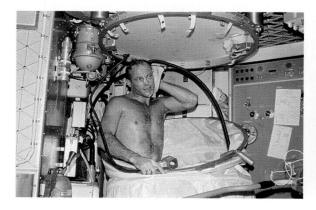

Skylab was the first U.S. space station. Astronauts staying in space for a long time need to eat, sleep, and bathe the same as on Earth. What do you think it would be like to take a shower in orbit? (Courtesy of NASA.)

T E C H N O F A C T

Technofact 96
How many sunsets and sunrises do you see when you're in orbit? Because the shuttle goes around the earth about every hour and a half, you can catch about sixteen in a 24-hour period!

Then in 1986, the Soviet Union launched the *Mir 1* space station. It had a computer-controlled system that took over some of the tiresome jobs people had to do. *Mir* was the first to have docking ports for other vehicles. Cosmonauts on *Mir* proved that people can stay in a microgravity environment for over a year!

The manned *Skylab, Salyut,* and *Mir* missions gave us lots of information about ways to make life better for people living and working in space. We want to use this information in building future space stations.

Building the space station of the future will be an engineering challenge. We've learned that building large structures in space is possible. The shuttle flights show we can move materials back and forth from Earth. It seems the next step is to build a large space station that can be a stepping stone to other far away places in space.

Besides the United States and the former Soviet Union, other countries including Japan are working together to build a space station. One plan calls for twenty shuttle trips over a 3-year period to assemble, outfit, and get a space station started. The modules (separate compartments) would be taken up separately and attached to a special set of trusses made of strong but lightweight struts. Power would come from solar cells and additional battery power. The modules would be like rooms in your own house, only smaller. The idea is for you to be as comfortable as possible during your time in space. Here's a quick look at life in space as you might find it today.

▶ Do you want to be taller? If you stay in a microgravity or zero-g environment very long you'll get that chance! You will actually grow an inch or two, because gravity isn't **compressing** (pushing together) the spongy disks between your **vertebrae** (the bones in your backbone). Sometimes your face will get puffy because gravity doesn't pull the blood away from your head. Your leg

The next space station will provide a platform for conducting more experiments with long-term exposure to microgravity, materials testing, and recycling of water. It may even serve as a building site for making a Mars exploration vehicle. (Courtesy of NASA.)

muscles will also become smaller because you don't have to work them as much to do things in a microgravity environment. Doctors have also found that you will lose calcium from your bones and they will become weak.

Researchers are working on medicines to help solve some of these problems. The best thing is exercise, and lots of it. Spacecraft now have exercise equipment such as treadmills and exercise bicycles on board. The astronauts are scheduled to use it for certain amounts of time daily to keep in shape.

Show NASA video of astronauts exercising in orbit.

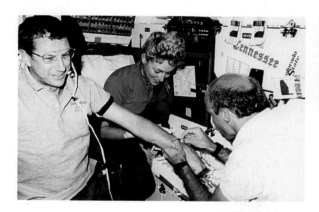

The crew of any space flight needs to stay in good health. During the flight, exercise is important to maintain muscle strength. Medical tests are a routine part of space flight. (Courtesy of NASA.)

▶ What's for dinner? In the early days, astronauts just squeezed food tubes and ate precooked things. Today, the astronauts have a diet especially planned for them to make sure they don't lose too many important vitamins and minerals such as sodium, calcium, and nitrogen. Breakfasts, lunches, and dinners are packed as meals in containers. Many foods have been vacuum-packed, dehydrated, and deep-frozen. Some, like spaghetti, cornflakes, bananas, and scrambled eggs, have been dried to save weight and space. Most drinks come in powdered form except orange juice and grapefruit juice. Luckily some things such as peanut butter, nuts, and chewing gum stay in natural form. Preparing the food takes special equipment. For example, a punching machine inserts hot or cold water into dehydrated food packs. Magnets or Velcro tape holds the food onto your tray. Even your fork is held down by a magnet. Most of the food is in a thick gravy or jelly-like substance, so it sticks to your spoon or fork. Yum!

Work with the home economics program to design space meals.

▶ In a space station, you can grow many of your vegetables. Any waste is recycled as fertilizer. Roots and stems have to be separated by a special plastic form or they will get all tangled up. The plants use the carbon dioxide breathed out by people, and they return oxygen in its place, just as they do on Earth.

▶ Sleeping in space is a little different from sleeping on Earth. You don't even need a bed. You just have to tie your sleeping bag someplace so that you don't float off when you turn over in your sleep! Even your breathing can make you move! In the space sta-

Show NASA video of day-to-day life in orbit.

Once you get over the possible motion sickness that weightless conditions can cause, it is fun to eat in space. (Courtesy of NASA.)

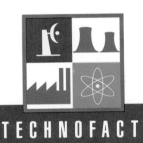

TECHNOFACT

Technofact 97

Like to play darts? In space, playing darts is a bit different than on Earth. When you aim for the target on Earth, you aim a little higher because you know gravity will pull the dart down during its flight. In space, the dart goes in a straight line from where you throw it! And you don't even have to throw it hard. In fact, you can just give it a little push and it will actually float toward the bull's-eye. The big challenge is to try playing darts in space without attaching your feet to something!

tion, there may be small sleeping compartments if you want some privacy.

▶ If you want to take a shower in space, you have to hook your feet to the floor or you might end up upside down from how you started. Suction pumps pull the water through holes in the floor so that the droplets don't float around. Even to use the toilet you have to strap yourself down with a seat belt!

▶ Did you know that even the clothes astronauts wear are made to be as safe as possible? Loose clothing is a problem, especially when you're in microgravity. You might catch it on something. There are also lots of pockets to carry things!

▶ Spacesuits contain enough water and oxygen to last about 7 hours. Also, the suit has to be able to withstand temperatures from $-157°$ C (centigrade) to $+121°$ C. Spacesuits must also be pressurized, because there is hardly any atmosphere outside the spacecraft or space station. The visor on the helmet is made to protect your eyes from the ultraviolet rays and micrometeorites.

▶ A pen, a fork, or any kind of food can't just be put down unless you fasten it to something. If you bite into a cracker, the crumbs will float around. The same is true if you spill water. The droplets would drift around. Try putting pepper on your salad, and what do you think might happen? The pepper grains could cause serious problems by clogging air filters.

▶ From a space station, the astronauts can study the Earth from above using very complex equipment. Then they can send that information to Earth for use. We can watch for changes in the atmosphere such as the hole in the ozone layer above the South Pole. There are many other areas of research better handled in space than on Earth. Some of these include making medicines, new alloys, and discovering new things about the universe.

If possible show live coverage of current space activities on television or satellite.

THINGS TO SEE AND DO:
Build a Space Station

Introduction:

Space stations have been a part of science fiction writing for many years. Today, they are a reality. Space stations provide a place for doing experiments on materials, astronomy, and testing the effects of microgravity on people. Space stations provide a place for careful observation of the Earth. They might even help us locate and fight the harmful effects of air or water pollution.

In this activity, you and your teacher will work as engineers, drafters, and technicians to design and build a space station simulator.

Design Brief:

Design and build a space station simulator that is large enough to hold four student astronauts. Design and build a mission control center and a communications system that will connect to the space station.

Materials:
▶ Plywood, prefinished hardboard
▶ Electrical components: wire, switches, and so on
▶ Fasteners: nails, wood screws
▶ Panel adhesive

Equipment:
▶ Band saw, scroll saw, saber saw
▶ Electric soldering gun

Older spacesuits were custom-made for each astronaut. (Courtesy of NASA.)

Procedure:

❶ This is a long-term activity that will require the help of all technology students in each class. You may work in groups within your class to make the parts of the space station.

❷ Brainstorm ideas for the shape or your space station. The room available in your classroom and the availability of materials may be factors to consider. Here are some ideas to consider in your design:

• Strength: Your space station must be strong enough to support the weight of the building materials used and the equipment you put in it.

• Structural Safety: The design of your space station must provide for easy entry and exit in an emergency. At least two exits must be provided. The exits should not be able to be locked or blocked.

Testing on Earth is often conducted under water to see how tiring it is for astronauts to work in spacesuits for many hours. (Courtesy of NASA.)

Have students research the design of space stations using NASA materials.

- Electrical Safety: The use of extension cords must be approved by your teacher. No electrical connections should be made without the approval of your teacher.
- Simulation: The space station will be used to simulate experiments that astronauts would perform in space. Plan what kind of experiments you could do, and make a list of the equipment that you might put into the space station.

❸ Use CAD software to design the shape of your space station. Keep the design specifications in mind.

❹ Choose the design that best suits your needs. Plan the structural skeleton that will support the station. Your design might look like this:

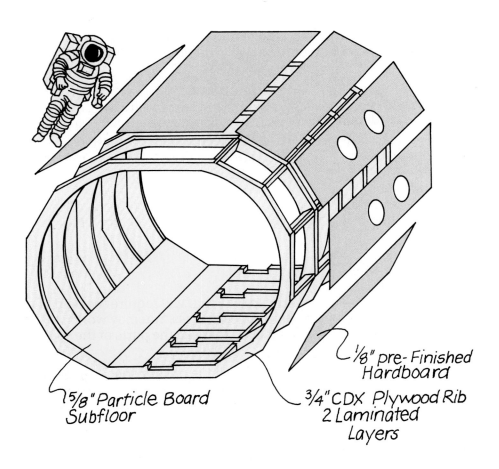

⅝" Particle Board Subfloor

⅛" pre-Finished Hardboard

¾" CDX Plywood Rib 2 Laminated Layers

❺ When the skeleton is complete, install the inside walls of the space station. Leave the outside uncovered for now so that you can easily run wires and connect equipment.

6 Design the electrical circuits so that they will run on 12 volts DC or less. Put a fuse in-line with each circuit to prevent possible fires. Ask your teacher for help.

7 Complete the outside "skin" of your space station. Keep in mind that you may want to remodel your design or add more equipment in the future. Each outside panel should be removable to give easy access.

8 Make a list of the equipment and its use in the space station. Your list might include:

- Computer: Can your computer be connected to a modem so that you could communicate with other student astronauts?
- Laser: Could a laser be used to simulate the alignment of your space station with the space shuttle for docking?
- Video phone: Maybe you and your teacher can use video cameras and monitors to let student astronauts see and talk with mission control in another classroom.
- Exercise equipment: The muscles of your body begin to weaken in the microgravity of space. Astronauts must exercise in space to keep their muscle tone. How could you simulate exercise activities in your space station?
- Cooking appliances: How would food preparation be different in space? What would it be like to eat in microgravity?
- Other? Use your imagination to think of ways other equipment might be used in the simulation of space station experiments.

9 Work in groups to perform space-related experiments in the space station. Switch groups so that everyone has a chance to be at mission control and "in space."

Evaluation:

1 List the features that your space station simulator has in common with a real space station.

2 If you could add one more piece of equipment to your space station, what would it be and how would it be used?

3 Ask other teachers to have their students help with your space station simulation activity.

4 How could space station research help us on a mission to Mars?

5 Why would it be important to recycle materials on a space station? How would recycling in space be different from recycling on Earth?

There aren't enough spacesuits taken on some shuttle flights for every astronaut. One possible solution is to use a personal rescue enclosure called a rescue ball. Each rescue ball would have life support and communications equipment for one astronaut. (Courtesy of NASA.)

Ask students to describe the conditions they might find on Mars.

Challenges:

1 Research space stations in science fiction and in reality. What do the two have in common? How are they different?

2 Contact other schools that have space station simulators. Communicate with them using a modem or FAX machine while you are simulating a space mission.

3 Write to NASA requesting information on space stations. Ask NASA if you could set up a phone call to an astronaut or mission specialist.

4 Download information from SpaceLink (available from NASA) or other space-related bulletin board services. Downlink information from a satellite dish receiving system related to space, such as NASA Select television.

5 Research the design and construction of simulations conducted by NASA to prepare for the building of a space station. Research the former Soviet Union's *Mir* space station.

Living in space would be a challenge just learning how to move and work in microgravity! Exploring new places is often dangerous as well as exciting. Many safety measures are taken to make space travel safe. Just like driving a car, however, space travel will never be totally risk-free.

Emergency evacuation of the Space Shuttle is possible on the ground using a sliding basket. Astronauts or technicians can jump into these baskets to quickly slide down a wire to safety. (Courtesy of NASA.)

The Moon, Mars, and Me! People are thinking of developing space colonies on the moon or Mars. They are looking at these space colonies as permanent homes. That's different from a short trip in a space shuttle or a longer time on a space station. The space colony will be built from materials made both in space and brought from the Earth. It will have an artificial environment that can be controlled for maximum comfort. The colony may have its own agricultural areas, artificial gravity system and even shopping centers. Space colonies will have to become almost **self-sufficient**. That means they can produce what they need to survive.

The moon is the most logical place for the first space colony. It is close to Earth and has many of the materials needed for construction. Mars is also in the running for a space colony in the next century. A space village there would probably be built underneath the planet's surface. A big problem is the time needed to get to Mars. It would take 14 months. Just think about how many supplies you'd need for a 14-month flight on the spacecraft!

Special Report
A Moon Base

The mission to Mars has been in the minds of space scientists for many years. It is now within reach thanks to advancements in technology. There is a good chance it will happen in the next 20 years. (Courtesy of NASA.)

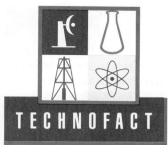

T E C H N O F A C T

Technofact 98
Some space futurists, who predict what space will be like in the future, think space colonies will someday be 20 miles long and about 4 miles wide. There will be room for 100,000 to 1,000,000 people in the large cylindrical places. People will be able to arrange the landscape to be anything they want, from shopping centers to open fields with trees, meadows, hills, and even small rivers.

Space Spinoffs and the Future

What are the benefits of space travel for you right now? **Space spinoffs** are new technologies and products that are developed during a space project but can be used for different purposes. Space is an ideal place for making certain products that are very difficult to make or can't be manufactured on earth. Some medicines, such as drugs for diabetes and blood diseases, are easier to make in space. These drugs use materials that are easier to separate with a process using electrical currents. On Earth, gravity gets in the way of making large quantities of these drugs.

New alloys are easier to make in space, too. This process involves melting two metals and then mixing them to make an **alloy**. It works best in space because the metals float together. On Earth, the lighter one floats on top of the denser one. Once the new alloys are made, they can be used anywhere—even back on Earth.

THINGS TO SEE AND DO:
Laser Guidance System

Techno Teasers
Space Spin-Offs

Introduction:
The docking of spacecraft is necessary to bring supplies to space stations and to unload building materials. Docking in space can be very tricky. The speeds of both vehicles are over 17,000 miles per hour. One mistake could mean disaster for both vehicles. Lasers might be used to help

Challenge students to research NASA spinoff books.

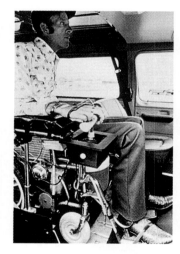

There have been many beneficial spinoffs of the space program. Space technology is used to make it possible for physically challenged people to lead a productive life. Here, a paralyzed man is able to drive a wheelchair using a controller called a Unistik. (Courtesy of NASA.)

An unexpected spinoff of space technology has been used to measure skin temperatures to test the effectiveness of moisturizers. This type of technology is called thermographic imaging. It detects temperature differences. (Courtesy of NASA.)

the docking of spacecraft. In this activity, you will work as a researcher experimenting with possible laser-guided docking of spacecraft.

Design Brief:
Build a laser guidance system that could be used to give feedback to astronauts during docking procedures.

Materials:
- Light-sensitive buzzer circuit kit
- Acrylic plastic (Plexiglas)
- Drinking straw
- Protractor

Equipment:
- Electronic soldering gun
- Plastics strip heater
- HeNe laser
- Drill press

Light-sensitive buzzer kits are available from science and technology suppliers. See Teacher's Resource Guide.

Procedure:

1 Work in a small group to assemble and design your guidance system. The following tasks can be divided among your group members:

- Assemble and solder the light-sensitive buzzer circuit kit.
- Design and build a clear plastic holder for the circuit board and battery.

Students may need to design and cut a paper model before cutting the plastic.

2 Follow the directions that come with the circuit kit. Be sure to place the transistor properly. Solder the components to the printed circuit board. Be careful not to overheat the components or run solder between the traces (conducting part) of the printed circuit board. Ask your teacher to inspect your work. Add a switch in the positive battery wire to make it easier to turn off the buzzer.

3 Cut and bend the plastic to fit around the circuit board. Your design might look like this:

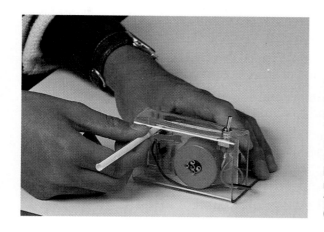

The light-sensitive buzzer is used with a laser to check the alignment of spacecraft as they dock. This simple circuit can be soldered together in one class period. A clear plastic shield makes it easy to attach a straw to direct the laser and a switch to turn off the battery power. (Photograph © Pam Benham.)

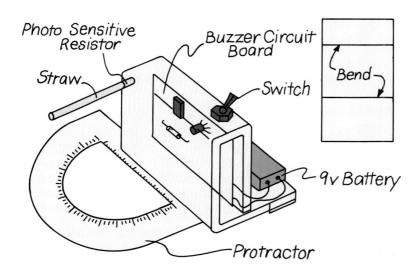

Photo Sensitive Resistor

Buzzer Circuit Board

Bend

Straw

Switch

9v Battery

Protractor

Drill a hole in the plastic cover for a 3½" long piece of plastic drinking straw over the light-sensitive resistor on the circuit board. This will serve as an alignment tunnel that the laser must enter to turn off the alarm. Glue or tape a protractor to the bottom of your buzzer circuit as shown. The zero point of the protractor should be under the photoresistor.

④ Align the laser and buzzer about 3′ apart. Tape a piece of paper under the buzzer circuit. Trace the outline of the protractor, and mark the 90° point of alignment when the laser goes straight down the straw and the buzzer stops.

SAFETY NOTE:
Do not look at the laser beam or directly at a mirror reflection. Warn others in the area that laser light is being used. Lasers operate at high voltages. Keep the area

around the laser dry. Do not place extension cords in aisles or where someone could trip over them.

❺ Turn the buzzer in both directions, and mark the point where the buzzer starts again.

❻ Repeat the experiment with a distance of 12′ between the laser and the buzzer.

❼ Remove the 3½″ length of straw and replace it with a 7″- long straw. Repeat the experiment.

Evaluation:

❶ What was the range of alignment in the first test?

❷ What effect did moving the laser farther away have on the range of alignment? Explain.

❸ What was the effect of lengthening the straw on the range of alignment? Explain.

❹ Would this type of laser guidance system work in space? How could you improve the design?

Challenges:

❶ Design your guidance system so it could be supported by a helium-filled balloon. Add or subtract weight from your guidance system until it is **neutrally buoyant**. That means it floats in one place because the lifting of the helium is offset by its weight. What new problems are there when you use your guidance system now?

❷ Test the guidance system for the maximum distance between the laser and the buzzer. How could this distance be an advantage to an actual space vehicle docking?

❸ Why would a normal white light be difficult to use instead of a laser? Explain.

❹ How could this guidance system be used on Earth? Explain your idea in words, or make a sketch of this spinoff.

There are many other spinoffs. A new material fiber called PBI, used to make spacesuits safer from fire, is now used in airline seat cushions. Battery-powered tools such as drills and screwdrivers came from the Apollo moon program. Even a special forest fire mapping system using computers, a satellite, and scanners to detect fires is a spinoff from NASA and the Jet Propulsion Laboratory.

The Future: How would you like to take a vacation to the moon or Mars? Aren't you curious just to see how things work in space and how people live? What kinds of sports do you think you could play on the moon? The big question facing the future of space travel and space colonization now is money.

TECHNOFACT

Technofact 99

The only major part of the space shuttle that can't be reused is the huge external fuel tank. It is built of aluminum and steel alloys and holds 145,000 gallons of liquid oxygen and 390,000 gallons of liquid hydrogen. It takes a lot of power to propel a 4.5 million pound space shuttle from liftoff to a velocity of 17,000 miles per hour in less than 10 minutes! The external tank is jettisoned (released) 8 minutes into the flight and tumbles back toward Earth. Most of it burns up in the atmosphere over the Indian or Pacific Ocean.

Challenge students to brainstorm future space activities.

Space travel costs a lot. Planning and building things in space costs lots of money and takes time. Many people feel the space program is worth the price. They can see the benefits to medicine and other industries here on Earth. They see space programs as being valuable in helping us to learn more about Earth from a different viewpoint. Other people do not want their tax dollars spent on space travel when there are many other problems on Earth to solve. What do you think the future of space travel should be?

What will the future of space exploration hold? Will your children live on Mars? Will you be able to work and live on the moon? The future is very exciting for the exploration of the "final frontier." (Courtesy of Westinghouse Electric Corporation.)

Summary

Space is often called the "final frontier." Since early times, people have dreamed about traveling in space. Now space travel is a reality. There are many important events in space history, from the launch of *Sputnik*, to the first men on the moon, to the Space Shuttle missions.

The most important problem to solve first was how to get spacecraft off the Earth. Early space pioneers like Konstantin Tsiolkovsky, Robert H. Goddard, and Hermann Oberth believed that rockets would power spacecraft, and they were right. The launching of *Sputnik* marked the beginning of the Space Age.

Until 1981 most space vehicles were launched only once. The development of the Space Shuttle changed many things. Now there is a reusable spacecraft that takes off like a rocket but lands like a glider. The Soviets built *Buran*, which works just like the U.S. shuttle. But even though the space shuttle was the answer to many dreams, the

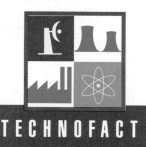

TECHNOFACT

Technofact 100

No time to exercise even in space? There's a new exercise device that astronauts might be able to use! It's called ROM The Time Machine. In 4 minutes, the designer claims you can stretch, strengthen, and tone every major muscle group. The faster you spin the flywheel, the more friction is created, and that means your muscles work harder.

TECHNOFACT

Technofact 101

Where will you be in 2019? If a plan being designed by some former Soviet and U.S. space engineers works, you might be one of the astronauts on Mars. A special Soviet rocket would lift a 60-foot ship into space for a trip to Mars. A crew of six would spend a year on Mars. But even before they got there, prefabricated structures and supplies would be landed at a Martian base site.

Techno Talk
Beyond Our Solar System

Challenger disaster in 1986 made everyone realize that space travel was still a risk.

Space is a near vacuum, which means it has very little air. There is also microgravity, or very little gravity. Working with things in space is different because of these two conditions. Gravity is a force of attraction between objects. The Earth's gravity pulls you and everything toward the center of the Earth.

In order to get spacecraft off the Earth, rockets have to use lots of power to lift off. Newton's laws of motion explain how rockets work on Earth and why they work even better in space. Launchers give the spacecraft extra power, or thrust.

All orbiting spacecraft are actually "falling around" the Earth. They don't fall straight down to Earth. That's because of the spacecraft's forward velocity and the fact that the Earth's surface is curved. The spacecraft is traveling forward fast enough that it falls toward the Earth just as fast as the Earth curves away from it.

Living in space is different because of microgravity. The way you eat, the way you sleep, the way you dress, how you bathe, and the way you move are all affected by microgravity. Researchers have found that people who stay in space for long periods of time have some changes in their muscles and bones. For that reason, all astronauts must exercise daily to keep fit.

In order to find out more about space, people are designing space stations and space colonies for the future. These would provide places where people would stay for months or even years. There are many challenges in building these future structures. One of the main problems is money. Right now there are many people who see how we all benefit from space exploration and space spinoffs. Others would rather spend tax dollars helping solve problems on Earth.

Challengers:

1 Do you think life exists beyond the Earth? Write a paragraph explaining your answer.

2 Research space achievements. Make a chart of what you think are major advancements in space technology.

3 Make a model of a space station. Include a living module, lab module, and any other modules you think are important.

4 Design and conduct three experiments that demonstrate Newton's laws of motion.

5 Explain why spacecraft are said to be "falling" toward the Earth, when they are actually orbiting the Earth.

6 Make a vector diagram that shows football players blocking each other.

7 Research different types of rocket engines. Make a drawing of each type.

8 Assemble and launch a model rocket. Use the altitude gauge to measure its height.

9 Design and prepare a space meal that would be tasty, lightweight, and easy to eat in microgravity.

10 Design a space colony or lunar city of the future. Include living quarters, power sources, and so on.

See Teacher's Resource Guide.

Chapter 15

Biotechnology

Things to Explore

When you finish this chapter, you will know that:

▶ Biotechnology raises ethical questions on how these technologies affect people and the environment.

▶ Foods of the future might come from new materials.

▶ Hydroponics is one method to raise plants in areas where land is limited.

▶ Distillation is a process that changes salt water into fresh water.

▶ Fermentation can be used to change biomass into biofuels such as ethanol.

▶ Bioengineering is a process that changes the gene structure of a plant or animal to improve it.

▶ Bionics, or biomechanics, is the design of artificial parts for the human body.

▶ The future of biotechnology is wide open to new ways to benefit people, plants, and the environment.

TechnoTerms

bioengi- neering	ethical	engineering
biomass	fermentation	genetics
biomechanics	fractional distillation	growth hormones
bionics	gasohol	hereditary
biotechnology	gene	hydroponics
degenerative	genetic	reverse osmosis
distillation	genetic	

Careers in Technology

Biotechnology is an exciting field with many different career opportunities. One area of biotechnology deals with how we manage the growth of plants. Horticulturists (plant growers) today use very sophisticated technologies to precisely control greenhouse temperature and humidity. In the United States, China, Greece, the Commonwealth of Independent States (the former Soviet Union), the Netherlands, and other world markets, horticulture technicians use special computer control systems to manage the quality of plants and the amount of energy used to make them grow. These systems allow technicians remote monitoring and programming of greenhouse temperature and humidity to keep the growing environment the best for the plants. You could really blossom in this job!

What is **biotechnology**? Biotechnology is technology related to living things. Many new advancements in biotechnology will hopefully improve health care and food. Vaccines will be produced in the next few years to protect against many diseases. Many **genetic** or **hereditary** (carried from parents to children) diseases are being studied in hopes of finding a way to reduce or change their effects on animals and plants. Helping to slow the aging process is part of biotechnology. Even turning sea water into fresh water uses biotechnology. Today, we are also using biotechnology to make new fuel sources.

Ask students to brainstorm examples of biotechnology.

Biotechnology can improve many things, but it raises some **ethical** questions about whether these changes are good or bad.

Biotechnology is related to living things. There have been many advancements in biotechnology in recent years. Many people think biotechnology will become a larger industry than the electronics industry is today. Can you think of ways biotechnology affects you? (Courtesy of GLAXO.)

You Are What You Eat

What's your favorite food? Bet you didn't say a nice big piece of bark! Foods of the future might well come from materials that you never thought about eating. Chemists are trying to break down foods into the basic materials that most of them are made of: carbon, hydrogen, and oxygen. Then these elements can be put together into new combinations to form artificial foods. Granted, they might not sound as good to you as a salad or a hamburger, but they will provide good nutrition. In countries where there is not enough food to eat, artificial foods might be the answer.

In order to make more food, farmers will use **growth hormones**. These hormones make cows fatter and help them produce more milk than they normally would. You can even have "custom-made calves." Today, scientists can create a line of leaner, stronger, healthier cattle. How would you like a turkey with more than two drumsticks? It might soon be possible!

In the future, our food may come from different sources. For example, biotechnologists are experimenting with a process to make tree bark into food. How would you like a juicy bark-burger? (Courtesy of United States Department of Agriculture.)

Growth hormones given to cattle can produce healthier, stronger animals. The testing of biotechnology-produced products is an ongoing process. (Courtesy of NASA.)

Challenge students to research and make a world map that shows desert climates that may someday be used for agricultural purposes.

Have you ever tried to grow plants without soil? Circulating water provides the nutrients for rapid plant growth. Hydroponics is sometimes used where space is limited, such as in a greenhouse or, someday, in a space station. (Courtesy of United States Department of Energy.)

Many vegetables can be improved, too. Corn has been improved to make it higher in nutritional value, more resistant to disease, and able to grow in colder climates. Scientists hope to try the same processes on other key crops such as wheat and rice. In France, a potato and a tomato were crossed to make a new vegetable called a "pomato."

Researchers are also trying to find nutritional foods that are easy and inexpensive to grow. One idea that has been developed is saltwater plants, called **halophytes**, that can be grown in the desert areas of the world. These crops would be used to make other processed foods. Already, foods based on soybeans are replacing meat and milk.

Visit a local hydroponics greenhouse or set up your own experimental hydroponics research station.

Finding ways to improve plants is an important research area in biotechnology.

Hydroponics, growing plants without soil, has potential for plant production in areas where land is limited. You can grow more plants in a limited amount of space using hydroponics. In addition, the crops grow faster and produce more. You can even reuse the water and fertilizer. In some places, scientists are growing salad greens outside soil on a sheet of plastic.

Sometimes biotechnology develops artificial substances to take the place of things like sugar, butter, and natural flavoring and coloring. You've probably heard of Nutrasweet™. It is used in many diet foods as a substitute for sugar.

Experiments with plants have increased the yield of many crops. Scientists are trying to "design" plants to resist insects and diseases. (Courtesy of NASA.)

Biotechnology is sometimes used to make substitutes for food products. Many of the foods you eat contain artificial sweeteners, flavors, and colors. (Courtesy of Nutrasweet.)

What Are Distillation and Fermentation?

Distillation: Did you know it's just as easy to die from thirst in the middle of the ocean as it is in the middle of a desert? That's because you can't drink salt water. A biotechnology process called distillation is important in changing salt water into fresh water.

Distillation is a way of separating substances by their different boiling points. In this process, as salt water is boiled in a container, the water in it will change to a gas. It then moves to a tube where it is cooled and again becomes a liquid you can drink. The salt's boiling point is well over 1000° C, so it stays in the original container. This method takes lots of energy to produce fresh water.

Distillation is used to change sea water into fresh water for drinking. The distillation process is also used to change crude oil into useful products such as gasoline, jet fuel, and heating oil. (Courtesy of American Petroleum Institute.)

New methods are being tested that would use less energy. One method forces sea water through filters to remove the salt. This method is called **reverse osmosis**. A high-pressure pump forces the sea water through microscopic openings in special filtering materials. This method may one day bring inexpensive water to desert nations.

Challenge students to research desalination processes and make a chart.

THINGS TO SEE AND DO:
Making Salt Water into Fresh Water

Special Report
A Cleaner Environment

Introduction:

It seems strange that some countries are running out of drinking water when they are right next to an ocean of water. Sea water contains salt and other minerals that make it impossible to drink or to use to irrigate most crops. The method most commonly used to desalinate (take the salt out) sea water is distillation. The distillation process boils water and condenses the steam into pure water. In this activity, you will be working as a biotechnology engineer. You are going to distill salt water into fresh water.

Design Brief:

Purify salt water into fresh water using the process of distillation.

Work with the science teacher in the demonstration of distillation.

Materials:
- Flask, rubber stopper, glass tubing
- Plastic tubing, ring stand, clamps
- Beaker, condenser, salt, water

Equipment:
- Electric hot plate or Bunsen burner

Procedure:

1 This activity will involve the entire class. Your teacher will ask you to help with the experiment. Teams of students should complete the following tasks:

- Mix 50 cc of water with one teaspoon of salt
- Set up the ring stand, condenser, and hot plate as illustrated
- Connect the plastic tubing to the flask and condenser

SAFETY NOTE:
All the equipment used in this experiment should be clean and free of any other chemicals. Food-grade plastic tubing should be used. Do not start this experiment until your teacher has checked the setup.

The completed setup should look like this:

Inspect all lab equipment for cleanliness before starting the experiment so students can taste the distillate safely.

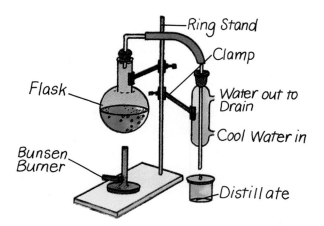

Salt water can be distilled into fresh water in the technology lab on a small scale. Can you imagine the size of the equipment needed to supply a large city with fresh water?

. . . .
Be sure that only steam passes through the tube to the condenser, not liquid.
. . . .

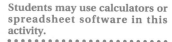

. . . .
Students may use calculators or spreadsheet software in this activity.
. . . .

❷ When everything is ready, pour the salt water into the flask. Turn on the hot plate or burner. Watch for the first signs of boiling and condensation.

❸ Collect the distillate (condensed vapor) in a clean beaker or cup. Allow the distillate to cool.

❹ Dip your finger in the distillate and taste the water.

Evaluation:

❶ Did the distillate taste salty? Explain.

❷ Why is it expensive to make fresh water this way?

❸ Design a solar-powered desalination plant. How do you think the output of the solar-powered plant would compare with the distillation process?

❹ Calculate the number of gallons of fresh water you use in one day. Use the following table to make your calculation.

Event	Gallons
Bath or shower	25
Toilet	5
Dish washing	7
Laundry	40
Garbage disposal	2
Other	?

How many gallons of water do you use in one year?

Challenges:

❶ Calculate the output of your distillation experiment. How long would it take to make the fresh water you use in one day using the equipment in this distillation experiment?

2 Make a list of five ways your family could save water.

3 Research the amount of fresh water used in your city or town every day. Sketch a plan for a large desalination plant that could supply the fresh water needs of your city or town.

4 Can you filter out the salt in sea water using filter paper? Explain.

Distillation is also used to make gasoline from crude oil. In this process, called **fractional distillation**, the crude oil is heated in a column, or distillation tower. As the vapor rises up the tower, it cools. The parts of crude oil, called **fractions**, condense out and become liquids again at different boiling points in the distillation tower. The substance left is revaporized many times as it flows through the tower. The fractions with the lowest boiling points are at the top. Gasoline has the lowest boiling point, followed by jet fuel. Fuel oil and lubricating oil are near the bottom of the tower because they have high boiling points.

The processing of crude oil into gasoline adds to the cost of gasoline when you buy it at a gas station.

Fermentation: Fermentation is a **biomass process** that uses microorganisms to turn grain into alcohols. It is a promising source of biofuels. Almost any kind of biomass (vegetation or animal wastes) that contains starch or cellulose, from cornstalks to cow dung, can be changed by fermentation and distillation into ethanol or grain alcohol.

Ethanol can be used as an alternative to gasoline when mixed with gasoline to produce **gasohol**. In some countries such as Brazil, over half of the automobile fuel used is gasohol. Making gasohol is an energy-saving step because these countries have so much extra sugar cane to use in fermentation. While ethanol is presently somewhat more expensive to produce than gasoline, it can be produced with resources countries already have.

Challenge students to make a chart or model of a fractional distillation tower used to refine petroleum.

TECHNOFACT

Technofact 104
If we replaced all gasoline in the United States with gasohol, we would cut use of gasoline by 10 percent and cut oil imports by 20 percent. Considering that we use about 15.7 million barrels of oil per day in this country, that's a lot!

Challenge students to contact local service stations. Ask if their gasoline contains alcohol.

Alcohol made from the fermentation process can help reduce our dependency on fossil fuels. Here, 10% of the "gas" is alcohol making a product called gasohol. (Courtesy of Archer Daniels Midland Company.)

Biofuels, like ethanol, do help control air pollution better than gasoline does. Methanol, another product made from fermentation of wood-product waste, is not as good an alternative as ethanol because it pollutes more. However, it still does not put as much carbon monoxide into the air as gasoline does. Many car manufacturers are planning to produce cars that burn alternative fuels in the future.

Designing New Life: Bioengineering and Medicine

Bioengineering, or **genetic engineering**, is a process of changing the gene structure of a plant or animal to improve it. A **gene** is a basic unit of heredity that carries certain information about the development of a plant or animal. **Genetics** is the study of how **traits**, or characteristics, of plants and animals are passed on from generation to generation. What color is your hair? Do you have blue eyes? Are you color-blind? All these traits are determined by your genes.

Genetic engineering research may make it possible for biotechnologists to design and produce new medicines, plants, or even animals. (Courtesy of United States Deparment of Agriculture.)

Genetic engineering tries to change, or **alter**, genes to make an improvement in the way a plant or animal develops. In most cases, a sample of the genetic material is taken from a plant or animal subject and then grown in cell cultures for use. This genetic engineering process is used in the manufacture of antibiotics such as penicillin and streptomycin. Even the production of insulin, a drug to treat diabetes, uses genetic engineering techniques.

Genetic engineering can produce plants with new genes that make them resistant to disease or able to grow faster than the original plant. Researchers are looking for a way to put a gene into plant cells so that they would not need nitrogen fertilizer. The **Environmental Protection Agency** (EPA) just approved the release of a genetically changed plant virus that destroys an insect that eats cabbage. All genetic engineering

Genetic engineering is being used to develop plants that are more resistant to the harmful effects of mildew, insects, and diseases. Disease-resistant plants help reduce the use of chemical pesticides or fungicides. (Courtesy of United States Department of Agriculture.)

depends on using cell culture techniques, where the cells multiply to produce more of the same kind of cells.

Researchers are trying to **isolate** (locate alone) genes that cause hereditary diseases such as muscular dystrophy. They want to be able to change the gene structure early in an **embryo**'s (beginning of a life form) development. That might change how the disease affects the person.

TECHNOFACT

THINGS TO SEE AND DO:
Designer Genes

Introduction:

Today's geneticists are trying to find ways to change the genetic makeup of plants to improve them for our use. The plants can be made more resistant to disease or they can be changed to produce more food than the original plant. To do so, geneticists must find out which plant characteristics, or **traits** determined by genes, are **dominant** or **recessive**. Each trait requires two genes to express it, one from each parent.

Gregor Mendel (1822–1884) , an Austrian monk, was the first person to study how traits are carried from one generation of living things to the next generation. After many experiments with pea plants, Mendel stated the **law of dominance**. A dominant gene will always mask, or hide, a recessive gene when they are together.

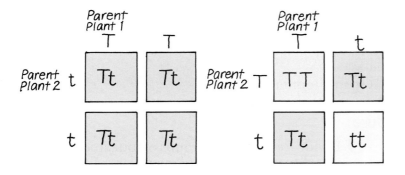

In this activity you are going to see how dominant and recessive genes make up the **phenotype** (what an organism looks like) and the **genotype** (what actual genes the organism has). Geneticists can use this information to alter life forms.

You will use a box grid called a **Punnett square** to show the offspring of mating two plants. Figure A shows the offspring from mating two parent plants, each pure or having identical genes for one trait. The four boxes show the genotypes and phenotypes of the offspring trait for tallness. T is the dominant gene for tallness and t is the recessive gene for shortness in plants. Will the plants be tall or short? You can tell just by looking to see if a dominant T appears in the box. If either one gene or both are T, then the offspring will be tall. Are there any short plants in this generation? The second Punnett square shows the possible plant sizes if two parent plants with different genes (Tt) are crossed. Are all the plants tall?

Technofact 106

Now you can have true coffee flavor without the caffeine. Some people don't like to drink decaffinated coffees on the market today because they don't taste like real coffee. A research team is currently working on altering the genetic structure of the coffee bean. The new genetic structure will block the caffeine gene from manufacturing caffeine in the plant, but the real coffee taste will still be there!

This grid of boxes is called a Punnet square. The symbols stand for:
- T=dominant gene for tall plants
- t=recessive gene for short plants

Ask the life science teacher to help in this activity.

Challenge students to make a chart of Mendel's genetics pea research in dominance and recessives.

You are going to find the possible combinations for two different traits. You are going to cross two parent plants that might be tall and have smooth seeds (TW), tall and have wrinkled seeds (Tw), short and have smooth seeds (tW), or short and have wrinkled seeds (tw). Look at the chart below.

This Punnett square has two different characteristics. The two characteristics make 16 possible combinations. The symbols stand for:
- T=dominant gene for tall plants
- t=recessive gene for short plants
- W=dominant gene for smooth peas
- w=recessive gene for wrinkled peas

Can you complete the chart to find out what the plants will be like?

	TW	Tw	tW	tw
TW	TTWW			
Tw				
tW				
tw				

Materials:
- Pencil
- 3″ × 5″ cards (at least 64)
- Paper
- Cardboard or tagboard
- Colored paper or markers
- PTC paper (optional)

Equipment:
- Computer with graphics software (optional)

PTC paper is available from science suppliers. See Teacher's Resource Guide.

Procedure:

1. Divide the class into groups of four students. Assign each group to make up the following sixty-four 3″ × 5″ cards to represent genes:

    - Sixteen cards with a large T to stand for dominant tallness
    - Sixteen cards with a small t for recessive shortness
    - Sixteen cards with a large W for dominant smooth pea seed
    - Sixteen cards with a small w for recessive wrinkled pea seed

2. Put all the cards together in a deck.

3. Make a large desktop sixteen-box grid or Punnett square to show plant combinations from a cross of two different traits. Have each student draw a card from the deck and place it in the appropriate box on the Punnett square. Refer to the chart above.

Students may use computer software in this activity.

④ Determine how many plants are:
- Tall and Smooth (TW) ____
- Tall and wrinkled (Tw) ____
- Short and smooth (tW) ____
- Short and wrinkled (tw) ____

Evaluation:

① Did you get a 9:3:3:1 ratio? If not, try again.

② Is a dominant gene necessarily better than a recessive gene? What is the real difference?

③ Check your classmates for the following dominant and recessive human traits:

- Tongue curl or twist into an S shape. Only individuals who have both recessive genes show this rare ability!
- Widow's peak. This is a dominant gene causing the hairline to drop downward in the center of the forehead.
- Tongue rolling. Another dominant gene gives some people the ability to roll their tongues into a distinct U shape.
- Ability to taste PTC. Another dominant gene allows you to taste PTC.

④ If you wanted to develop a plant that would resist a certain disease, would you try to make that trait for resistance dominant or recessive? Why?

⑤ Research some genetic diseases such as sickle-cell anemia, cystic fibrosis, or Tay-Sachs disease. Find out if they result from dominant genes or recessive genes.

Do you remember the huge oil spill in Alaska? Certain microorganisms that feed on petroleum can be **modified**, or changed, by genetic engineering to help fight this kind of pollution. Researchers don't know what might happen if they are able to modify these microorganisms. Maybe they will end up multiplying too fast and become another pollution problem themselves!

Many people worry that genetic engineering might produce dangerous bacteria and viruses that could be used in war. This problem is one that many countries have tried to prevent through special agreements with each other.

Genetic engineering when it involves people raises many questions. When it is used in the treatment of genetic diseases or helps produce an antibiotic, you can see its benefits. Whether technologists and scientists should change or improve humans by trying to **clone**, or identically duplicate, a person is another issue that genetic engineering has raised. What if we could "design" the perfect person by genetic engineering?

Challenge students to make a bulletin board display showing the impacts of the Alaska oil spill.

Genetic engineering might help us clean up the harmful effects of technology. Here, genetically engineered bacteria are used to "eat" an oil spill. (Courtesy of InterBio Inc.)

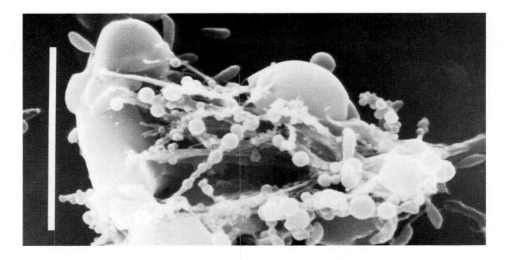

Assign students to write a paragraph expressing their opinions of genetic engineering.

Who would decide what traits are the best? Now that you know something about the possibilities of genetic engineering, how do you think we should use it to benefit people and the environment?

THINGS TO SEE AND DO:
Designer Foods

Introduction:
The ability to design plants or animals using genetic engineering has increased rapidly in the past few years. The first patented animal is already for sale. Scientists have "designed" a genetically engineered mouse that is ideal for laboratory experiments in cancer research. Many other possibilities exist for genetically engineered plants and animals. In this activity, you will work as a scientist to design a plant or an animal. Many people predict the future of biotechnology to be bigger than the electronics industry today. Some people fear what might happen if the technology is used strictly for profit or without thinking about how it could change our world.

Design Brief:
Design a plant or animal that would be of benefit to people.

Materials:
▶ Paper, pencil

Equipment:
▶ Computer with word-processing and graphics software (optional)

Procedure:
❶ Some of the most important experiments in the history of science and technology have taken place only in someone's mind. This activity will require you to think about the possibilities of genetic engineering and the benefits they could bring. You will work individually or in small groups.

2 Brainstorm ideas about how food crops or animals could be altered to make food production better, or combine the good qualities of existing plants or animals. Some possibilities might include:

Idea	Benefit
• A cross between a tomato and a water melon	More tomato to eat
• A rabbit that tastes like bacon	Higher meat production
• A plastic-eating rodent	Reduce landfill garbage
• Oil-eating bacteria	Clean up oil spills
• Biting insects that spread vitamins or vaccines	Improve world health
• Wheat that grows in salt water	Improve food production

3 The plant or animal you design must be beneficial. You should consider its impact on the environment and its resistance to diseases and insects. If the plant or animal were released into the wild, would it kill other beneficial plants or animals?

4 Write a paragraph explaining your idea. Use a computer and word-processing software if possible.

5 Make a drawing of your idea. Use graphics software if possible.

6 Make a bulletin board display of all of the ideas in the class.

Evaluation:

1 Do you think genetic engineering can be beneficial to people? Explain.

2 What are the dangers associated with uncontrolled genetic engineering? How should they be controlled?

3 Which of the ideas in your class was the most promising? Why?

4 Why were the *Apollo* astronauts kept away from other people for 2 weeks after their return to Earth? Was this precaution needed?

Challenges:

1 What effects do you think genetic engineering will have on the future? List three possible changes in your lifetime.

Ask a doctor or physical therapist about prosthetic devices.

② Research the development of genetic engineering. What other genetically engineered products are waiting for approval?

③ Research early experiments in genetic theory. What scientist is considered the father of genetics?

④ Research the genome project or the discovery of the structure of DNA. Write a short report on either subject.

Bionics and the future of Biotechnology

Bionics: Did you ever think technology would be able to make a human from almost totally artificial parts? **Bionics**, or **biomechanics**, is a field where researchers and scientists are designing prostheses (artifical parts such as arms, legs, and hips), and teeth to rebuild people who need replacement parts. Bionics also includes computers, new wheelchair designs, and other technologies that make it possible for people with birth defects, accidental injuries, or **degenerative** (progessively get worse) diseases to lead more productive lives.

Thousands of ordinary people, handicapped by disease or genetic disorders, rely on computerized devices that make up for the body's weakness. Computers keep time in heart pacemakers, read books to the blind, provide sound with hearing aids, and help paralyzed people communicate. Many of these computerized devices are hidden or so small that no one knows they're being used.

The design of a reliable artificial heart is still a problem that biotechnologists are working to solve. Electronic devices such as pacemakers help thousands of people lead healthy lives. (Photograph by Doug Persons.)

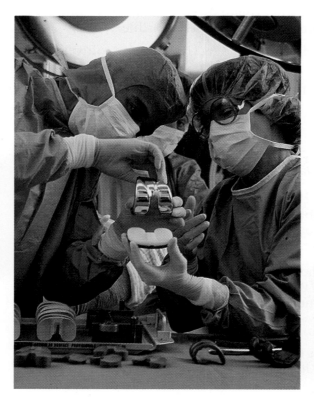

There are hundreds of artificial body parts available today. The branch of biotechnology called bionics or biomechanics designs new body parts. Here, an artificial knee joint is being prepared for an operation. (Courtesy of Brystol-Meyers Squibb Company.)

THINGS TO SEE AND DO:
The Artificial Heart

Introduction:
There are over 1,000 artificial body parts available today. Artificial hearts have been around for almost 10 years. It has been very difficult to perfect a device that can replace the human heart. Some people think that because it is just a pump it should be easy to replace. The problem is much more complicated. One problem is that the white blood cells in the blood try to attack any foreign material in the body. Drugs have been developed to offset this problem, but there are many other complications that have prevented doctors from making a perfect replacement for the heart. In this activity, you will make a working model of the human heart. You will begin to appreciate the complex problems that must be solved to make a functioning and reliable artificial heart.

Design Brief:
Design, build, and test an artificial heart. The model must be able to pump water and its parts must be visible.

Plastic heart model kits are available from science supply houses. See Teacher's Resource Guide.

Materials:
- Acrylic plastic (Plexiglas), ⅛″ and ½″
- Solvent cement for acrylic plastic
- Red food coloring, water, plastic tubing
- Rubberbands, ¹⁄₁₆″ welding rod
- Siphon pump, beaker

Equipment:
- Scroll saw
- Drill press
- Wire cutters
- Medicine dropper
- Disk or belt sander

Procedure:

1 In this activity, you will work in a small group to build an artificial heart model. Study the diagram on page 410 to see what tasks are needed to complete the assignment.

2 Make a list of the tasks needed. Assign a task to each group member. Place a ½″-thick clear acrylic plastic sheet over the diagram on page 410. Trace the design to mark the place to cut. Cut the plastic using a scroll saw.

SAFETY NOTE:
Remember to wear eye protection at all times. Follow the general safety rules and the specific safety rules for the machines you are using. Do not talk to others while you are using a machine.

3 Trace the outline of the plastic heart model onto two pieces of ⅛″ clear acrylic. Cut the shapes using a scroll saw.

Two pieces of plastic may be cut together if they are temporarily attached with rubber cement.

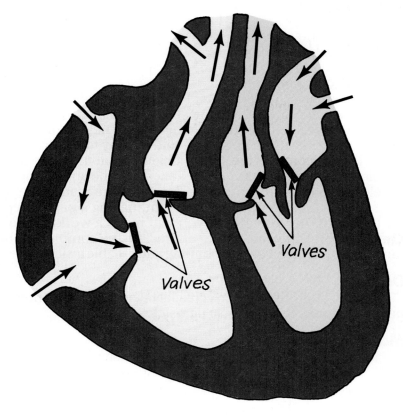

The plastic heart model will give you an idea of the difficulty of designing and making an artificial heart.

Demonstrate to students the difference between chemical goggles and those used for other general purposes.

❹ Make a valve mechanism similar to the illustration, or design your own. Drill the holes needed for the valves.

❺ Wipe off the plastic pieces, and assemble them using rubberbands to hold them together. Use a medicine dropper to apply a small amount of solvent cement to the edges of the plastic.

SAFETY NOTE:
Ask your teacher for help before attempting to use the solvent cement. Wear chemical goggles and follow the directions on the label carefully. Wash your hands after use. Use in a well ventilated area. Keep the solvent cement away from heat or flame.

Capillary action draws the cement into the space between the layers. You should avoid letting the solvent cement drip on the flat surfaces of the plastic. If it does, do not wipe it off. It will evaporate quickly. Allow the cement to dry.

❻ Attach the pump and hoses to the heart model. Mix red food coloring and water in a beaker to simulate blood.

❼ Put the ends of the hoses in the beaker. Pump the "blood" through the heart model. Check for leaks, and patch them with silicone sealant if needed.

Evaluation:

1 Measure the flow rate of your heart model. Squeeze the pump about once a second to match your resting pulse. How much "blood" can your model pump in one minute at a resting pulse rate?

2 Double the pulse rate to 120 beats per minute. Measure the flow rate. What is the amount of "blood" pumped in one minute?

3 Research the amount of blood a human heart pumps in one minute. How did your model compare?

4 Why is it an advantage for some doctors to be skilled in both medicine and engineering?

Challenges:

1 Research heart transplants. Why are they more reliable than artificial hearts at this time? What part of an artificial heart is the most difficult to design and make?

2 Use a vacuum pump and a bell jar to make a model of an artificial lung.

3 Research the operation of a heart-lung machine. How is it used during a transplant operation?

4 Research the Jarvik artificial heart. Make a sketch showing how it was constructed. Find out how it was used to keep patients alive.

Bionics, besides trying to find replacement parts for the human body, also tries to find ways to use electronics to give motion to paralyzed people directly. One system, called **FES** (functional electrical stimulation), uses low levels of electrical current to stimulate movement in paralyzed arms and legs. Scientists have known for a long time that muscles can be made to move by electricity. The problem is to develop a system that can handle the complex motions of even a simple body movement. How many muscles do you think you use just taking a single step? A computer trying to help a person take one step needs almost 1,500 commands to move thirty-two muscles. That's just one step!

The Future of Biotechnology: Advancements in biotechnology are rapidly reshaping the field of biology and medicine. What we can do with biotechnology to produce more nutritional foods and substitute foods is changing the way the world eats and how people take care of themselves. The energy alternatives provided by biomass processes such as fermentation will provide a more economical energy source and clean up pollution at the same time.

Genetic engineering will make it possible to create different crops that are resistant to disease and can grow in unusual places. Genetic engineering holds great possibilities in attacking genetic diseases and producing new cures for diseases. Bionics, by putting electronics and

Ask the life science teacher for help with this project.

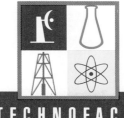

TECHNOFACT

Technofact 108

Watch out for those genetically engineered honey bees! Insect geneticists are trying to develop a strain of bees that would be protected from the Africanized ("killer") honey bees (AHBs). Because the U.S. diet is directly or indirectly dependent on crops that are pollinated by honey bees, we need to control what happens to our honey bees. If there is genetic mixing, our honey bees might not continue to pollinate crops as efficiently.

Genetic engineering might even make it possible to grow square trees. If trees were square instead of round, it would be possible to make more usable lumber from a tree and reduce waste. (Courtesy of Georgia Pacific.)

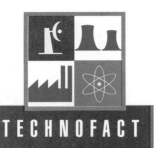

TECHNOFACT

Technofact 109

The first human gene procedure aimed at correcting a hereditary disease happened in 1990. A 4-year-old girl was born with adenosine deaminase (ADA) deficiency. She didn't have an important immunity cell chemical in her white blood cells. Those are the cells that fight diseases. The girl was treated with a combination of her own white blood cells mixed with a specially engineered virus. These new cells were then put back into her bloodstream, where they would gradually make her own immune system stronger.

the human body together, will bring many exciting advances toward helping physically challenged people.

In the 1970s, there was a television show called *The Six Million Dollar Man*. In that show, a former astronaut who had been injured in a plane crash was totally rebuilt with bionic parts. His new bionic arms, legs, muscles, eyes, and ears worked better than the real things. He was stronger and could move faster, see better, and hear better than an ordinary human. What was science fiction then is fast becoming a reality!

Summary

Biotechnology is technology related to living things. Biotechnology is rapidly changing the fields of biology and medicine. Through biotechnolgy processes, new and more nutritional foods are being developed. New artificial foods are helping to feed people in countries where food is scarce.

Much of the world's water is sea water, which people can't use. The process of distillation is making it possible to change sea water into fresh water. Alternative fuel sources that don't pollute the air as much, like gasohol, are produced using a biotechnology technique called fermentation.

Genetic engineering is helping us make new crops that are more nutritious and able to grow in different areas. Many of these crops also are disease-resistant, which means more of the crop will make it to harvest time. It is now possible to make a new plant from a combination of two other plants. When it comes to genetic engineering and humans, there are many questions still to be answered. Genetic engineering can help to get rid of some genetic diseases and bring cures for other diseases. But there are ethical questions about cloning people or designing people using genetic engineering.

Bionics is replacing body parts, or bringing electronics and the human body together. Today, many body parts can be replaced by artificial parts. In other instances, electronics is being used to help the human body work when it can't by itself. These developments allow physically challenged people to live more normal and productive lives. Computerized devices have greatly helped the field of bionics.

The future of biotechnology is wide open. There are many ways it can benefit people, plants, and the environment if we use it right.

Challengers:

1 Research two genetic diseases, and tell how they affect people.

2 Make a model illustrating fractional distillation of crude oil in a refinery.

3 Ask a physician to talk to your class about how the heart works and how pacemakers can help an unhealthy heart.

4 Research the number of bionic parts there are available to be used in the human body.

5 Design an experiment using drosophila (fruit flies) that would show how traits are passed from one generation to another.

6 Ask a doctor or geneticist to talk with your class about genetic engineering and how it can benefit you.

7 Design a "perfect" wheelchair for a physically challenged person who likes to be very active in sports.

8 Research the many uses of soybeans as food substitutes for meats and other products. Prepare a dish with soybeans, and see if your class can tell the difference.

9 Grow some plants hydroponically. Compare their growth with plants grown in soil.

10 Research the structure of DNA, and build a model.

See Teacher's Resource Guide.

Chapter 16

Technology and Your Future

Things to Explore

When you finish this chapter, you will know that:

▶ People determine how a technology should be used.

▶ Technology changes very rapidly.

▶ Decisions you make will determine whether technologies are harmful or helpful.

▶ The future of technology is open to many ideas.

▶ Technology must fit the needs of people and the environment.

Chapter Opener
Technology & Your Future

(Art by Patricia A. Hutchinson.)

Careers in Technology

New developments in technology come from people who find a need to solve a problem. Nicholas Ristow, a second-grade student in Texas, wanted a way to keep rain from hitting him in the face while riding his bike. So he put his imagination to work and came up with this simple design. He uses a clear plastic rolled shade placed into a hollow bar that goes between his handlebars. A plastic pipe extends above the back of his bicycle seat. When it starts to rain, he unrolls the shade from his handlebars and hooks it to the pipe behind him. For his design he won national recognition and a U.S. Savings Bond. Maybe you'll develop a new technology to help people.

People decide whether technology is good or bad. The way technology is used depends on people like you! In this book, you have learned how quickly technology is growing and changing. Those changes affect all of us in many different ways. Think how communication and transportation have changed in the last century. You can communicate with people anywhere in the world in a matter of seconds through computers and satellites. You can travel across the Atlantic Ocean in about 3 hours or travel into space at speeds of 17,000 miles per hour.

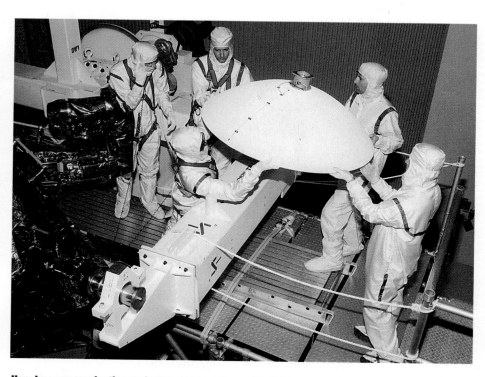

How has communication technology changed your life? What do you think will happen in the future? (Courtesy of NASA.)

Einstein had a vision of the future. His theories are still being studied today. Maybe you will have even a bigger effect on the future of technology. (Courtesy of The Bettmann Archive.)

The important thing is that you understand how technology affects you today and in the future. You will make decisions about whether technologies that are developed now and in the future are beneficial or harmful to people and the environment.

Now that you've completed this technology book, your idea of what technology is may have changed since you first started. What is your definition of technology now? Did you remember to include problem-solving skills in your definition? You need to be able to think through problems associated with technology as well as any other subject.

The future of technology is wide open. The only limit is your imagination. As Albert Einstein said: "Imagination is more important than knowledge, for knowledge is limited to all we know and understand, while imagination embraces the entire world and all there ever will be to know and understand."

Many courageous people who are risk takers, who are curious about how things work, and who believe in trying, have led the way in creating technologies that make your life easier. Technology must continue to fit the needs of people and the environment. That means you have to plan how new technologies can help make life better for people around the world.

Technology can make your life easier and healthier. You make decisions about your future based on your understanding of technology. (Courtesy of Hosmer Dorrance Corporation.)

You are on the road to becoming "technoliterate." One thing is certain. Technology will keep changing quickly! It will be your responsibility to keep up with the future changes in technology so that you will know how your world works.

THINGS TO SEE AND DO: *Technology in Your Future*

Introduction:
In this book, you have been learning about how technology is used today and how it might be used tomorrow. It's now time for you to express your idea of what the future might bring. In this activity, you will work as a futurist to give us a glimpse of what you think the future might hold.

Design Brief:
Create a story that demonstrates what you think the future holds.

Ask a social studies teacher for help on this futuring project.

Materials:
▶ Pencil, paper

Equipment:
▶ Brain, imagination
▶ Computer with word-processing software, camcorder, and so on (optional)

Procedure:

❶ Work individually or in small groups to complete this design brief. Remember the steps in problem solving? They will help you solve this problem and others you are faced with when you are finished with this book.

❷ Brainstorm ideas of what the future might be like 20, 50, or 100 years from now. Make a list of the things your group thinks of. Consider the following topics in your discussion of the future of technology:

- Transportation
- Energy
- Production
- Communications
- Computers
- Space colonization
- Biotechnology

Other topics to consider in the discussion of the future might include music, art, food, entertainment, sports, etc.

Which of the technologies represented in these photos do you think can help make your future better? (Figure A Courtesy of US West/John Blaustein Photography. Figure B Courtesy of the Mayo Clinic. Figure C Courtesy of Dow Corporation. Figure D Courtesy of EG&G.)

A

B

C

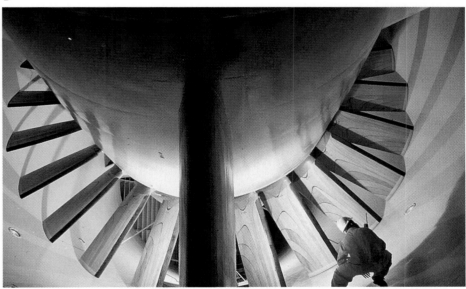

D

3 Decide how your idea of the future will be presented to the class. Choose one of the following presentation methods:

- Written story
- Video production
- Drawing
- Other?

4 Gather all the materials and information you think you need. Complete your production.

5 Present your idea to the class.

Challenge students to present their ideas about the future to other classes or elementary grades.

Evaluation:

1 Which idea of the future did you like best? Why?

2 In your opinion, which idea seemed the least likely to occur? Explain.

3 What is the most important problem facing technology in the future? Explain.

4 How can the understanding of technology help you in your future? Explain.

Challengers:

❶ Read a science fiction book about the future. Make a list of the things you think are likely and those that seem completely impossible. Report your findings.

❷ Make a prediction of what you will be doing 5, 10, 20, and 50 years from now. Will technology play a part in your future?

❸ Research the technology-related classes that are available to you next year. Are you interested in learning more about technology? What other classes in technology will be available in high school or college?

❹ What area of technology do you find most interesting? Why? How could your interest be developed into a career?

Ask students to give their definition of technology. Compare it to what they thought at the beginning of the course.

Techno Talk
Decisions, Decisions

Techno Talk
Answer Segment–1

Techno Talk
Answer Segment–2

APPENDIX

Table of Trigonometric Ratios

Angle Measure	sin	cos	tan	Angle Measure	sin	cos	tan
1°	0.0175	0.9998	0.0175	46°	0.7193	0.6947	1.0355
2°	0.0349	0.9994	0.0349	47°	0.7314	0.6820	1.0724
3°	0.0523	0.9986	0.0524	48°	0.7431	0.6691	1.1106
4°	0.0698	0.9976	0.0699	49°	0.7547	0.6561	1.1504
5°	0.0872	0.9962	0.0875	50°	0.7660	0.6428	1.1918
6°	0.1045	0.9945	0.1051	51°	0.7771	0.6293	1.2349
7°	0.1219	0.9925	0.1228	52°	0.7880	0.6157	1.2799
8°	0.1392	0.9903	0.1405	53°	0.7986	0.6018	1.3270
9°	0.1564	0.9877	0.1584	54°	0.8090	0.5878	1.3764
10°	0.1736	0.9848	0.1763	55°	0.8192	0.5736	1.4281
11°	0.1908	0.9816	0.1944	56°	0.8290	0.5592	1.4826
12°	0.2079	0.9781	0.2126	57°	0.8387	0.5446	1.5399
13°	0.2250	0.9744	0.2309	58°	0.8480	0.5299	1.6003
14°	0.2419	0.9703	0.2493	59°	0.8572	0.5150	1.6643
15°	0.2588	0.9659	0.2679	60°	0.8660	0.5000	1.7321
16°	0.2756	0.9613	0.2867	61°	0.8746	0.4848	1.8040
17°	0.2924	0.9563	0.3057	62°	0.8829	0.4695	1.8807
18°	0.3090	0.9511	0.3249	63°	0.8910	0.4540	1.9626
19°	0.3256	0.9455	0.3443	64°	0.8988	0.4384	2.0503
20°	0.3420	0.9397	0.3640	65°	0.9063	0.4226	2.1445
21°	0.3584	0.9336	0.3839	66°	0.9135	0.4067	2.2460
22°	0.3746	0.9272	0.4040	67°	0.9205	0.3907	2.3559
23°	0.3907	0.9205	0.4245	68°	0.9272	0.3746	2.4751
24°	0.4067	0.9135	0.4452	69°	0.9336	0.3584	2.6051
25°	0.4226	0.9063	0.4663	70°	0.9397	0.3420	2.7475
26°	0.4384	0.8968	0.4877	71°	0.9455	0.3256	2.9042
27°	0.4540	0.8910	0.5095	72°	0.9511	0.3090	3.0777
28°	0.4695	0.8829	0.5317	73°	0.9563	0.2924	3.2709
29°	0.4848	0.8746	0.5543	74°	0.9613	0.2756	3.4874
30°	0.5000	0.8660	0.5774	75°	0.9659	0.2588	3.7321
31°	0.5150	0.8572	0.6009	76°	0.9703	0.2419	4.0108
32°	0.5299	0.8480	0.6249	77°	0.9744	0.2250	4.3315
33°	0.5446	0.8387	0.6494	78°	0.9781	0.2079	4.7046
34°	0.5592	0.8290	0.6745	79°	0.9816	0.1908	5.1446
35°	0.5736	0.8192	0.7002	80°	0.9848	0.1736	5.6713
36°	0.5878	0.8090	0.7265	81°	0.9877	0.1564	6.3138
37°	0.6018	0.7986	0.7536	82°	0.9903	0.1392	7.1154
38°	0.6157	0.7880	0.7813	83°	0.9925	0.1219	8.1443
39°	0.6293	0.7771	0.8098	84°	0.9945	0.1045	9.5144
40°	0.6428	0.7660	0.8391	85°	0.9962	0.0872	11.4301
41°	0.6561	0.7547	0.8693	86°	0.9976	0.0698	14.3007
42°	0.6691	0.7431	0.9004	87°	0.9986	0.0523	19.0811
43°	0.6820	0.7314	0.9325	88°	0.9994	0.0349	28.6363
44°	0.6947	0.7193	0.9657	89°	0.9998	0.0175	57.2900
45°	0.7071	0.7071	1.0000				

GLOSSARY

acid rain   Rain that damages the environment. It contains chemicals from cars and factories.

aerodynamics   The study or science of moving air.

alloy   A material made when two or more metals are mixed together.

anthropometry   The science of measuring people.

applied research   Research done to solve a particular problem.

artificial intelligence   Programming a computer to recognize problem situations and make the right decisions.

ASCII (ask-key)   American Standard Code for Information Interchange.

assembly line   A manufacturing method where each machine or worker does a job and the product being made moves from worker to worker.

automation   The automatic control of a process by a machine.

basic research   Gathering information to use when it is needed.

binary system (base 2)   The number system using only zero and one as digits. It is used in computers.

bioengineering   A process of changing the genetic structure of a plant or animal to improve it.

biomass   Living or dead plant or animal matter used to make energy.

biomechanics (bionics)   Research and design of artificial body parts and devices that help people.

biotechnology   Using living organisms to make products.

bit   An acronym for binary digit; it is the smallest unit of information used by a computer.

broadcast   Sending a message to many receivers at the same time.

bronze   A metal made from combining copper and tin.

byte   Eight bits together that represents information used by a computer.

CAD   Computer Aided Design. The use of a computer to help in designing a part, building, etc.

CAD/CAM   The technology that hooks computer aided design with computer aided manufacturing.

CAM   Computer Aided Manufacturing. Using computers to control a manufacturing process.

CD-ROM   Compact Disk–Read Only Memory. A thin round disk with information that is scanned by a laser and then displayed on the computer.

CIM   Computer Integrated Manufacturing. The technology combining manufacturing, design, and business systems together.

cam   A mechanical part that changes rotary (turning) motion into up-and-down motion.

capacitor   An electronic component that can temporarily hold an electric charge.

capital   Money used as a resource in a company.

catalysts   Substances that make chemical reactions go faster without becoming part of the reaction.

characters   Different fonts or kinds of letters used in word processing software.

circuit   Electronic components put together to make a complete pathway for electricity to flow.

Clarke Belt   The orbit of communication satellites in geosynchronous orbit 22,300 miles above the Earth's equator.

communication   The process of exchanging information either by sending it or receiving it.

compression   A force that squeezes materials.

compression strength   The ability of a material to resist being smashed.

component   A part that does a certain job.

composite material   Made up of more than one material.

conductor   A material that lets electrons easily pass from one atom to another.

conveyor   A transportation device that moves parts on an assembly line.

corporation   A company organized and owned by stockholders.

data   Figures and facts that become information.

decompose   To break down naturally.

design brief   The definition of a problem; also called a problem statement.

design elements   Graphics (symbols and pictures) that add emphasis to a printed page. They include unity, rhythm, and the rule of thirds.

desktop publishing (DTP)   Text and graphics used together to make a report, newsletter, or newspaper.

digitizer   A device that changes video images into computer images.

diode   An electronic component that lets electrons flow in only one direction.

distillation   A method of separating substances by their different boiling points.

downtime   A stop in production of a product.

dynamic loads   Forces on a structure that change rapidly, like wind or an earthquake.

ecology   The study of how things interact with the environment.

electricity   The flow of electrons through a conductor.

electromechanical   A machine with a combination of electrical and mechanical parts.

electron   The negatively charged part of an atom.

electronic components   Parts put into circuits that control the flow of electrons. Resistors, capacitors, diodes, and transistors are electronic components.

electronics   A part of technology that deals with the movement of electrons through different materials such as conductors, insulators, semiconductors, and superconductors.

energy   The ability to do work. Energy has many forms (mechanical, electrical, etc.) and comes from many sources (nuclear, chemical, solar, etc.).

ergonomics   The study of how the human body relates to things around it.

exponential rate of change   Increasing or decreasing at a changing rate.

fatigue strength   The ability to withstand many cycles of bending or flexing.

FAX (Facsimile)   A way of sending graphics or text electronically using a telephone line.

fermentation   A process where microorganisms digest starch and sugar, producing alcohol.

fiber optics   Light pulses moving through a flexible glass or plastic cable for communication.

flowchart   Major steps in a process shown as they happen in sequence.

gears   Mechanical devices that transmit forces. They can change the speed of spinning parts, or change direction.

genetic engineering (bioengineering)   Designing new life forms.

genetics   The study of how traits of plants and animals are passed on from generation to generation.

geostationary (geosynchronous orbit)   Satellites that stay in the same place above the Earth at all times. They move just fast enough to turn with the Earth.

greenhouse effect   The natural buildup of heat trapped by atmospheric gases, mainly carbon dioxide, that lets visible light in but keeps infrared radiation from leaving the earth's surface.

hereditary   Passed on from one generation to another.

hologram   A three-dimensional picture made by a laser.

hovercraft (air cushion vehicles)   Vehicles that ride on a cushion of air.

hydraulic   Using a liquid such as oil under pressure to create a force.

hydroponics   Growing plants without soil.

insulators   Materials that do not allow electrons to flow easily from one atom to another.

Integrated Circuit (IC or microchip)   A combination of microscopic components and circuits.

Just-in-time manufacturing (JIT)   A system scheduled by computer where all materials and ordered parts get to the factory just in time to be used in production.

kinetic energy   Energy of motion.

knowledge base   All the facts known to people today. It is growing exponentially.

laser   Light Amplification by Stimulated Emission of Radiation. A laser is a monochromatic (one-colored) focused light source that can be very strong.

lever   A simple machine used to change the direction and the size of a movement or force.

magnetic levitation (Maglev)   Floating above a magnetic field. A Maglev train will float above the magnetic field instead of using wheels on a track.

manufacturing   The building of products in a factory or business.

mass production   Manufacturing many of the same item at one time, usually using assembly lines and interchangeable parts.

microchip (integrated circuit)   A small silicon chip containing thousands of tiny electronic circuits.

microgravity   Sometimes called zero gravity or zero-g. It means very little gravity.

MIDI   Musical Instrument Digital Interface. A device that lets you put music into a computer from an electronic keyboard or other instrument.

modem   A device that lets your computer communicate with another computer using the telephone line.

molecule   A combination of atoms.

multimeter   A device that measures electricity.

NASA   National Aeronautics and Space Administration.

optical processing   Gathering and comparing thousands of images in a few seconds.

orthographic-projection   Drawings that show three views of an object in three dimensions—from the top, from one side, and looking directly at the front.

partnership   A business owned by two or more people.

patent   A special government license that protects an invention.

petrochemicals   Products made from oil or petroleum.

plan view   A view looking down on a structure or building site.

pneumatic   Using a gas such as air under pressure to create a force.

potential energy   Energy at rest waiting to do work.

power   The amount of work done in a certain amount of time.

problem statement   The definition of a problem; also called the design brief.

processes   Special machines or operations that change a raw material into a finished product. Machining, forming, fastening, and finishing are major processes.

proprietorship   A business owned by just one person.

prototype   A model of a product.

quality control   Inspecting or testing for faults in a product and then correcting the causes.

RAM   Random Access Memory. The computer memory used by software programs and individual documents.

ROM   Read Only Memory. It is built into a computer's circuits and cannot be changed by the user.

Research and Development (R&D)   A company division that investigates resources in order to make new products or improve existing products.

resistor   An electronic component that resists electron flow.

resources   Anything used in the production of a product. Resources can be people, machines, information, raw materials, energy, time, or money (capital).

robotic vision   Vision system using television camera eyes in robots.

SI   The International System of Weights and Measures used internationally for trade.

sales forecast   A prediction of how many sales a company will make of a product.

scanner   A machine that copies text and graphics from paper to the computer.

schematic diagram   A drawing using a set of symbols to represent real parts.

scientific method   The process of scientific problem solving developed by Galileo and Bacon in the 1600s.

semiconductor   A group of materials that conduct electricity under certain conditions.

serendipity   The process of discovery by accident.

shear   A sliding force made by a pair of forces acting on an object in opposite directions (the blades on a pair of scissors.)

simulation   A method of modeling a real product or process without using the real things.

smart house   A house where many things are controlled by computer.

software   The coded instructions for a computer to use.

static loads   Forces that act on a structure. They are unchanging or slowly changing.

stepper motors   Electric motors that turn in small steps. Often used in automation.

strategy   A step-by-step plan.

superconductor   A material that conducts electricity perfectly, usually at low temperatures.

system   A combination of parts that work together as a whole. A system model has input, process, output, and feedback.

technology   Using critical thinking skills, resources, and the devices people have invented to solve problems.

technologically literate   The ability to understand technology and evaluate the effects of technology on people and the environment.

telemetry   The process of sending and receiving information from a distance.

tensile strength   The ability of a material to resist stretching or pulling apart.

theory   Ideas about how nature and the world works.

thermocouple   An electrical temperature sensor.

thermoplastic   A plastic that can be melted and remelted many times using heat.

thrust   A force or push exerted on an object to make it move.

transistor   An electronic component used to switch or amplify electrical circuits.

transportation   The movement of people or goods from one place to another.

trends   Current needs people have.

videoconferencing   A meeting between people at different locations using television video and audio signals beamed by satellite from one place to the other.

voice recognition   A computer's ability to react to a person's voice instead of commands from a keyboard or other input device.

work envelope   The maximum distance that each part of a robotic arm can move in any direction.

INDEX

Abstracts, 73
Acceptance sampling, 220
Acronyms, 16
    for computers, 61
Acrylic, 154
Administration, 206
Advertising, 227–31
Aerodynamics, 313–17
Ailerons, 318
Air compressor, 130
Air cushion vehicles, 305
Airfoil, 315–18
Airplanes, 311–12
    flight of, 318–19
Air transportation, 310–20
Alloy, 387
Aluminum, 154
Amp, 145
Amperage, 145
AMTRAK, 296
Analogy, 130
Angle of attack, 316
Animation, 337
Anthropometric data, 150
Application, 60
Applied research, 18
Artificial heart, 409–11
Artificial intelligence, 47
ASCII, 56
Audio, 39
Automated storage and retrieval system
    (AS/RS), 177
Automated transit systems, 302
Automatic guided vehicle system
    (AGVS), 177
Automation, 172–76
    computer control and, 190–96
    factory of the future, 196–99
    materials handling, 176–82
    robots, 182–90
Automobiles, 294–95

Barges, 306
Basic research, 18

Baud rate, 73
Beam-break system, 195
Bernoulli's principle, 312
Bibliography, 72
Bill of materials, 139
Bimetallic strip, 136
Binary system, 54–55
Biodegradable, 166
Bioengineering, 402–408
Biomass, 283
Biomass process, 401
Biomechanics, 408–11
Bionics, 408–11
Biotechnology, 45
    bioengineering and, 402–408
    bionics and, 408–11
    definition of, 396
    distillation and, 398–401
    fermentation and, 401–402
    foods of future, 396–98
    future of, 411–12
Bit, 56
Blimps, 310
Board of directors, 206
Boat hull design, 306–10
Boats, 303–304
Boroscope, 220
Brainstorm, 21–22
Brand names, 228
Break-even point, 230
Broadcasting, 344
Bronze Age, 5
Burn in, 220
Buses, 302
Business
    companies and, 204–12
    future of, 231–32
    marketing and advertising in,
        227–31
    mass production and, 212–19
    product packaging and,
        223–27
    quality control and, 219–23
Byte, 56

CAD/CAM, 190
CAD station, 67
Cam, 123
Capacitors, 127
Capillary action, 410
Capital, 204, 205
Career clusters, 37
Cargo ships, 305
Carpal tunnel syndrome, 151
Catalysts, 137
CD-ROM (compact disk-read only
        memory), 72
Cellular phones, 347
Central processing unit (CPU), 57
Challenger, 364
Characters, 62
Chemical systems, 118, 136–37
Chord, 316
Circuit, 125
Clearance hole, 132
Clip art, 33, 65
Closed system, 111
Columbia, 364
Commercial zones, 255
Commission, 228
Communication, 328–29
    electronic, 342–46
    future of, 356–58
    pictures and symbols, 334–42
    writing and drawing, 329–34
Community design, 252–56
Compact disk (CD), 72
Company, 204–12
Components, 53, 125
    electronic, 127
Composite materials, 154, 157, 320
Composition, 335
Compression, 248–52
Compression strength, 160
Computer-aided design (CAD), 65–67
Computer-aided manufacturing
        (CAM), 67, 190
Computer-integrated manufacturing
        (CIM), 67, 190–91

Computers, 51
    acronyms for, 61
    basics of, 52–54
    binary system and, 54–55
    bits and bytes and, 56
    computer-aided design and, 64–67
    databases and, 68–69
    finding information with, 72–75
    future of, 75
    memory in, 58
    peripheral devices and, 59
    software for, 60
    spreadsheets and, 70–72
    system parts, 57
    transportation and, 320
    using, 60
    word processing and, 62–64
Conductors, 124
Consensus, 42
Conserve, 104
Construction, 236–38
    communities, designing, 252–56
    design and, 236–38
    future of, 257–61
    structural design and, 245–47
    tension and compression and,
        248–52
Consumer, 210
Consumer surveys, 208
Control, 167
Conveyors, 176–82, 303
Cordless telephones, 347
Corporation, 204–208
Corpus collosum, 88
Cost-effective, 190
Cranes, 177
Cranks, 121
Crude oil, 137
Cylinders
    double-acting, 130
    single-acting, 130

Data, 53
Database, 68–69
Dead loads, 247
Decimal system, 55, 144
Decoding, 73
Decompose, 111
Degenerative diseases, 408

Depth of field, 195
Desalination, 399–401
Design, 238–45
    of communities, 252–56
    materials for, 155–62
    measurement and, 144–50
    models and prototypes for, 162–65
    product ecology and, 165–68
    products for people, 150–55
    structural, 245–47
Design brief, 12, 14
Design elements, 334
Desktop publishing, 64
Destructive testing, 161
Digital image processor, 342
Digitizer, 59
Diode, 128
Direct sales, 228
Dirigibles, 310
Distillation, 398–401
Distribution, 229
Dividend, 205
Downlink, 349
Downloading, 73, 333
Downtime, 208
DPDT (double-pole, double-throw)
        switch, 129–30
Drag, 316
Drawings
    making, 332–34
    orthographic, 239, 330
    pictorial, 242, 330
    structural, 245–46
    technical, 330
Dry cell, 136
Dynamic loads, 247

Ecology, 165–68
Electrical circuits, 125, 127
Electrical systems, 118, 124–30
Electricity, 124, 269
    measuring, 145
    source of, 270–74
Electrodes, 136
Electrolytes, 136
Electromechanical machines, 124
Electronic bulletin board, 354
Electronic communication, 342–46
Electronic noise, 329

Electrons, 36, 124
End effector, 185
Energy, 270–74
    alternative sources of, 282–88,
        320
    conservation of, 256, 274–82
    conventional forms of, 269
    definition of, 266
    kinds of, 267
    sources of, 267–68
Energy-efficient car, 280–82
English system, 146
Environmental Protection Agency
        (EPA), 402
Ergonomics, 150–55
Escape velocity, 377
Experimenting, 19
Exponential rate of change, 10–11
Extravehicular mobility units (EMU),
        151

Fads, 86
Fatigue strength, 160
FAX (facsimile), 354
Feedback, 90, 116, 210, 328
Fermentation, 401–402
Ferrocement, 159
Ferrous metals, 157
FES (functional electrical stimulation),
        411
Fiberglass, 154
Fiber-optic communication, 348,
        350–52
Fifth wheel, 296
Fission, 269
Fixtures, 216
Floor plan, 65, 238
Flowchart, 31–32, 188
Fluid systems, 118, 130–33
Flywheel, 124
Fonts, 62
Foods, designer, 396–98, 406–408
Footprint, 349
Force, 121
Forklift, 177
Fossil fuels, 103, 269
Fractional distillation, 401
Frequency, 343
Fringe benefits, 207

Fulcrum, 121
Fusion, 269

Gasohol, 401
Gears, 123
Gene, 402
    dominant, 403
    recessive, 403
Generator, 269
Genetic diseases, 396
Genetic engineering, 45, 402–408
Genetics, 402
Genotype, 403
Geographic information systems
        analyst, 48
Geostationary, 348
Geosynchronous orbit, 348
Geothermal energy, 103, 284
Go-no-go gauge, 222
Graphic arts layout, 341–42
Gravity, 372, 374, 377
Greenhouse effect, 164
Growth hormones, 396

Halophytes, 397
Ham operators, 344
Hard copy, 333
Hardness, 160
Hardware, 57
Hardwoods, 156
Heavier-than-air vehicles, 310
Helicopters, 319
Hereditary diseases, 396
Hertz, 343
Hoists, 177
Hologram, 357
Hovercrafts, 305, 321–23
Human resources, 207
Hydraulic pump, 130
Hydraulic systems, 130
Hydroelectric power, 269
Hydrofoils, 304
Hydroponics, 398
Hypothesis, 6

Icons, 43
Illustrations, 330, 331
Incentive, 228
Industrial Revolution, 5

Industrial robots, 183
Industrial zones, 254
Innovations/inventions, 80–86
    definition of, 80
    history of, 91–93
    ideas for, 86–91
    patents for, 93–97
Input, 116
Insulation, 275, 278–80
Insulators, 124
Integrated circuit (IC), 53
Interchangeable parts, 212
Interference, 328
International System of Units, 144
Inventory, 213
Iron Age, 5

Jigs, 216
Joystick, 59
Just-in-time (JIT) manufacturing, 213

Kelvin, 126
Kinetic energy, 267
Knowledge base, 10

Laminar flow, 315
Landfills, 166
Land transportation, 294–303
Laptop computer, 64
Laser guidance system, 387–90
Lasers, 353
Launchers (rocket), 377
Law of dominance, 403
Leading edge, 316
Levers, 121
Lift, 316
Lighter-than-air vehicles, 310
Line of golden proportion, 335
Linkages, 121
Liquidation, 231
Live loads, 247
Loads, 247
Logo, 66

Maglev trains, 297–302
Magnetic levitation, 47, 125
Management, 206
Market, 209
Marketing, 227–31

Marketing department, 208
Marketing plan, 208
Mass, 372
Mass production, 173, 212–19
Mass transit rail vehicles, 302
Materials
    classification of, 156–57
    natural, 155
    properties of, 160
    synthetic, 155
    testing, 159
Materials handling, 176–82
Measurement, 144–50
Mechanical systems, 118, 120–24
Medical researcher, 46
Metals, 157
Metric system, 144, 146
Microchip, 18, 53
MIDI (musical instrument digital
        interface), 59
Models, 162–65
Modem, 59, 73, 354
Modular, 257
Molecules, 36
Monorails, 177
Mother board, 57
MRI (magnetic resonance imaging),
        126
Multimeters, 144

NASA (National Aeronautics and
        Space Administration), 81,
        364, 386
Natural materials, 155
Negative, 340
Networking, 67, 354
Neutrally buoyant, 390
Newton's laws of motion, 372, 373
Nichrome wire, 134
Nonferrous metals, 157
Nonrenewable resources, 269
Nuclear energy, 269
Nucleus, 36
Number crunching, 70

Ocean liners, 304
Ohm, 145
Ohm's law, 145
On line, 208

Optical processing, 196
Orthographic projection, 239, 330
Output, 116

Packaging products, 223–27
Pallets, 177
Parallel circuits, 125
Parameters, 14
Partnership, 204
Passenger ships, 304
Pasteurize, 163
Patent, 93, 95–97
    combination of materials, 94
    design, 94
    living cell, 94
    process/method/system, 94
    structure, 94
Payload, 183
Peripherals, 59
Persistence of vision, 337
Petrochemicals, 137
Petroleum, 137
Phenotype, 403
Photoelectric sensors, 195
Photograph, 340–41
Photography, 334
Photon, 353
Photovoltaic cells, 282
Physically challenged, 82
Pick-and-place maneuver, 195
Pictorial drawing, 242
    isometric, 330
    oblique, 330
    perspective, 330
Piggybacked trailers, 296
Pilot hole, 132
Pipelines, 302
Piston, 130
Pixels, 65
Plastics, 157
Plotter, 65
Pneumatic systems, 130
Point-to-point transmission, 344
Polycarbonate, 154
Potential energy, 267
Power, 266
Precipitation, 268
Pre-Industrial Revolution, 5
Preliminary sketch, 238

Print, 340
Printed circuit (PC) board, 57
Printer, 59
Problem-solving loop, 15
Problem-solving strategy, 12–17
Problem statement, 12
Process, 116
Processes, 106–11
Production department, 208
Products
    marketing and advertising, 227–31
    packaging, 223–27
Program, 60
Properties, 158
Proprietorship, 204
Prosthetic devices, 155
Protocol, 73
Prototype, 89, 162–65
Public areas, 254
Punnett square, 403

Quality control, 106, 219–23

Rack and pinion, 123
Radio, 343–44, 346
RAM (random access memory), 58
Reciprocating motion, 123
Recycling, 104, 165
Refining, 137
Remote manipulator system (RMS),
    183
Renewable resources, 282
Research and development (R&D),
    17–18, 81, 207
Residential zones, 254
Resistance, 121
Resistors, 127
Resources
    energy, 103
    information, 102
    machines, 102
    money, 103
    nonrenewable, 103
    people, 102
    processes used to make products,
        106–11
    raw materials, 103
    renewable, 103
    time, 103

using, 104–106
    wasting of, 111–12
Retail sales, 228
Retrieving, 18
Reverse osmosis, 399
Rhythm, 335
Robotic vision, 191
Robots, 182–99
Rocket launch system, 366–71
ROM (read only memory), 58
Rotational motion, 123
RPMs (revolutions per minute),
    123
"Rube Goldberg" invention, 83–86
Rudder, 318
Rule of thirds, 335

Safe lights, 341
Safety rules, 8
Sales forecast, 227
Salyut, 379–80
Satellite communications, 348, 352
Scale, 147
Scanner, 59
Schematic diagrams, 129
Scientific method, 6
Self-sufficient, 386
Semiconductors, 125
Serendipity, 80
Series circuits, 125
Series-parallel circuits, 125
Shear, 249
Shelter, 237
Ships, 303–304
Short circuit, 127
Simulations, 165
Site, 238
Skylab, 364, 379–80
Smart house, 257
Software, 60
Softwoods, 156
Solar array, 286
Solar battery charger, 285–88
Solar cells, 282
Solar energy, 256, 282
Sound effects, 40
Space colonies, 386–87
Space spinoffs, 387
Space stations, 379–86

Space technology
    history of, 362–65
    living and working in space,
        379–87
    physics of, 372–77
    spinoffs and the future, 387–91
    travel in space, 377–79
Spreadsheets, 70–72
Springs, 124
Sprocket, 123
Sputnik, 363
Standardized parts, 212
Standards, 144
Standard sizes, 121
Static loads, 247
Statistical process control, 220
Stepper motors, 184
Stereoscopic vision, 195
Stockholders, 206
Stone Age, 5
Structural design, 245–47
Structural drawings, 245–46
Submarines, 305
Subsystems, 118
Superconductivity, 126
Superconductor, 45, 125
Survey, 210
Synchronized production, 213
Synonyms, 62
Synthetic materials, 155
Systematic, 117
Systems
    chemical, 136–37
    combining, 137–39
    electronic, 124–30
    fluid, 130–33
    mechanical, 120–24
    model of, 116–17
    thermal, 133–36
    types of, 118

Teach pendant, 188

Tech, 2
Technical illustrators, 330
Technical writers, 329
Technologically literate, 2
Technology
    careers in, 43–48
    definition of, 2
    effect of, 30–31
    future of, 45, 47, 416–20
    growth of, 3
    health/P.E. and, 42–43
    history of, 5–7
    language arts and, 39–40
    math and, 40–42
    science and, 35–36
    social studies and, 36–39
Technology teacher, 44
Telemetry, 191–95
Telephone, 347–48
Television, 344, 347–48
Tensile strength, 160
Tension, 248–52
Theories, 35
Thermal systems, 118, 133–36
Thermocouple, 136
Thermoplastics, 157
Thermosetting plastics, 157
Thesaurus, 62
Thrust, 377
Tidal energy, 284
Timeline, 37–39
Tolerance, 222
Trailing edge, 316
Trains, 296–302
Traits, 402, 403
Transistors, 128
Transportation, 293–94
    air, 310–20
    definition of, 292
    future of, 320
    land, 294–303
    water, 303–10

Trends, 81
Trigonometric ratios (table), 423
Troubleshooting, 120
Trucks, 295–96
Tugboats, 306
Turbine, 269, 270–74
Turbulence, 307

Unity, 335
Uplink, 349
Uploading, 73, 333

Vacuum, 378–79
Variable, 167
Vectors, 374–77
Vessels, 303
Video, 39
Videoconferencing, 357
Video effects, 40
Video image, 355–56
Video photography, 338
Visualizing, 89
Voice recognition, 45
Volt, 145
Voltage, 145
Voyager, 364

Water transportation, 303–10
Wet cell, 136
Wholesale sales, 228
Wind bracing, 247
Wind energy, 283
Wind loads, 247
Wind tunnel, 313–18
Woods, 156
Word processing, 62–64
Work envelope, 186
WYSIWYG, 60

Yaw, 318

Zero gravity (zero-g), 372